MATHEMATICAL ANALYSIS OF PHYSICAL SYSTEMS

Ronald E. Mickens

Atlanta University

D1457858

VAN NOSTRAND REINHOLD COMPANY

New York

Copyright © 1985 by **Van Nostrand Reinhold Company Inc.**
Library of Congress Catalog Card Number: 84-25712
ISBN: 0-442-26077-6

Manufactured in the United States of America.

Published by Van Nostrand Reinhold Company Inc.
135 West 50th Street
New York, New York 10020

Van Nostrand Reinhold Company Limited
Molly Millars Lane
Wokingham, Berkshire RG11 2PY, England

Van Nostrand Reinhold
480 Latrobe Street
Melbourne, Victoria 3000, Australia

Macmillan of Canada
Division of Gage Publishing Limited
164 Commander Boulevard
Agincourt, Ontario MIS 3C7, Canada

15 14 13 12 11 10 9 8 7 6 5 4 3 2 1

Library of Congress Cataloging in Publication Data
Mickens, Ronald E., 1943–
  Mathematical analysis of physical systems.
  Bibliography: p.
  Includes index.
  1. Mathematical physics.   I. Title.
QC20.M48   1985   530.1′5   84-25712
ISBN 0-442-26077-6

Contents

Preface

An epistemological problem of great interest to natural scientists is why mathematics is so effective a tool in the formulation and analysis of physical phenomena.[1] Most scientists and mathematicians would accept as correct that mathematics is not a part of natural science itself. Its axioms may derive from the processes of pure reason or they may be suggested by experience; however, once obtained, they form a basis upon which the entire structure of mathematics can be developed independently of experience. One can certainly do meaningful mathematics without being concerned with the happenings of the physical world.[2]

There have always been, in each generation, a few great mathematicians and scientists who have tried to resolve the difficulties inherent in this problem. Within the past century, the names of H. Poincaré, Albert Einstein, Hermann Weyl, Niels Bohr, Eugene P. Wigner, and Raymond L. Wilder appear prominently.

The physicist Eugene P. Wigner looked at this problem and concluded that "the unreasonable effectiveness of mathematics in the natural sciences" is a mystery not yet solved.[1] He ends his article with the following statements.

> The miracle of the appropriateness of the language of mathematics for the formulation of the laws of physics is a wonderful gift which we neither understand nor deserve. We should be grateful for it and hope that it will remain valid in future research and that it will extend, for better or for worse, to our pleasure even though perhaps also to our bafflement, to wide branches of learning. (p. 14)

I shall not attempt to either bring forth or discuss the multitude of arguments that a variety of authors have presented to deal with the problem of "the unreasonable effectiveness of mathematics in the natural sciences." The references given will easily lead the reader to the relevant literature and the important names.[3-8] However, from my viewpoint, there are two points of importance that should be stated. First, it is clear, in a deep sense, that mathematics is the language of science. Also, at the present time, there is a powerful, often unstated assumption that underlies most investigations in the natural sciences.[9] "Understanding means understanding through mathematics" (p. 51). Second, a reasonable and thoughtful solution of the explanation of the applicability of mathematics to the natural sciences has been provided by Wilder. In summary, he concludes that mathematics, as well as the natural sciences, is a cultural system. Therefore[10]

> If the problems worked on by mathematicians, physicists and chemists are imposed on them by the cultural forces of their respective sub cultures, and if there are sufficiently many contacts between the mathematicians and the scientists . . . then it is to be expected that mathematics will be applicable to the sciences. . . . (p. 12)

This explanation and the philosophy behind it is particularly satisfactory for me since it provides a human framework for the basis and direction of scientific and mathematical activities. In addition, it allows us to understand and compare the diversity of scientific and mathematical activities in other (non-Western) cultures without being chauvinistic.

The basic thrust of this book is to show through examples how mathematics can be applied to the formulation and analysis of physical systems. The first chapter consists of reprints of two thought-provoking articles on the "unreasonable effectiveness of mathematics in the natural sciences." Six chapters follow, each of which examines in detail a particular physical system or concept. The authors of these chapters are experts in their respective fields of endeavor.

The underlying theme of the book is that the unity of nature is revealed through its mathematical expression. The various chapter contributions allow the reader to readily comprehend the fact that a diversity of phenomena can be understood within the framework of a relatively small number of physical concepts and mathematical relations.

In order of their appearance, the contributors and their topics are as follows: Sylvester J. Gates, Jr. discuses the concept of gauge invariance so that nonspecialists can gain an understanding of its importance in modern physics. Stanley R. Deans presents some of the fundamental properties of the Radon transform and shows how it can be applied to problems in a number of scientific areas. James. V. Lindesay and Harry L. Morrison indicate how a general set of equations can be developed for a stationary physical system with an internal local guage symmetry. William P. Reinhardt illustrates the basic concepts relating to the onset of chaos in deterministic dynamics by using conservative Hamiltonian dynamics in two degrees of freedom. Vladimir Z. Kresin and William A. Lester, Jr. present the adiabatic approach as a method in quantum many-body theory and describe recent advances in the technique. Robert Gilmore writes on the subject of catastrophe theory, whose objective is to study the qualitative properties of the solutions of equations and, in particular, to determine how these qualitative properties change as the parameter in the equations change.

I would like to thank Eric Rosen who, as a senior editor at Van Nostrand Reinhold, first encouraged me to bring this book into existence. My gratitude also extends to Charles Hutchinson, Publisher, Van Nostrand Reinhold, who helped me to complete this project through his understanding and patience.

RONALD E. MICKENS

REFERENCES

1. E. P. Wigner, "The Unreasonable Effectiveness of Mathematics in the Natural Sciences," *Communications on Pure and Applied Mathematics* 13 (1960): 1–14.
2. S. Bochner, *The Role of Mathematics in the Rise of Science* (Princeton, N.J.: Princeton University Press, 1966).
3. H. Weyl, *Philosophy of Mathematics and Natural Science* (Princeton, N.J.: Princeton University Press, 1949).
4. D. A. T. Gasking, "Mathematics and the World," *Australasian Journal of Philosophy* 8 (1940): 97–116.

5. M. Kline, *Mathematics and the Physical World* (New York: Crowell, 1959).
6. F. J. Dyson, "Mathematics in the Physical Sciences," *Scientific American* 211 (1964): 127–146.
7. M. Kline, *Mathematics: The Loss of Certainty* (New York: Oxford University Press, 1980).
8. P. J. Davis and R. Hersh, *The Mathematical Experience* (Boston: Birkhauser, 1981).
9. R. L. Wilder, *Mathematics as a Cultural System* (Oxford: Pergamon Press, 1981).
10. C. Smorynski, "Mathematics as a Cultural System," *The Mathematical Intelligencer* 5 (1983): 9–15.

Contributors

Stanley R. Deans
565 Willow Road, Apt. 3, Menlo Park, California 94025

Sylvester James Gates
Department of Physics and Astronomy, University of Maryland, College Park, Maryland 20742

Robert Gilmore
Department of Physics and Atmospheric Sciences, Drexel University, Philadelphia, Pennsylvania 19104

Vladimir Z. Kresin
Materials and Molecular Research Division, Lawrence Berkeley Laboratory, University of California, Berkeley, California 94720

William A. Lester
Materials and Molecular Research Division, Lawrence Berkeley Laboratory, University of California, Berkeley, California 94720

James V. Lindesay
Department of Physics, University of California, Berkeley, California 94703

Harry L. Morrison
Department of Physics, University of California, Berkeley, California 94703

William P. Reinhardt
Department of Chemistry, University of Pennsylvania, Philadelphia, Pennsylvania 19104

1-1

The Unreasonable Effectiveness of Mathematics in the Natural Sciences*

Eugene P. Wigner

Princeton University

> and it is probable that there is some secret here which
> remains to be discovered. (C. S. Pierce)

There is a story about two friends, who were classmates in high school, talking about their jobs. One of them became a statistician and was working on population trends. He showed a reprint to his former classmate. The reprint started, as usual, with the Gaussian distribution and the statistician explained to his former classmate the meaning of the symbols for the actual population, for the average population, and so on. His classmate was a bit incredulous and was not quite sure whether the statistician was pulling his leg. "How can you know that?" was his query. "And what is this symbol here?" "Oh," said the statistician, "this is π." "What is that?" "The ratio of the circumference of the circle to its diameter." "Well, now you are pushing your joke too far," said the classmate, "surely the population has nothing to do with the circumference of the circle."

Naturally, we are inclined to smile about the simplicity of the classmate's approach. Nevertheless, when I heard this story, I had to admit to an eerie feeling because, surely, the reaction of the classmate betrayed only plain

*Reproduced from *Commmunications on Pure and Applied Mathematics* **13**:1–14 (1960). Copyright © 1960 by John Wiley & Sons, Inc.*

common sense. I was even more confused when, not many days later, some-one came to me and expressed his bewilderment[1] with the fact that we make a rather narrow selection when choosing the data on which we test our theories.

How do we know that, if we made a theory which focusses its attention on phenom-ena we disregard and disregards some of the phenomena now commanding our at-tention, that we could not build another theory which has little in common with the present one but which, nevertheless, explains just as many phenomena as the present theory.

It has to be admitted that we have no definite evidence that there is no such theory.

The preceding two stories illustrate the two main points which are the subjects of the present discourse. The first point is that mathematical con-cepts turn up in entirely unexpected connections. Moreover, they often permit an unexpectedly close and accurate description of the phenomena in these connections. Secondly, just because of this circumstance, and because we do not understand the reasons for their usefulness, we cannot know whether a theory formulated in terms of mathematical concepts is uniquely appropriate. We are in a position similar to that of a man who was provided with a bunch of keys and who, having to open several doors in succession, always hit on the right key on the first or second trial. He became skeptical concerning the uniqueness of the coordination between keys and doors.

Most of what will be said on these questions will not be new; it has prob-ably occurred to most scientists in one form or another. My principal aim is to illuminate it from several sides. The first point is that the enormous use-fulness of mathematics in the natural sciences is something bordering on the mysterious and that there is no rational explanation for it. Second, it is just this uncanny usefulness of mathematical concepts that raises the question of the uniqueness of our physical theories. In order to establish the first point, that mathematics plays an unreasonably important role in physics, it will be useful to say a few words on the question "What is mathematics?," then, "What is physics?,"then, how mathematics enters physical theories, and last, why the success of mathematics in its role in physics appears so baffling. Much less will be said on the second point: the uniqueness of the theories of physics. A proper answer to this question would require elaborate exper-imental and theoretical work which has not been undertaken to date.

WHAT IS MATHEMATICS?

Somebody once said that philosophy is the misuse of a terminology which was invented just for this purpose.[2] In the same vein, I would say that mathematics is the science of skillful operations with concepts and rules invented just for this purpose. The principal emphasis is on the invention of concepts. Mathematics would soon run out of interesting theorems if these had to be formulated in terms of the concepts which already appear in the axioms. Furthermore, whereas it is unquestionably true that the concepts of elementary mathematics and particularly elementary geometry were formulated to describe entities which are directly suggested by the actual world, the same does not seem to be true of the more advanced concepts, in particular the concepts which play such an important role in physics. Thus, the rules for operations with pairs of numbers are obviously designed to give the same results as the operations with fractions which we first learned without reference to "pairs of numbers." The rules for the operations with sequences, that is with irrational numbers, still belong to the category of rules which were determined so as to reproduce rules for the operations which were already known to us. Most more advanced mathematical concepts, such as complex numbers, algebras, linear operators, Borel sets—and this list could be continued almost indefinitely—were so devised that they are apt subjects on which the mathematician can demonstrate his ingenuity and sense of formal beauty. In fact, the definition of these concepts, with a realization that interesting and ingenious considerations could be applied to them, is the first demonstration of the ingeniousness of the mathematician who defines them. The depth of thought which goes into the formation of the mathematical concepts is later justified by the skill with which these concepts are used. The great mathematician fully, almost ruthlessly, exploits the domain of permissible reasoning and skirts the impermissible. That his recklessness does not lead him into a morass of contradictions is a miracle in itself: certainly it is hard to believe that our reasoning power was brought, by Darwin's process of natural selection, to the perfection which it seems to possess. However, this is not our present subject. The principal point which will have to be recalled later is that the mathematician could formulate only a handful of interesting theorems without defining concepts beyond those contained in the axioms and that the concepts outside those contained in the axioms are defined with a view of permitting ingenious logical operations which appeal to our aesthetic sense both as operations and also in their results of great generality and simplicity.[3]

The complex numbers provide a particularly striking example of the foregoing. Certainly, nothing in our experience suggests the introduction of these quantities. Indeed, if a mathematician is asked to justify his interest in complex numbers, he will point, with some indignation, to the many beautiful theorems in the theory of equations, of power series and of analytic functions in general, which owe their origin to the introduction of complex numbers. The mathematician is not willing to give up his interest in these most beautiful accomplishments of his genius.[4]

WHAT IS PHYSICS?

The physicist is interested in discovering the laws of inanimate nature. In order to understand this statement, it is necessary to analyze the concept of "law of nature."

The world around us is of baffling complexity and the most obvious fact about it is that we cannot predict the future. Although the joke attributes only to the optimist the view that the future is uncertain, the optimist is right in this case: the future is unpredictable. It is, as Schrödinger has remarked, a miracle that in spite of the baffling complexity of the world, certain regularities in the events could be discovered [1]. One such regularity, discovered by Galileo, is that two rocks, dropped at the same time from the same height, reach the ground at the same time. The laws of nature are concerned with such regularities. Galileo's regularity is a prototype of a large class of regularities. It is a surprising regularity for three reasons.

The first reason that it is surprising is that it is true not only in Pisa, and in Galileo's time, it is true everywhere on the Earth, was always true, and will always be true. This property of the regularity is a recognized invariance property, and, as I had occasion to point out some time ago [2], without invariance principles similar to those implied in the preceding generalization of Galileo's observation, physics would not be possible. The second surprising feature is that the regularity which we are discussing is independent of so many conditions which could have an effect on it. It is valid no matter whether it rains or not, whether the experiment is carried out in a room or from the Leaning Tower, no matter whether the person who drops the rocks is a man or a woman. It is valid even if the two rocks are dropped, simultaneously and from the same height, by two different people. There are, obviously, innumerable other conditions which are all immaterial from the point of view of the validity of Galileo's regularity. The irrelevancy of so many circumstances which *could* play a role in the phenomenon observed, has also been called an invariance [2]. However, this invariance is of a different character than the preceding one since it cannot be formulated as a general principle. The exploration of the conditions which do, and which do not, influence a phenomenon is part of the early experimental exploration of a field. It is the skill and ingenuity of the experimenter which shows him phenomena which depend on a relatively narrow set of relatively easily realizable and reproducible conditions.[5] In the present case, Galileo's restriction of his observations to relatively heavy bodies was the most important step in this regard. Again, it is true that if there were no phenomena which are independent of all but a manageably small set of conditions, physics would be impossible.

The preceding two points, though highly significant from the point of view of the philosopher, are not the ones which surprised Galileo most, nor do they contain a specific law of nature. The law of nature is contained in the statement that the length of time which it takes for a heavy object to fall from a given height is independent of the size, material and shape of the body which drops. In the framework of Newton's second "law,"this amounts to the statement that the gravitational force which acts on the falling body is proportional to its mass but independent of the size, material and shape of the body which falls.

The preceding discussion is intended to remind, first, that it is not at all natural that "laws of nature" exist, much less that man is able to discover them.[6] The present writer had occasion, some time ago, to call attention to the succession of layers of "laws of nature," each layer containing more general and more encompassing laws than the previous one and its discovery constituting a deeper penetration into the structure of the universe than the layers recognized before [3]. However, the point which is most significant in the present context is that all these laws of nature contain, in even their remotest consequences, only a small part of our knowledge of the inanimate world. All the laws of nature are conditional statements which permit a prediction of some future events on the basis of the knowledge of the present, except that some aspects of the present state of the world, in practice the overwhelming majority of the determinants of the present state of the world, are irrelevant from the point of view of prediction. The irrelevancy is meant in the sense of the second point in the discussion of Galileo's theorem.[7]

As regards the present state of the world, such as the existence of the earth on which we live and on which Galileo's experiments were performed, the existence of the sun and of all our surroundings, the laws of nature are entirely silent. It is in consonance with this, first, that the laws of nature can be used to predict future events only under exceptional circumstances—when all the relevant determinants of the present state of the world are known. It is also in consonance with this that the construction of machines, the functioning of which he can foresee, constitutes the most spectacular accomplishment of the physicist. In these machines, the physicist creates a situation in which all the relevant coordinates are known so that the behavior of the machine can be predicted. Radars and nuclear reactors are examples of such machines.

The principal purpose of the preceding discussion is to point out that the laws of nature are all conditional statements and they relate only to a very small part of our knowledge of the world. Thus, classical mechanics, which is the best known prototype of a physical theory, gives the second derivatives of the positional coordinates of all bodies, on the basis of the knowledge of the positions, etc., of these bodies. It gives no information on the existence, the present positions, or velocities of these bodies. It should be mentioned, for the sake of accuracy, that we have learned about thirty years ago that even the conditional statements cannot be entirely precise: that the conditional statements are probability laws which enable us only to place intelligent bets on future properties of the inanimate world, based on the knowledge of the present state. They do not allow us to make categorical statements, not even categorical statements conditional on the present state of the world. The probabilistic nature of the "laws of nature" manifests itself in the case of machines also, and can be verified, at least in the case of nuclear reactors, if one runs them at very low power. However, the additional limitation of the scope of the laws of nature[8] which follows from their probabilistic nature, will play no role in the rest of the discussion.

THE ROLE OF MATHEMATICS IN
PHYSICAL THEORIES

Having refreshed our minds as to the essence of mathematics and physics, we should be in a better position to review the role of mathematics in physical theories.

Naturally, we do use mathematics in everyday physics to evaluate the results of the laws of nature, to apply the conditional statements to the particular conditions which happen to prevail or happen to interest us. In order that this be possible, the laws of nature must already be formulated in mathematical language. However, the role of evaluating the consequences of already established theories is not the most important role of mathematics in physics. Mathematics, or, rather, applied mathematics, is not so much the master of the situation in this function: it is merely serving as a tool.

Mathematics does play, however, also a more sovereign role in physics. This was already implied in the statement, made when discussing the role of applied mathematics, that the laws of nature must be already formulated in the language of mathematics to be an object for the use of applied mathematics. The statement that the laws of nature are written in the language of mathematics was properly made three hundred years ago;[9] it is now more true than ever before. In order to show the importance which mathematical concepts possess in the formulation of the laws of physics, let us recall, as an example, the axioms of quantum mechanics as formulated, explicitly, by the great mathematician, von Neumann, or, implicitly, by the great physicist, Dirac [4, 5]. There are two basic concepts in quantum mechanics: states and observables. The states are vectors in Hilbert space, the observables self-adjoint operators on these vectors. The possible values of the observations are the characteristic values of the operators—but we had better stop here lest we engage in a listing of the mathematical concepts developed in the theory of linear operators.

It is true, of course, that physics chooses certain mathematical concepts for the formulation of the laws of nature, and surely only a fraction of all mathematical concepts is used in physics. It is true also that the concepts which were chosen were not selected arbitrarily from a listing of mathematical terms but were developed, in many if not most cases, independently by the physicist and recognized then as having been conceived before by the mathematician. It is not true, however, as is so often stated, that this had to happen because mathematics uses the simplest possible concepts and these were bound to occur in any formalism. As we saw before, the concepts of mathematics are not chosen for their conceptual simplicity—even sequences of pairs of numbers are far from being the simplest concepts—but for the amenability to clever manipulations and to striking, brilliant arguments. Let us not forget that the Hilbert space of quantum mechanics is the complex Hilbert space, with a Hermitean scalar product. Surely to the unpreoccupied mind, complex numbers are far from natural or simple and they cannot be suggested by physical observations. Furthermore, the use of complex numbers is in this case not a calculational trick of applied mathematics but comes close to being a necessity in the formulation of the laws of quantum mechanics. Finally, it now begins to appear that not only numbers but so-called analytic functions are destined to play a decisive role in the formulation of quantum theory. I am referring to the rapidly developing theory of dispersion relations.

It is difficult to avoid the impression that a miracle confronts us here, quite comparable in its striking nature to the miracle that the human mind can string a thousand arguments together without getting itself into contradictions or to the two miracles of the existence of laws of nature and of the human mind's capacity to divine them. The observation which comes closest to an explanation for the mathematical concepts' cropping up in physics which I know is Einstein's statement that the only physical theories which we are willing to accept are the beautiful ones. It stands to argue that the concepts of mathematics, which invite the exercise of so much wit, have the quality of beauty. However, Einstein's observation can at best explain properties of theories which we are willing to believe and has no reference to the intrinsic accuracy of the theory. We shall, therefore, turn to this latter question.

IS THE SUCCESS OF PHYSICAL THEORIES TRULY SURPRISING?

A possible explanation of the physicist's use of mathematics to formulate his laws of nature is that he is a somewhat irresponsible person. As a result, when he finds a connection between two quantities which resembles a connection well-known from mathematics, he will jump at the conclusion that the connection *is* that discussed in mathematics simply because he does not know of any other similar connection. It is not the intention of the present discussion to refute the charge that the physicist is a somewhat irresponsible person. Perhaps he is. However, it is important to point out that the mathematical formulation of the physicist's often crude experience leads in an uncanny number of cases to an amazingly accurate description of a large class of phenomena. This shows that the mathematical language has more to commend it than being the only language which we can speak; it shows that it is, in a very real sense, the correct language. Let us consider a few examples.

The first example is the oft quoted one of planetary motion. The laws of falling bodies became rather well established as a result of experiments carried out principally in Italy. These experiments could not be very accurate in the sense in which we understand accuracy today partly because of the effect of air resistance and partly because of the impossibility, at that time, to measure short time intervals. Nevertheless, it is not surprising that as a result of their studies, the Italian natural scientists acquired a familiarity with the ways in which objects travel through the atmosphere. It was Newton who then brought the law of freely falling objects into relation with the motion of the moon, noted that the parabola of the thrown rock's path on the earth, and the circle of the moon's path in the sky, are particular cases of the same mathematical object of an ellipse and postulated the universal law of gravitation, on the basis of a single, and at that time very approximate, numerical coincidence. Philosophically, the law of gravitation as formulated by Newton was repugnant to his time and to himself. Empirically, it was based on very scanty observations. The mathematical language in which it was formulated contained the concept of a second derivative and those of us who have tried to draw an osculating circle to a curve know that the second derivative is not a very immediate concept. The law of gravity which Newton reluctantly established and which he could verify with an accuracy of about 4% has proved to be accurate to less than a ten thousandth of a per cent and became so closely associated with the idea of absolute accuracy that only recently did physicists become again bold enough to inquire into the limitations of its accuracy.[10] Certainly, the example of Newton's law, quoted over and over again, must be mentioned first as a monumental example of a law, formulated in terms which appear to be simple to the mathematician, which has proved accurate beyond all reasonable expectation. Let us just recapitulate our thesis on this example: first, the law, particularly since a second derivative appears in it, is simple only to the mathematician, not to common sense or to non-mathematically-minded freshmen; second, it is a conditional law of very limited scope. It explains nothing about the earth which attracts Galileo's rocks, or about the circular form of the moon's orbit, or about the planets of the sun. The explanation of these initial conditions is left to the geologist and the astronomer, and they have a hard time with them.

The second example is that of ordinary, elementary quantum mechanics. This originated when Max Born noticed that some rules of computation, given by Heisenberg, were formally identical with the rules of computation with matrices, established a long time before by mathematicians. Born, Jordan and Heisenberg then proposed to replace by matrices the position and momentum variables of the equations of classical mechanics [6]. They applied the rules of matrix mechanics to a few highly idealized problems and the results were quite satisfactory. However, there was, at that time, no rational evidence that their matrix mechanics would prove correct under more realistic conditions. Indeed, they say "if the mechanics as here proposed should already be correct in its essential traits." As a matter of fact, the first application of their mechanics to a realistic problem, that of the hydrogen atom, was given several months later, by Pauli. This application gave results in agreement with experience. This was satisfactory but still understandable because Heisenberg's rules of calculation were abstracted from problems which included the old theory of the hydrogen atom. The miracle occurred only when matrix mechanics, or a mathematically equivalent theory, was applied to problems for which Heisenberg's calculating rules were meaningless. Heisenberg's rules presupposed that the classical equations of motion had solutions with certain periodicity properties; and the equations of motion had solutions with certain periodicity properties; and the equations of motion of the two electrons of the helium atom, or of the even greater number of electrons of heavier atoms, simply do not have these properties, so that Heisenberg's rules cannot be applied to these cases. Nevertheless, the calculation of the lowest energy level of helium, as carried out a few months ago by Kinoshita at Cornell and by Bazley at the Bureau of Standards, agree with the experimental data within the accuracy of the observations, which is one part in ten millions. Surely in this case we "got something out" of the equations that we did not put in.

The same is true of the qualitative characteristics of the "complex spectra," that is the spectra of heavier atoms. I wish to recall a conversation with Jordan who told me, when the qualitative features of the spectra were derived, that a disagreement of the rules derived from quantum mechanical theory, and the rules established by empirical research, would have provided the last opportunity to make a change in the framework of matrix mechanics. In other words, Jordan felt that we would have been, at least temporarily, helpless had an unexpected disagreement occurred in the theory of the helium atom. This was, at that time, developed by Kellner and by Hilleraas. The mathematical formalism was too clear and unchangeable so that, had the miracle of helium which was mentioned before not occurred, a true crisis would have arisen. Surely, physics would have overcome that crisis in one way or another. It is true, on the other hand, that physics as we know it today would not be possible without a constant recurrence of miracles similar to the one of the helium atom which is perhaps the most striking miracle that has occurred in the course of the development of elementary quantum mechanics, but by far not the only one. In fact, the number of analogous miracles is limited, in our view, only by our willingness to go after more similar ones. Quantum mechanics had, nevertheless, many almost equally striking successes which gave us the firm conviction that it is, what we call, correct.

The last example is that of quantum electrodynamics, or the theory of the Lamb shift. Whereas Newton's theory of gravitation still had obvious connections with experience, experience entered the formulation of matrix mechanics only in the refined or sublimated form of Heisenberg's prescriptions. The quantum theory of the Lamb shift, as conceived by Bethe and established by Schwinger, is a purely mathematical theory and the only direct contribution of experiment was to show the existence of a measurable effect. The agreement with calculation is better than one part in a thousand.

The preceding three examples, which could be multiplied almost indefinitely, should illustrate the appropriateness and accuracy of the mathematical formulation of the laws of nature in terms of concepts chosen for their manipulability, the "laws of nature" being of almost fantastic accuracy but of strictly limited scope. I propose to refer to the observation which these examples illustrate as the empirical law of epistemology. Together with the laws of invariance of physical theories, it is an indispensable foundation of these theories. Without the laws of invariance the physical theories could have been given no foundation of fact; if the empirical law of epistemology were not correct, we would lack the encouragement and reassurance which are emotional necessities without which the "laws of nature" could not have been successfully explored. Dr. R. G. Sachs, with whom I discussed the empirical law of epistemology, called it an article of faith of the theoretical physicist, and it is surely that. However, what he called our article of faith can well be supported by actual examples—many examples in addition to the three which have been mentioned.

THE UNIQUENESS OF THE THEORIES OF PHYSICS

The empirical nature of the preceding observation seems to me to be self-evident. It surely is not a "necessity of thought" and it should not be necessary, in order to prove this, to point to the fact that it applies only to a very small part of our knowledge of the inanimate world. It is absurd to believe that the existence of mathematically simple expressions for the second derivative of the position is self-evident, when no similar expressions for the position itself or for the velocity exist. It is therefore surprising how readily the wonderful gift contained in the empirical law of epistemology was take for granted. The ability of the human mind to form a string of 1000 conclusions and still remain "right," which was mentioned before, is a similar gift.

Every empirical law has the disquieting quality that one does not know its limitations. We have seen that there are regularities in the events in the world around us which can be formulated in terms of mathematical concepts with an uncanny accuracy. There are, on the other hand, aspects of the world concerning which we do not believe in the existence of any accurate regularities. We call these initial conditions. The question which presents itself is whether the different regularities, that is the various laws of nature which will be discovered, will fuse into a single consistent unit, or at least asymptotically approach such a fusion. Alternately, it is possible that there always will be some laws of nature which have nothing in common with each other. At present, this is true, for instance, of the laws of heredity and of physics. It is even possible that some of the laws of nature will be in conflict with each other in their implications, but each convincing enough in its own domain so that we may not be willing to abandon any of them. We may resign ourselves to such a state of affairs or our interest in clearing up the conflict between the various theories may fade out. We may lose interest in the "ultimate truth," that is in a picture which is a consistent fusion into a single unit of the little pictures, formed on the various aspects of nature.

It may be useful to illustrate the alternatives by an example. We now have, in physics, two theories of great power and interest: the theory of quantum phenomena and the theory of relativity. These two theories have their roots in mutually exclusive groups of phenomena. Relativity theory applies to macroscopic bodies, such as stars. The event of coincidence, that is in ultimate analysis of collision, is the primitive event in the theory of relativity and defines a point in space-time, or at least would define a point if the colliding particles were infinitely small. Quantum theory has its roots in the microscopic world, and from its point of view, the event of coincidence, or of collision, even if it takes place between particles of no spatial extent, is not primitive and not at all sharply isolated in space-time. The two theories operate with different mathematical concepts—the four dimensional Riemann space and the infinite dimensional Hilbert space, respectively. So far, the two theories could not be united, that is, no mathematical formulation exists to which both of these theories are approximations. All physicists believe that a union of the two theories is inherently possible and that we shall find it. Nevertheless, it is possible also to imagine that no union of the two theories can be found. This example illustrates the two possibilities, of union and of conflict, mentioned before, both of which are conceivable.

In order to obtain an indication as to which alternative to expect ultimately, we can pretend to be a little more ignorant than we are and place ourselves at a lower level of knowledge than we actually possess. If we can find a fusion of our theories on this lower level of intelligence, we can confidently expect that we will find a fusion of our theories also at our real level of intelligence. On the other hand, if we would arrive at mutually contradictory theories at a somewhat lower level of knowledge, the possibility of the permanence of conflicting theories cannot be excluded for ourselves either. The level of knowledge and ingenuity is a continuous variable and it is unlikely that a relatively small variation of this continuous variable changes the attainable picture of the world from inconsistent to consistent.[11]

Considered from this point of view, the fact that some of the theories which we know to be false give such amazingly accurate results, is an adverse factor. Had we somewhat less knowledge, the group of phenomena which these "false" theories explain, would appear to us to be large enough to "prove" these theories. However, these theories are considered to be "false" by us just for the reason that they are, in ultimate analysis, incompatible with more encompassing pictures and, if sufficiently many such false theories are discovered, they are bound to prove also to be in conflict with each other. Similarly, it is possible that the theories, which we consider to be "proved" by a number of numerical agreements which appears to be large enough for us, are false because they are in conflict with a possible more encompassing theory which is beyond our means of discovery. If this were true, we would have to expect conflict between our theories as soon as their number grows beyond a certain point and as soon as they cover a sufficiently large number of groups of phenomena. In contrast to the article of faith of the theoretical physicist mentioned before, this is the nightmare of the theorist.

Let us consider a few examples of "false" theories which give, in view of their falseness, alarmingly accurate descriptions of groups of phenomena. With some goodwill, one can dismiss some of the evidence which these examples provide. The success of Bohr's early and pioneering ideas on the atom was always a rather narrow one and the same applies to Ptolemy's epicycles. Our present vantage point gives an accurate description of all phenomena which these more primitive theories can describe. The same is not true any more of the so-called free-electron theory which gives a marvelously accurate picture of many, if not most, properties of metals, semiconductors and insulators. In particular, it explains the fact, never properly understood on the basis of the "real theory," that insulators show a specific resistance to electricity which may be 1,026 times greater than that of metals. In fact, there is no experimental evidence to show that the resistance is not infinite under the conditions under which the free-electron theory would lead us to expect an infinite resistance. Nevertheless, we are convinced that the free-electron theory is a crude approximation which should be replaced, in the description of all phenomena concerning solids, by a more accurate picture.

If viewed form our real vantage point, the situation presented by the free-electron theory is irritating but is not likely to forebode any inconsistencies which are unsurmountable for us. The free-electron theory raises doubts as to how much we should trust numerical agreement between theory and experiment as evidence for the correctness of the theory. We are used to such doubts.

A much more difficult and confusing situation would arise if we could, some day, establish a theory of the phenomena of consciousness, or of biology, which would be as coherent and convincing as our present theories of the inanimate world. Mendel's laws of inheritance and the subsequent work on genes may well form the beginning of such a theory as far as biology is concerned. Furthermore, it is quite possible that an abstract argument can be found which shows that there is a conflict between such a theory and the accepted principles of physics. The argument could be of such abstract nature that it might not be possible to resolve the conflict, in favor of one or of the other theory, by an experiment. Such a situation would put a heavy strain on our faith in our theories and on our belief in the reality of the concepts which we form. It would give us a deep sense of frustration in our search for what I called the "ultimate truth." The reason that such a situation is conceivable is that, fundamentally, we do not know why our theories work so well. Hence their accuracy may not prove their truth and consistency. Indeed, it is this writer's belief that something rather akin to the situation which was described above exists if the present laws of heredity and of physics are confronted.

Let me end on a more cheerful note. The miracle of the appropriateness of the language of mathematics for the formulation of the laws of physics is a wonderful gift which we neither understand nor deserve. We should be grateful for it and hope that it will remain valid in future research and that it will extend, for better or for worse, to our pleasure even though perhaps also to our bafflement, to wide branches of learning.

The writer wishes to record here his indebtedness to Dr. M. Polanyi who, many years ago, deeply influenced his thinking on problems of epistemology, and to V. Bargmann whose friendly criticism was material in achieving whatever clarity was achieved. He is also greatly indebted to A. Shimony for reviewing the present article and calling his attention to C. S. Peirce's papers.

NOTES

1. The remark to be quoted was made by F. Werner when he was a student at Princeton.
2. This statement is quoted here from *W. Dubislav's Die Philosophie der Mathematik in der Gegenwart.* Junker und Dunnhaupt Verlag, Berlin, 1932, p. 1.

3. Polanyi, in his *Personal Knowledge*, University of Chicago Press, 1958 says: "All these difficulties are but consequences of our refusal to see that mathematics cannot be defined without acknowledging its most obvious feature: namely, that it is interesting," (page 188).

4. The reader may be interested, in this connection, in Hilbert's rather testy remarks about intuitionism which "seeks to break up and to disfigure mathematics," Abh. Math. Sem. Univ. Hamburg, Vol. 158, 1922, or Gesammelte Werke, Springer, Berlin, 1935, page 188.

5. See, in this connection, the graphic essay of M. Deutsch, Daedalus, Vol. 87, 1958, page 86. A. Shimony has called my attention to a similar passage in C. S. Peirce's *Essays in the Philosophy of Science*, The Leberal Arts Press, New York, 1957 (page 237).

6. E. Schrödinger, in his *What is Life*, Cambridge University Press, 1945, says that this second miracle may well be beyond human understanding, (page 31).

7. The writer feels sure that it is unneccessary to mention that Galileo's theorem, as given in the text, does not exhaust the content of Galileo's observations in connection with the laws of freely falling bodies.

8. See, for instance, E. Schrödinger, reference [1].

9. It is attributed to Galileo.

10. See, for instance, R. H. Dicke, American Scientist, Vol. 25, 1959.

11. This passage was written after a great deal of hesitation. The writer is convinced that it is useful, in epistemological discussions, to abandon the idealization that the level of human intelligence has a singular position on an absolute scale. In some cases it may even be useful to consider the attainment which is possible at the level of the intelligence of some other species. However, the writer also realizes that his thinking along the lines indicated in the text was too brief and not subject to sufficient critical appraisal to be reliable.

REFERENCES

[1] Schrödinger, E. *Über Indeterminismus in der Physik*, J. A. Barth, Leipzig, 1932; also Dubislav, W., *Naturphilosophie*, Junker und Dunnhaupt, Berlin, 1933, Chap. 4.

[2] Wigner, E. P., *Invariance in physical theory*, Proc. Amer. Philos. Soc., Vol. 93, 1949, pp. 521–526.

[3] Wigner, E. P., *The limits of science*, Proc. Amer. Philos. Soc. Vol. 94, 1950, p. 442. See also Margenau, H., *The Nature of Physical Reality*, McGraw-Hill, New York, 1950, Chap. 8.

[4] Dirac, P. A. M., *Quantum Mechanics*, 3rd Edit., Clarendon Press, Oxford, 1947.

[5] von Neumann, J., *Mathematische Grundlagen der Quantenmechanik*, Springer, Berlin, 1932. English translation, Princeton Univ. Press, 1955.

[6] Born, M., and Jordan, P., *On quantum mechanics*, Zeits. f. Physik, No. 34, 1925, pp. 858–888. Born, M., Heisenberg, W., and Jordan, P., *On quantum mechanics, Part II*, Zeits. f. Physik. No. 35, 1926, pp. 557–615. (The quoted sentence occurs in the latter article, page 558.)

1-2

The Unreasonable Effectiveness of Mathematics*

R. W. Hamming

PROLOGUE

It is evident from the title that this is a philosophical discussion. I shall not apologize for the philosophy, though I am well aware that most scientists, engineers, and mathematicians have little regard for it; instead, I shall give this short prologue to justify the approach.

Man, so far as we know, has always wondered about himself, the world around him, and what life is all about. We have many myths from the past that tell how and why God, or the gods, made man and the universe. These I shall call *theological explanations*. They have one principal characteristic in common—there is little point in asking why things are the way they are, since we are given mainly a description of the creation as the gods chose to do it.

Philosophy started when man began to wonder about the world outside of this theological framework. An early example is the description by the philosophers that the world is made of earth, fire, water, and air. No doubt they were told at the time that the gods made things that way and to stop worrying about it.

*This article is reproduced with permission from *American Mathematical Monthly* 87 (1980): 81–90.

From these early attempts to explain things slowly came philosophy as well as our present science. Not that science explains "why" things are as they are—gravitation does not explain why things fall—but science gives so many details of "how" that we have the feeling we understand "why." Let us be clear about this point; it is by the sea of interrelated details that science seems to say "why" the universe is as it is.

Our main tool for carrying out the long chains of tight reasoning required by science is mathematics. Indeed, mathematics might be defined as being the mental tool designed for this purpose. Many people through the ages have asked the question I am effectively asking in the title, "Why is mathematics so unreasonably effective?" In asking this we are merely looking more at the logical side and less at the material side of what the universe is and how it works.

Mathematicians working in the foundations of mathematics are concerned mainly with the self consistency and limitations of the system. They seem not to concern themselves with why the world apparently admits of a logical explanation. In a sense I am in the position of the early Greek philosophers who wondered about the material side, and my answers on the logical side are probably not much better than their answers were in their time. But we must begin somewhere and sometime to explain the phenomenon that the world seems to be organized in a logical pattern that parallels much of mathematics, that mathematics is the language of science and engineering.

Once I had organized the main outline, I had then to consider how best to communicate my ideas and opinions to others. Experience shows that I am not always successful in this matter. It finally occurred to me that the following preliminary remarks would help.

In some respects this discussion is highly theoretical. I have to mention, at least slightly, various theories of the general activity called mathematics, as well as touch on selected parts of it. Furthermore, there are various theories of applications. Thus, to some extent, this leads to a theory of theories. What may surprise you is that I shall take the experimentalist's approach in discussing things. Never mind what the theories are supposed to be, or what you think they should be, or even what the experts in the field assert they are; let us take the scientific attitude and look at what they are. I am well aware that much of what I say, especially about the nature of mathematics, will annoy many mathematicians. My experimental approach is quite foreign to their mentality and preconceived beliefs. So be it!

The inspiration for this article came from the similarly entitled article, "The Unreasonable Effectiveness of Mathematics in the Natural Sciences" [1], by E. P. Wigner. It will be noticed by those who have already read it that I have left out part of the title, and that I do not duplicate much of his material (I do not feel I can improve on his presentation). On the other hand, I shall spend relatively more time trying to explain the implied question of the title. But when all my explanations are over, the residue is still so large as to leave the question essentially unanswered.

THE EFFECTIVENESS OF MATHEMATICS

In his paper, Wigner gives a large number of examples of the effectiveness of mathematics in physical sciences. Let me, therefore, draw on my own experiences that are closer to engineering. My first real experience in the use of mathematics to predict things in the real world was in connection with the design of atomic bombs during the Second World War. How was it that the numbers we so patiently computed on the primitive relay computers agreed so well with what happened on the first test shot at Almagordo? There were, and could be, no small-scale experiments to check the computations directly. Later experience with guided missiles showed me that this was not an isolated phenomenon—constantly what we predict from the manipulation of mathematical symbols is realized in the real world. Naturally, working as I did for the Bell System, I did many telephone computations and other mathematical work on such varied things as traveling wave tubes, the equalization of television lines, the stability of complex communication systems, the blocking of calls through a telephone central office, to name but a few. For glamour, I can cite transistor research, space flight, and computer design, but almost all of science and engineering has used extensive mathematical manipulations with remarkable successes.

Many of you know the story of Maxwell's equations, how to some extent for reasons of symmetry he put in a certain term, and in time the radio waves that the theory predicted were found by Hertz. Many other examples of successfully predicting unknown physical effects from a mathematical formulation are well known and need not be repeated here.

The fundamental role of *invariance* is stressed by Wigner. It is basic to much of mathematics as well as to science. It was the lack of invariance of Newton's equations (the need for an absolute frame of reference for velocities) that drove Lorentz, Fitzgerald, Poincaré, and Einstein to the special theory of relativity.

Wigner also observes that *the same mathematical concepts* turn up in entirely unexpected connections. For example, the trigonometric functions which occur in Ptolemy's astronomy turn out to be the functions which are invariant with respect to translation (time invariance). They are also the appropriate functions for linear systems. The enormous usefulness of the same pieces of mathematics in widely different situations has no rational explanation (as yet).

Furthermore, the *simplicity* of mathematics has long been held to be the key to applications in physics. Einstein is the most famous exponent of this belief. But even in mathematics itself the simplicity is remarkable, at least to me; the simplest algebraic equations, linear and quadratic, correspond to the simplest geometric entities, straight lines, circles, and conics. This makes analytic geometry possible in a practical way. How can it be that simple mathematics, being after all a product of the human mind, can be so remarkably useful in so many widely different situations?

Because of these successes of mathematics, there is at present a strong trend toward making each of the sciences mathematical. It is usually regarded as a goal to be achieved, if not today, then tomorrow. For this audience I will stick to physics and astronomy for further examples.

Pythagoras is the first man to be recorded who clearly stated that "Mathematics is the easy way to understand the universe." He said it both loudly and clearly, "Number is the measure of all things."

Kepler is another famous example of this attitude. He passionately believed that God's handiwork could be understood only through mathematics. After twenty years of tedious computations, he found his three laws of planetary motion—three comparatively simple mathematical expressions that described the apparently complex motions of the planets.

It was Galileo who said, "The laws of Nature are written in the language of mathematics." Newton used the results of both Kepler and Galileo to deduce the famous Newtonian laws of motion, which together with the law of gravitation are perhaps the most famous example of the unreasonable effectiveness of mathematics in science. They not only predicted where the known planets would be but successfully predicted the positions of unknown planets, the motions of distant stars, tides, and so forth.

Science is composed of laws which were originally based on a small, carefully selected set of observations, often not very accurately measured originally; but the laws have later been found to apply over much wider ranges of observations and much more accurately than the original data justified. Not always, to be sure, but often enough to require explanation.

During my thirty years of practicing mathematics in industry, I often worried about the predictions I made. From the mathematics that I did in my office I confidently (at least to others) predicted some future events—if you do so and so, you will see such and such—and it usually turned out that I was right. How could the phenomena know what I had predicted (based on human-made mathematics) so that it could support my predictions? It is ridiculous to think that is the way things go. No, it is that mathematics provides, somehow, a reliable model for much of what happens in the universe. And since I am able to do only comparatively simple mathematics, how can it be that simple mathematics suffices to predict so much?

I could go on citing more examples illustrating the unreasonable effectiveness of mathematics, but it would only be boring. Indeed, I suspect that many of you know examples that I do not. Let me, therefore, assume that you grant me a very long list of successes, many of them as spectacular as the prediction of a new planet, of a new physical phenomenon, of a new artifact. With limited time, I want to spend it attempting to do what I think Wigner evaded—to give at least some partial answers to the implied question of the title.

WHAT IS MATHEMATICS?

Having looked at the effectiveness of mathematics, we need to look at the question, *"What is Mathematics?"* This is the title of a famous book by Courant and Robbins [2]. In it they do not attempt to give a formal definition, rather they are content to show what mathematics is by giving many examples. Similarly, I shall not give a comprehensive definition. But I will come closer than they did to discussing certain salient features of mathematics as I see them.

Perhaps the best way to approach the question of what mathematics is, is to start at the beginning. In the far distant, prehistoric past, where we must look for the beginnings of mathematics, there were already four major faces of mathematics. First, there was the ability to carry on the *long chains of close reasoning* that to this day characterize much of mathematics. Second, there was *geometry*, leading through the concept of continuity to topology and beyond. Third, there was *number,* leading to arithmetic, algebra, and beyond. Finally there was *artistic taste,* which plays so large a role in modern mathematics. There are, of course, many different kinds of beauty in mathematics. In number theory it seems to be mainly the beauty of the almost infinite detail; in abstract algebra the beauty is mainly in the generality. Various areas of mathematics thus have various standards of aesthetics.

The earliest history of mathematics must, of course, be all speculation, since there is not now, nor does there ever seem likely to be, any actual, convincing evidence. It seems, however, that in the very foundations of primitive life there was built in, for survival purposes if for nothing else, an understanding of cause and effect. Once this trait is built up beyond a single observation to a sequence of "If this, then that, and then it follows still further that . . . ," we are on the path of the first feature of mathematics I mentioned, long chains of close reasoning. But it is hard for me to see how simple Darwinian survival of the fittest would select for the ability to do the long chains that mathematicians and science seem to require.

Geometry seems to have arisen from the problems of decorating the human body for various purposes, such as religious rites, social affairs, and attracting the opposite sex, as well as from the problems of decorating the surfaces of walls, pots, utensils, and clothing. This also implies the fourth aspect I mentioned, aesthetic taste, and this is one of the deep foundations of mathematics. Most textbooks repeat the Greeks and say that geometry arose from the needs of the Egyptians to survey the land after each flooding by the Nile River, but I attribute much more to aesthetics than do most historians of mathematics and correspondingly less to immediate utility.

The third aspect of mathematics, numbers, arose from counting. So basic are numbers that a famous mathematician once said, "God made the integers, man did the rest" [3]. The integers seem to us to be so fundamental that we expect to find them wherever we find intelligent life in the universe. I have tried, with little success, to get some of my friends to understand my amazement that the abstraction of integers for counting is both possible and useful. Is it not remarkable that 6 sheep plus 7 sheep make 13 sheep; that 6 stones plus 7 stones make 13 stones? Is it not a miracle that the universe is so constructed that such a simple abstraction as a number is possible? To me this is one of the strongest examples of the unreasonable effectiveness of mathematics. Indeed, I find it both strange and unexplainable.

In development of numbers, we next come to the fact that these counting numbers, the integers, were used successfully in measuring how many times a standard length can be used to exhaust the desired length that is being measured. But it must have soon happened, comparatively speaking, that a whole number of units did not exactly fit the length being measured, and the measurers were driven to the fractions—the extra piece that was left over was used to measure the standard length. Fractions are not counting numbers, they are measuring numbers. Because of their common use in measuring, the fractions were, by a suitable extension of ideas, soon found to obey the same rules for manipulations as did the integers, with the added benefit that they made division possible in all cases (I have not yet come to the number zero). Some acquaintance with the fractions soon reveals that between any two fractions you can put as many more as you please and that in some sense they are homogeneously dense everywhere. But when we extend the concept of number to include the fractions, we have to give up the idea of the next number.

This bridge brings us again to Pythagoras, who is reputed to be the first man to prove that the diagonal of a square and the side of a square have no common measure—that they are irrationally related. The observation apparently produced upheaval in Greek mathematics. Up to that time the discrete number system and the continuous geometry flourished side by side with little conflict. The crisis of incommensurability tripped off the Euclidean approach to mathematics. It is a curious fact that the early Greeks attempted to make mathematics rigorous by replacing the uncertainties of numbers by what they felt was the more certain geometry (due to Eudoxus). It was a major event to Euclid, and as a result you find in *The Elements* [4] a lot of what we now consider number theory and algebra cast in the form of geometry. Opposed to the early Greeks, who doubted the existence of the real number system, we have decided that there should be a number that measures the length of the diagonal of a unit square (though we need not do so), and that is more or less how we extend the rational number system to include the algebraic numbers. It was the simple desire to measure lengths that did it. How can anyone deny that there is a number to measure the length of any straight line segment?

The algebraic numbers, which are roots of polynomials with integer, fractional, and, as was later proved, even algebraic numbers as coefficients, were soon under control by simply extending the same operations that were used on the simpler system of numbers.

However, the measurement of the circumference of a circle with respect to its diameter soon forced us to consider the ratio called pi. This is not an algebraic number, since no linear combination of the powers of pi with integer coefficients will exactly vanish. One length, the circumference, being a curved line, and the other length, the diameter, being a straight line, make the existence of the ratio less certain than is the ratio of the diagonal of a square to its side; but since it seems that there ought to be such a number, the trancendental numbers gradually got into the number system. Thus by a further suitable extension of the earlier ideas of numbers, the transcendental numbers were admitted consistently into the number system, though few students are at all comfortable with the technical apparatus we conventionally use to show the consistency.

Further tinkering with the number system brought both the number zero and the negative numbers. This time the extension required that we abandon the division for the single number zero. This seems to round out the real number system for us (as long as we confine ourselves to the processes of taking limits of sequences of numbers and do not admit still further operations)—not that we have to this day a firm, logical simple, foundation for them; but they say that familiarity breeds contempt, and we are all more or less familiar with the real number system. Very few of us in our saner moments believe that the particular postulates that some logicians have dreamed up create the numbers—no, most of us believe that the real numbers are simply there and that it has been an interesting, amusing, and important game to try to find a nice set of postulates to account for them. But let us not confuse ourselves—Zeno's paradoxes are still, even after 2,000 years, too fresh in our minds to delude ourselves that we understand all that we wish we did about the relationship between the discrete number system and the continuous line we want to model. We know, from nonstandard analysis if from no other place, that logicians can make postulates that put still further entities on the real line, but so far few of us have wanted to go down that path. It is only fair to mention that there are some mathematicians who doubt the existence of the conventional real number system. A few computer theoreticians admit the existence of only "the computable numbers."

The next step in the discussion is the complex number system. As I read history, it was Cardan who was the first to understand them in any real sense. In his *The Great Art or Rules of Algebra* [5] he says, "Putting aside the mental tortures involved mulitply $5 + \sqrt{-15}$ by $5 - \sqrt{-15}$ making $25 - (-15) \ldots$." Thus he clearly recognized that the same formal operations on the symbols for complex numbers would give meaningful results. In this way the real number system was gradually extended to the complex number system, except that this time the extension required giving up the property of ordering the numbers—the complex numbers cannot be ordered in the usual sense.

Cauchy was apparently led to the theory of complex variables by the problem of integrating real functions along the real line. He found that by bending the path of integration into the complex plane he could solve real integegration problems.

A few years ago I had the pleasure of teaching a course in complex variables. As always happens when I become involved in the topic, I again came away with the feeling that "God made the universe out of complex numbers." Clearly, they play a central role in quantum mechanics. They are a natural tool in many other areas of application, such as electric circuits, fields, and so on.

To summarize, from simple counting using the God-given integers, we made various extensions of the ideas of numbers to include more things. Sometimes the extensions were made for what amounted to aesthetic reasons, and often we gave up some property of the earlier number system. Thus we came to a number system that is unreasonably effective even in mathematics itself; witness the way we have solved many number theory problems of the original highly discrete counting system by using a complex variable.

From the above we see that one of the main strands of mathematics is the extension, the generalization, the abstraction—they are all more or less the same thing—of well-known concepts to new situations. But note that in the very process the definitions themselves are subtly altered. Therefore, what is not so widely recognized, old proofs of theorems may become false proofs. The old proofs no longer cover the newly defined things. The miracle is that almost always the same theorems are still true; it is merely a matter of fixing up the proofs. The classic example of this fixing up is Euclid's *The Elements* [4]. We have found it necessary to add quite a few new postulates (or axioms, if you wish, since we no longer care to distinguish between them) in order to meet current standards of proof. Yet how does it happen that no theorem in all the thirteen books is now false? Not one theorem has been found to be false, though often the proofs given by Euclid seem to be false. And this phenomenon is not confined to the past. It is claimed that an ex-editor of *Mathematical Reviews* once said that over half of the new theorems published these days are essentially true though the published proofs are false. How can this be if mathematics is the rigorous deduction of theorems from assumed postulates and earlier results? Well, it is obvious to anyone who is not blinded by authority that mathematics is not what the elementary teachers said it was. It is clearly something else.

What is this "else"? Once you start to look you find that if you were confined to the axioms and postulates then you could deduce very little. The first major step is to introduce new concepts derived from the assumptions, concepts such as triangles. The search for proper concepts and definitions is one of the main features of doing great mathematics.

While on the topic of proofs, classical geometry begins with the theorem and tries to find a proof. Apparently it was only in the 1850's or so that it was clearly recognized that the opposite approach is also valid (it must have been occcasionally used before then). Often it is the proof that generates the theorem. We see what we can prove and then examine the proof to see what we have proved! These are often called "proof generated theorems" [6]. A classic example is the concept of uniform convergence. Cauchy had proved that a convergent series of terms, each of which is continuous, converges to a continuous function. At the same time there were known to be Fourier series of continuous functions that converged to a discontinuous limit. By a careful examination of Cauchy's proof, the error was found and fixed up by changing the hypothesis of the theorem to read, "a uniformly convergent series."

More recently, we have had an intense study of what is called the foundations of mathematics—which in my opinion should be regarded as the top battlements of mathematics and not the foundations. It is an interesting field, but the main results of mathematics are impervious to what is found there— we simply will not abandon much of mathematics no matter how illogical it is made to appear by research in the foundations.

I hope that I have shown that mathematics is not the thing it is often assumed to be, that mathematics is constantly changing and hence even if I did succeed in defining it today the definition would not be appropriate tomorrow. Similarly with the idea of rigor—we have a changing standard. The dominant attitude in science is that we are not the center of the universe, that we are not uniquely placed, etc., and similarly it is difficult for me to believe that we have now reached the ultimate of rigor. Thus we cannot be sure of the current proofs of our theorems. Indeed it seems to me:

> The Postulates of Mathematics Were Not
> on the Stone Tablets that Moses Brought
> Down from Mt. Sinai.

It is necessary to emphasize this. We begin with a vague concept in our minds, then we create various sets of postulates, and gradually we settle down to one particular set. In the rigorous postulational approach the original concept is now replaced by what the postulates define. This makes further evolution of the concept rather difficult and as a result tends to slow down the evolution of mathematics. It is not that the postulation approach is wrong, only that its arbitrariness should be clearly recognized, and we should be prepared to change postulates when the need becomes apparent.

Mathematics has been made by man and therefore is apt to be altered rather continuously by him. Perhaps the original sources of mathematics were forced on us, but as in the example I have used we see that in the development of so simple a concept as number we have made choices for the extensions that were only partly controlled by necessity and often, it seems to me, more by aesthetics. We have tried to make mathematics a consistent, beautiful thing, and by so doing we have had an amazing number of successful applications to the real world.

The idea that theorems follow from the postulates does not correspond to simple observation. If the Pythagorean theorem were found to not follow from the postulates, we would again search for a way to alter the postulates until it was true. Euclid's postulates came from the Pythagorean theorem, not the other way. For over thirty years I have been making the remark that if you came into my office and showed me a proof that Cauchy's theorem was false I would be very interested, but I believe that in the final analysis we would alter the assumptions until the theorem was true. Thus there are many results in mathematics that are independent of the assumptions and the proof.

How do we decide in a "crisis" what parts of mathematics to keep and what parts to abandon? Usefulness is one main criterion, but often it is usefulness in creating more mathematics rather than in the application to the real world! So much for my discussion of mathematics.

SOME PARTIAL EXPLANATIONS

I will arrange my explanations of the unreasonable effectiveness of mathematics under four headings.

1. *We see what we look for.* No one is surprised if after putting on blue tinted glasses the world appears bluish. I propose to show some examples of how much this is true in current science. To do this I am again going to violate a lot of widely, passionately held beliefs. But hear me out.

I picked the example of scientists in the earlier part for a very good reason. Pythagoras is to my mind the first great physicist. It was he who found that we live in what the mathematicians call L_2—The sum of the squares of the two sides of a right triangle gives the square of the hypotenuse. As I said before, this is not a result of the postulates of geometry—this is one of the results that shaped the postulates.

Let us next consider Galileo. Not too long ago I was trying to put myself in Galileo's shoes, as it were, so that I might feel how he came to discover the law of falling bodies. I try to do this kind of thing so that I can learn to think like the masters did—I deliberately try to think as they might have done.

Well, Galileo was a well-educated man and a master of scholastic arguments. He well knew how to argue the number of angels on the head of a pin, how to argue both sides of any question. He was trained in these arts far better than any of us these days. I picture him sitting one day with a light and a heavy ball, one in each hand, tossing them gently. He says, hefting them, "It is obvious to anyone that heavy objects fall faster than light ones—and anyway, Aristotle says so." "But suppose," he says to himself, having that kind of a mind," that in falling the body broke into two pieces. Of course the two pieces would immediately slow down to their appropriate speeds. But suppose further that one piece happened to touch the other one. Would they now be one piece and both speed up? Suppose I tied the two pieces together. How tightly must I do it to make them one piece? A light string? A rope? Glue? When are two pieces one?

The more he thought about it—and the more you think about it—the more unreasonable becomes the question of when two bodies are one. There is simply no reasonable answer to the question of how a body knows how heavy it is—if it is one piece, or two, or many. Since falling bodies do something, the only possible thing is that they all fall at the same speed—unless interfered with by other forces. There is nothing else they can do. He may have later made some experiments, but I strongly suspect that something like what I imagined actually happened. I later found a similar story in a book by Pólya [7]. Galileo found his law not by experimenting but by simple, plain thinking, by scholastic reasoning.

I know that the textbooks often present the falling body law as an experimental observation; I am claiming that it is a logical law, a consequence of how we tend to think.

Newton, as you read in books, deduced the inverse square law from Kepler's laws, though they often present it the other way; from the inverse square law the textbooks deduce Kepler's laws. But if you believe in anything like the conservation of energy and think that we live in a three-dimensional Euclidean space, then how else could a symmetric central-force field fall off? Measurements of the exponent by doing experiments are to a great extent attempts to find out if we live in a Euclidean space, and not a test of the inverse square law at all.

But if you do not like these two examples, let me turn to the most highly touted law of recent times, the uncertainty principle. It happens that recently I became involved in writing a book on *Digital Filters* [8] when I knew very little about the topic. As a result I early asked the question, "Why should I do all the analysis in terms of Fourier integrals? Why are they the natural tools for the problem?" I soon found out, as many of you already know, that the eigenfunctions of translation are the complex exponentials. If you want the time invariance, and certainly physicists and engineers do (so that an experiment done today or tomorrow will give the same results), then you are led to these functions. Similarly, if you believe in linearity then they are again the eigenfunctions. In quantum mechanics the quantum states are absolutely additive; they are not just a convenient linear approximation. Thus the trigonometric functions are the eigenfunctions one needs in both digital filter theory and quantum mechanics, to name but two places.

Now when you use these eigenfunctions you are naturally led to representing various functions, first as a countable number and then as a non-countable number of them—namely, the Fourier series and the Fourier integral. Well, it is a theorem of Fourier integrals that the variability of the function multiplied by the variability of its transform exceeds a fixed constant, in one notation $1/2\pi$. This says to me that in any linear, time invariant system you must find an uncertainty principle. The size of Planck's constant is a matter of the detailed identification of the variables with integrals, but the inequality must occur.

As another example of what has often been thought to be a physical discovery but which turns out to have been put in there by ourselves, I turn to the well-known fact that the distribution of physical constraints is not uniform; rather the probability of a random physical constant having a leading digit of 1, 2, or 3 is approximately 60%, and of course the leading digits of 5, 6, 7, 8, and 9 occur in total only aobut 40% of the time. This distribution applies to many types of numbers, including the distribution of the coefficients of a power series having only one singularity on the circle of convergence. A close examination of the phenomenon shows that it is mainly an artifact of the way we use numbers.

Having given four widely different examples of nontrivial situations where it turns out that the original phenomenon arises from the mathematical tools we use and not from the real world, I am ready to strongly suggest that a lot of what we see comes from the glasses we put on. Of course this goes against much of what you have been taught, but consider the arguments carefully. You can say that it was the experiment that forced the model on us, but I suggest that the more you think about the four examples the more uncomfortable you are apt to become. They are not the arbitrary theories that I have selected, but ones which are central to physics.

In recent years it was Einstein who most loudly proclaimed the simplicity of the laws of physics, who used mathematics so extensively as to be popularly known as a mathematician. When examining his special theory of relativity paper [9] one has the feeling that one is dealing with a scholastic philosopher's approach. He knew in advance what the theory should look like, and he explored the theories with mathematical tools, not actual experiments. He was so confident of the rightness of the relativity theories that, when experiments were done to check them, he was not much interested in the outcomes, saying that they had to come out that way or else the experiments were wrong. And many people believe that the two relativity theories rest more on philosophical grounds than on actual experiments.

Thus my first answer to the implied question about the unreasonable effectiveness of mathematics is that we approach the situations with an intellectual apparatus so that we can only find what we do in many cases. It is both that simple, and that awful. What we were taught about the basis of science being experiments in the real world is only partially true. Eddington went further than this; he claimed that a sufficiently wise mind could deduce all of physics. I am only suggesting that a surprising amount can be so deduced. Eddington gave a lovely parable to illustrate this point. He said, "Some men went fishing in the sea with a net, and upon examining what they caught they concluded that there was a minimum size to the fish in the sea."

2. *We select the kind of mathematics to use.* Mathematics does not always work. When we found that scalars did not work for forces, we invented a new mathematics, vectors. And going further we have invented tensors. In a book I have recently written [10] conventional integers are used for labels, and real numbers are used for probabilities; but otherwise all the arithmetic and algebra that occurs in the book, and there is a lot of both, has the rule that

$$1 + 1 = 0 .$$

Thus my second explanation is that we select the mathematics to fit the situation, and it is simply not true that the same mathematics works every place.

3. *Science in fact answers comparatively few problems.* We have the illusion that science has answers to most of our questions, but this is not so. From the earliest of times man must have pondered over what Truth, Beauty, and Justice are. But so far as I can see science has contributed nothing to the answers, nor does it seem to me that science will do much in the near future. So long as we use a mathematics in which the whole is the sum of the parts we are not likely to have mathematics as a major tool in examining these famous three questions.

Indeed, to generalize, almost all of our experiences in this world do not fall under the domain of science or mathematics. Furthermore, we know (at least we think we do) that from Godel's theorem there are definite limits to what pure logical manipulation of symbols can do, there are limits to the domain of mathematics. It has been an act of faith on the part of the scientists that the world can be explained in the simple terms that mathematics handles. When you consider how much science has not answered then you see that our successes are not so impressive as they might otherwise appear.

4. *The evolution of man provided the model.* I have already touched on the matter of the evolution of man. I remarked that in the earliest forms of life there must have been the seeds of our current ability to create and follow long chains of close reasoning. Some people [11] have further claimed that Darwinian evolution would naturally select for survival those competing forms of life which had the best models of reality in their minds—"best" meaning best for surviving and propagating. There is no doubt that there is some truth in this. We find, for example, that we can cope with thinking about the world when it is of comparable size to ourselves and our raw un-aided senses, but that when we go to the very small or the very large then our thinking has great trouble. We seem not to be able to think appropriately about the extremes beyond normal size.

Just as there are odors that dogs can smell and we cannot, as well as sounds that dogs can hear and we cannot, so too there are wavelengths of light we cannot see and flavors we cannot taste. Why, then, given our brains wired the way they are, does the remark, "Perhaps there are thoughts we cannot think," surprise you? Evolution, so far, may possibly have blocked us from being able to think in some directions; there could be unthinkable thoughts.

If you recall that modern science is only about 400 years old, and that there have been from 3 to 5 generations per century, then there have been at most 20 generations since Newton and Galileo. If you pick 4,000 years for the age of science, generally, then you get an upper bound of 200 generations. Considering the effects of evolution we are looking for via selection of small chance variations, it does not seem to me that evolution can explain more than a small part of the unreasonable effectiveness of mathematics.

CONCLUSION

From all of this I am forced to conclude both that mathematics is unreasonably effective and that all of the explanations I have given when added together simply are not enough to explain what I set out to account for. I think that we—meaning you, mainly—must continue to try to explain why the logical side of science—meaning mathematics, mainly—is the proper tool for exploring the universe as we perceive it at present. I suspect that my explanations are hardly as good as those of the early Greeks, who said for the material side of the question that the nature of the universe is earth, fire, water, and air. The logical side of the nature of the universe requires further exploration.

REFERENCES

1. E. P. Wigner, "The unreasonable effectiveness of mathematics in the natural sciences,"*Communications on Pure and Applied Mathematics,* 13 (1960).
2. R. Courant and H. Robbins, *What is Mathematics?* (England: Oxford University Press, 1941).
3. L. Kronecker, Item 1634, in *On Mathematics and Mathematicians,* by R. E. Moritz (New York: Dover Publications, 1958).
4. Euclid, *Euclid's Elements,* T. L. Heath (New York: Dover Publications, 1956).
5. G. Cardano, *The Great Art or Rules of Algebra,* T. R. Witmer, trans. (Cambridge, Mass.: MIT Press, 1968), pp. 219–220.
6. Imre Lakatos, *Proofs and Refutations* (Cambridge, England: Cambridge University Press, 1976), p. 33.
7. G. Pólya, *Mathematical Methods in Science* (Mathematical Association of America, 1963), pp. 83–85.
8. R. W. Hamming, *Digital Filters,* (Englewood Cliffs, N.J.: Prentice-Hall, 1977).
9. G. Holton, *Thematic Origins of Scientific Thought, Kepler to Einstein* (Cambridge, Mass.: Harvard University Press, 1973).
10. R. W. Hamming, *Coding and Information Theory* (Englewood Cliffs, N.J.: Prentice-Hall, 1980).

2

Gauge Invariance in Nature: A Simple View

Sylvester J. Gates, Jr.

Massachusetts Institute of Technology

ABSTRACT: An introductory discussion of the concept of gauge invariance is presented so that nonspecialists can gain an understanding of its importance in elementary-particle and high-energy physics. The discussion is completely self-contained and assumes the reader has some knowledge of calculus, differential equations, introductory electromagnetism, and quantum mechanics. The concept of gauge invariance is illustrated first in the realm of classical electromagnetism. The action functional is introduced and used to show the relation of gauge invariance to charge conservation. The conservation of charge is also discussed from the point of view of a single electron. Next gauge invariance is illuminated within the realm of non-relativistic quantum mechanics. The illustration proceeds, after an introductory discussion about the philosophy of quantum mechanics, to show how gauge transformations enter in a more robust manner, affecting both the gauge fields and their sources. The concepts of action functionals, minimal coupling, and covariant derivatives are presented. Gauge invariance achieved through the use of integrals of the gauge fields is shown within some physical contexts. Section 2-4 begins with a general discussion of the theory of special relativity. The necessary concepts of four-vector, Lorentz invariance, co- and contravariant vector components are developed and utilized to show the relativistic covariance of classical electromagnetism and the nonrelativistic nature of Newton's second law and Schrödinger's equation. The Klein-Gordon and Dirac equations are presented as relativistic generalizations of Schrödinger's equation. An introductory description of the strong, weak, electromagnetic, and gravitational forces begins Section 2-5. The fundamental building blocks of matter, quarks and leptons, are introduced. The "chemical" properties of the fundamental blocks, usually referred to as color and flavor, are discussed. The structure of hadronic matter, baryons and mesons, is explained in terms of the quark model. Lepton-quark symmetry and the notion of family structure are discussed.

In the final section the actual construction of Yang-Mills gauge theories is explained. The gauging of color and flavor are shown to lead to quantum chromodynamics (QCD) and quantum flavor dynamics (QFD), respectively. The proposition of infrared slavery is explained in relation to the screening of color and the nonobservability of quarks. Next the spontaneous breaking of Yang-Mills theories is discussed. The Glashow-Salam-Weinberg (G-S-W) unified theory of the electromagnetic and weak interactions is discussed. Some modifications to Maxwell's equations implied by the G-S-W theory are noted. Finally, grand unified theory models (GUT models) are discussed in relation to the possibility of proton decay.

2-1   INTRODUCTION

In the quest for the resolution of a problem, sometimes an important clue is noticed that acts as a guide toward finding the solution. Any reader who has attempted to solve an intricate technical problem within the confines of any activity has probably had this experience. After noticing the clue, one initially only possesses an intuition, a "feeling" that one is proceeding in the right direction to solve the problem. The actual resolution may require a lot of hard work, including trying out dead ends. But that first big clue acts as a Rosetta Stone pointing toward an answer.

In particle physics, certain principles play the roles of important clues. For example, when data became available in the 1930s on the nuclear recoil produced during β-decay (see Section 2-5), it appeared by calculating the energies of the recoiling nucleus and the ejected electron that there was a violation of the principle of conservation of energy. In order to explain this apparent violation, Pauli proposed the existence of a new chargeless and massless particle, the neutrino. The violation was explained by assuming that the missing energy was simply being "carried" away by the neutrino that was not being detected.

Sometimes the principles themselves actually turn out to be wrong. There is the famous example of the principle of conservation of parity, which is essentially a way of saying that nature, at the level of fundamental particles, does not make a distinction between right- and left-handedness. (See Section 2-4 for the definition of handedness of a particle.) But nature does make such a distinction!

However, elementary particle physicists today seem to be in possession of another important clue: gauge invariance. With the present understanding it seems very likely that *all* of the fundamental forces in nature arise as an expression of gauge invariance. The author hopes that the reader will gain an understanding of this concept and come to appreciate its importance as the single most dominant theme in theoretical high-energy physics today.

In this chapter we will not discuss the existence of gravity as a result of gauge invariance. Unfortunately space forbids such a discussion. But it should be noted that gauge invariance also appears to play a role in gravitation. This relation can be shown in a completely mathematically rigorous manner.

2-2 CLASSICAL ELECTRODYNAMICS AND GAUGE INVARIANCE

One of the most important discoveries of nineteenth century physics was the fact that electricity and magnetism are not unrelated phenomena. Prior to this, electricity dealt with glass rods and cat fur, batteries, currents, and lightning; magnetism with magnets, compasses, iron filings, and the North Pole. The realization that electricity and magnetism are different aspects of one force was the consequence of the work of many people. But the person responsible for synthesizing the myth and lore into a concise, mathematically consistent theory was James Clerk Maxwell. In honor of his work, we now refer to the equations that describe the behavior of electromagnetic phenomena as Maxwell's equations. These remarkably simple equations underlie most of our modern technology, which relies heavily on our ability to manipulate electromagnetic phenomena efficiently.

Maxwell's equations are

$$\nabla \cdot \mathbf{D} = 4\pi\rho, \qquad \text{(Gauss's law)} \qquad \textbf{(2-1)}$$

$$\nabla \times \mathbf{H} - \frac{1}{c}\frac{\partial}{\partial t}\mathbf{D} = \frac{4\pi}{c}\mathbf{J}, \qquad \text{(Ampere-Maxwell law)} \qquad \textbf{(2-2)}$$

$$\nabla \cdot \mathbf{B} = 0, \qquad \textbf{(2-3)}$$

$$\nabla \times \mathbf{E} + \frac{1}{c}\frac{\partial}{\partial t}\mathbf{B} = 0, \qquad \text{(Faraday's law)} \qquad \textbf{(2-4)}$$

where $\mathbf{D} = \epsilon\mathbf{E}$ and $\mathbf{H} = \mu^{-1}\mathbf{B}$. (I beg the indulgence of my readers who are more familiar with these equations written in systems other than Gaussian units.) The first two equations state how the electric displacement field, $\mathbf{D}$, and the magnetic field intensity, $\mathbf{H}$, are created in response to the presence of distributions of charge densities, ρ, and current densities, $\mathbf{J}$, in matter. The third equation implies that there is no such thing as a magnetic charge. We will return to this equation later. Finally, the last equation states that the presence of a time-varying magnetic field must be accompanied by the presence of a spatially varying electric field. Finally, we have the "constituency equations" $\mathbf{D} = \epsilon\mathbf{E}$ and $\mathbf{H} = \mu^{-1}\mathbf{B}$, which relate the strengths of the electric and magnetic fields in the presence of matter, characterized by the dielectric constant ϵ and magnetic permeability μ, to the fields that would be present in the absence of such matter.

In the case of *statics* (i.e., all quantities independent of time) it is well known that solving Maxwell's equations is easier if we introduce a scalar potential V, which we relate to $\mathbf{E}$ via the equation $\mathbf{E} = -\nabla V$ and a vector potential $\mathbf{A}$ related to $\mathbf{B}$ by $\mathbf{B} = \nabla \times \mathbf{A}$. A virtue of introducing the potentials V and $\mathbf{A}$ lies in the fact that they automatically imply that $\mathbf{E}$ and $\mathbf{B}$ satisfy (2-3) and (2-4). In the case of dynamics, we must modify the relations between the fields and the potentials.

In this case we can easily show that (2-3) and (2-4) are satisfied if

$$\mathbf{B} = \nabla \times \mathbf{A}, \tag{2-5}$$

$$\mathbf{E} = -\nabla V - \frac{1}{c}\frac{\partial}{\partial t}\mathbf{A}. \tag{2-6}$$

These equations reduce to the prior relations in the case of the time-independent vector potentials.

Having introduced the scalar and vector potentials, we can use them to rewrite (2-1) and (2-2) as

$$\nabla^2 V + \frac{1}{c}\frac{\partial}{\partial t}\nabla \cdot \mathbf{A} = -4\pi\rho, \tag{2-7}$$

$$\left(\nabla^2 - \frac{1}{c}\frac{\partial^2}{\partial t^2}\right)\mathbf{A} - \nabla\left(\nabla \cdot \mathbf{A} + \frac{1}{c}\frac{\partial}{\partial t}V\right) = -\frac{4\pi}{c}\mathbf{J}, \tag{2-8}$$

where we have set the dielectric constant and the magnetic permeability to one. We now have a more economical description of electromagnetism (V and $\mathbf{A}$ constitute four fields compared to the six contained in $\mathbf{E}$ and $\mathbf{B}$) but the price we pay is that (2-7) and (2-8) are more complicated than (2-1)–(2-4).

In giving a description of electromagnetism in terms of potentials, something even more remarkable has occurred. The electric and magnetic fields are physically measurable. We can perform gedanken experiments with test charges and magnets to determine their configurations. We need only recall the Lorentz force law,

$$\mathbf{F} = q\left[\mathbf{E} + \frac{\mathbf{v}}{c} \times \mathbf{B}\right].$$

For the test charge we can use Newton's second law,

$$\frac{d}{dt}\mathbf{p} = q\left[\mathbf{E} + \frac{1}{c}\mathbf{v} \times \mathbf{B}\right]. \tag{2-9}$$

This equation tells us that by measuring the motion of the test charge we obtain information about the electric and magnetic fields. Similarly, for the test magnet of dipole moment m we measure the force exerted on the magnet field using the equation $\mathbf{F} = (\mathbf{m} \cdot \nabla)\mathbf{B}$, which also follows from the Lorentz force law. Now, for measured $\mathbf{E}$ and $\mathbf{B}$ fields, can we find a unique V and $\mathbf{A}$? The answer is no. To see this, consider two sets of potentials $(V, \mathbf{A})$ and $(V', \mathbf{A}')$, where

$$V' = V - \frac{1}{c}\frac{\partial}{\partial t}\Lambda, \tag{2-10}$$

$$\mathbf{A}' = \mathbf{A} + \nabla\Lambda, \tag{2-11}$$

and Λ is an arbitrary function of space and time. If the first set of potentials satisfies (2-5) and (2-6) for the measured values of $\mathbf{E}$ and $\mathbf{B}$, so will the second set. The lesson learned is that if we insist on a description of electromagnetism in terms of potentials (V and $\mathbf{A}$) there is no classical experiment that can determine these potentials uniquely.

Equations (2-10) and (2-11) may also be interpreted as a change (or transformation) of variables from $(V, \mathbf{A})$ to $(V', \mathbf{A}')$. In fact, these equations are referred to as a "gauge" transformation. Since the physical fields of classical electromagnetism are left invariant by this change of variables, we say that it is a gauge-invariant theory.

The gauge invariance of electromagnetism is a useful property of classical theory. We can use the freedom represented by $\Lambda(\mathbf{x}, t)$ to simplify some problems by making a particular choice for this function. Making such a choice is called picking a gauge. To see this procedure explicitly, we can return to (2-7) and (2-8). It is easy to show that these equations take the same form after the gauge transformation as before it, proving, by the way, that the charge and current densities are also gauge invariant. Now, consider only the divergence of $\mathbf{A}$. Under (2-11) we find

$$\nabla \cdot \mathbf{A}' = \nabla \cdot \mathbf{A} - \nabla^2\Lambda.$$

If $\nabla \cdot \mathbf{A}$ is nonzero, then we can force $\nabla \cdot \mathbf{A}'$ to vanish by solving the Poisson equation $\nabla^2\Lambda = \nabla \cdot \mathbf{A}$ to determine Λ. But if we can force $\nabla \cdot \mathbf{A}' = 0$ by a gauge transformation and the physics is gauge invariant, then we must be able to choose $\nabla \cdot \mathbf{A} = 0$ from the beginning. Imposing the condition $\nabla \cdot \mathbf{A} = 0$ is called the Coulomb gauge condition. In the Coulomb gauge (2-7) and (2-8) become

$$\nabla^2 V = -4\pi\rho,$$

$$\left(\nabla^2 - \frac{1}{c}\frac{\partial}{\partial t^2}\right)\mathbf{A} - \nabla(\nabla \cdot \mathbf{A}) = -\frac{4\pi}{c}\mathbf{J}.$$

Another well-known gauge is the Lorentz gauge defined by

$$\nabla \cdot \mathbf{A} + \frac{1}{c}\frac{\partial}{\partial t}V = 0.$$

In this gauge, (2-7) and (2-8) take the following forms:

$$-\Box V \equiv \left(\nabla^2 - \frac{1}{c^2}\frac{\partial^2}{\partial t^2}\right)V = -4\pi\rho,$$

$$-\Box \mathbf{A} \equiv \left(\nabla^2 - \frac{1}{c^2}\frac{\partial^2}{\partial t^2}\right)\mathbf{A} = -\frac{4\pi}{c}\mathbf{J},$$

where we have used the symbol $\Box$ for the d'Alembertian.

Once the electromagnetic potentials have been introduced, we can also rewrite the expression for the Lorentz force law. Direct substitution of (2-5) and (2-6) into (2-9), after some algebra, yields

$$\frac{D}{Dt}\left(\mathbf{p} + \frac{q}{c}\mathbf{A}\right) = -q\nabla\left(V - \frac{1}{c}\cdot\mathbf{A}\right), \tag{2-12}$$

where

$$\frac{D}{Dt} \equiv \frac{\partial}{\partial t} + \mathbf{v}\cdot\nabla$$

is the convective derivative. When acting on $\mathbf{p}$ we see

$$\frac{D}{Dt}\mathbf{p} = \frac{\partial\mathbf{p}}{\partial t} = \frac{d\mathbf{p}}{dt},$$

since $\mathbf{p}$ is only a function of time. Equation (2-12) can be written in a simpler form if we introduce the canonical momentum $\mathbf{P}$ and a velocity-dependent potential energy U, which are defined by

$$\mathbf{P} \equiv \mathbf{p} + \frac{q}{c}\mathbf{A}, \quad U \equiv q\left(V - \frac{1}{c}\mathbf{v}\cdot\mathbf{A}\right). \tag{2-13}$$

Equation (2-13) implies (2-12) can be written as

$$\frac{D}{Dt}\mathbf{P} = -\nabla U, \tag{2-14}$$

which is precisely the form of Newton's second law for a particle of momentum $\mathbf{P}$ in the presence of a potential energy function U. It is clear that (2-9) is gauge invariant since $\mathbf{E}$ and $\mathbf{B}$ are gauge invariant. But it is interesting to see how gauge invariance manifests itself in (2-14). We can define a new momentum $\mathbf{P}'$ and a new potential energy function U' by substituting (2-10) and (2-11) into the definitions of (2-13). Under this substitution (2-14) becomes

$$\frac{D}{Dt}\mathbf{P}' = -\nabla U' + \frac{q}{c}\left[\frac{D}{Dt}, \nabla\right]\Lambda,$$

$$\left[\frac{D}{Dt}, \nabla\right] \equiv \frac{D}{Dt}\nabla - \nabla\frac{D}{Dt}, \tag{2-15}$$

where we have introduced the notation $[D/Dt, \nabla]$ to denote the commutator as defined in (2-15). However, upon reflection we realize that the commutator vanishes because the order in which we apply the convective derivative and the gradient operator is irrelevant. So (2-14), like (2-7) and (2-8), is also gauge invariant.

We would now like to show that there is a connection between gauge invariance and the conservation of electric charge. In order to show this connection we need to briefly review the Lagrangian description of a point particle moving in the presence of a potential energy function.

Let the coordinates of a point-particle of mass m be denoted by $\mathbf{b}(t)$. Let $W(\mathbf{b})$ represent the potential energy function. The action (the time integral of the Lagrangian) is

$$S(\mathbf{b}) = \int_{t_1}^{t_2} dt \left[\frac{1}{2}m |\dot{\mathbf{b}}|^2 - W(\dot{\mathbf{b}})\right], \qquad \dot{\mathbf{b}} = \frac{d}{dt}\mathbf{b}. \tag{2-16}$$

As is well known, by varying $\mathbf{b}$ by an arbitrary infinitesimal amount $(\delta\mathbf{b})$ that vanishes at the endpoints of the integration, we are led to Newton's second law by requiring that the variation of $S(\mathbf{b})$ vanish. Explicitly, we have $\delta S \equiv S(\mathbf{b} + \delta\mathbf{b}) - S(\mathbf{b})$ and

$$\delta S = \int_{t_1}^{t_2} dt \, [m\delta \dot{\mathbf{b}} \cdot \dot{\mathbf{b}} - \delta \mathbf{b} \cdot \nabla_b W],$$

$$= \int_{t_1}^{t_2} dt \left[-\delta \mathbf{b} \cdot \left[m \frac{d}{dt} \dot{\mathbf{b}} + \nabla_b W \right] + \frac{d}{dt} (m\delta \dot{\mathbf{b}} \cdot \dot{\mathbf{b}}) \right].$$

$$\delta S = 0 \rightarrow m \frac{d}{dt} \dot{\mathbf{b}} = -\nabla_b W. \qquad \text{(2-17)}$$

The concept of an action as introduced in (2-16) is a useful device. It permits us to write a simple, concise expression like (2-16), which contains as much information about the dynamics of our particle as the equation of motion in (2-17). We will be making much use of actions in later discussions, so it is appropriate to make two observations here.

First, note that the action for a specified function $\mathbf{b}(t)$ assigns a number, $S(\mathbf{b})$, to this function. Such an object is called a *functional*. This name is to remind us that $S(\mathbf{b})$ is very much like a function. After all, a function $f(x)$, is simply a rule that, for a specified number x_0, assigns another number, $f(x_0)$, to it.

Second, we observe that demanding that the variation of $S(\mathbf{b})$ vanish is analogous to requiring the first derivative of a function vanish. In other words, we are looking for extremal functions that minimize the action. That the particle obeys Newton's second law (2-17) is equivalent to saying the particle will move along the path described by the $\mathbf{b}(t)$ that minimizes the action. (In the following we will not explicitly write the endpoints of the integration for the actions.)

In a similar manner we can find an action for V and $\mathbf{A}$ which upon variation yields Maxwell's equations in the form of (2-7) and (2-8). If we consider first the case where $\rho = \mathbf{J} = 0$, then variation of the following expression yields (2-7) and (2-8):

$$S_{\text{EM}}(V, \mathbf{A}) \equiv \frac{1}{4\pi} \int dt \, d^3x \, \frac{1}{2} \left\{ \left| \nabla V + \frac{1}{c} \frac{\partial}{\partial t} \mathbf{A} \right|^2 - |\nabla \times \mathbf{A}|^2 \right\}. \qquad \text{(2-18)}$$

Explicitly, the variation yields

$$S_{\text{EM}} = \frac{1}{4\pi} \int dt \, d^3x \left\{ -\delta V \left[\nabla^2 V + \frac{1}{c} \frac{\partial}{\partial t} \nabla \cdot \mathbf{A} \right] \right.$$

$$\left. - \delta \mathbf{A} \cdot \left[\Box \, \mathbf{A} + \nabla \left(\nabla \cdot \mathbf{A} + \frac{1}{c} \frac{\partial}{\partial t} V \right) \right] \right]$$

$$+ \nabla \cdot \left[\delta V \left(\nabla V + \frac{1}{c} \frac{\partial}{\partial t} \mathbf{A} \right) + (\mathbf{A} \cdot \nabla) \mathbf{A} - \frac{1}{2} \nabla |\mathbf{A}|^2 \right]$$

$$+ \frac{\partial}{\partial t} \left[\delta \mathbf{A} \cdot \left(\nabla V + \frac{1}{c} \frac{\partial}{\partial t} \mathbf{A} \right) \right] \Big\}.$$

where the surface terms (terms proportional to a total derivative) may be dropped. Now we can ask, "What terms must be added to S_{EM} in order to produce the source terms in (2-7) and (2-8)?" To answer this question, we first define $S_{EM\text{-matter}}$ by

$$S_{EM\text{-matter}} = \int dt \, d^3x \left[-V\rho + \frac{1}{c} \mathbf{A} \cdot \mathbf{J} \right]. \tag{2-19}$$

By starting from $(S_{EM} + S_{EM\text{-matter}})$ and again performing the variations δV and $\delta \mathbf{A}$ we find (2-7) and (2-8) but now with the source terms.

We can now show that gauge invariance of the action actually implies the conservation of electric charge. It is easy to see that S_{EM} is gauge-invariant. Explicitly we find

$$S_{EM}(V, \mathbf{A}) = S_{EM}(V', \mathbf{A}').$$

Now we define $S_{total}(V, \mathbf{A})$ by the relation

$$S_{total}(V, \mathbf{A}) \equiv S_{EM} + S_{EM\text{-matter}}$$

and impose the condition that S_{total} is gauge invariant. Substituting (2-10) and (2-11) into (2-18) and (2-19) yields

$$S_{total}(V, \mathbf{A}) = S_{total}(V', \mathbf{A}) - \frac{1}{c} \int dt \, d^3x \left[\rho \frac{\partial \Lambda}{\partial t} + \mathbf{J} \cdot \nabla \Lambda \right]. \tag{2-20}$$

Thus it appears that it is *not* possible to have a gauge-invariant S_{total} that satisfies the same relation as S_{EM} in (2-19). But let us investigate the last term in (2-20) more closely. Integrating by parts implies

$$\int dt \, d^3x \left[\rho \frac{\partial \Lambda}{\partial t} + \mathbf{J} \cdot \nabla \Lambda \right] = \int dt \, d^3x \left[\frac{\partial}{\partial t}(\rho\Lambda) + \nabla \cdot (\Lambda \mathbf{J}) \right]$$

$$- \int dt \, d^3x \left[\Lambda \left(\frac{\partial \rho}{\partial t} + \nabla \cdot \mathbf{J} \right) \right].$$

The first integral on the right-hand side yields only surface terms, which can be ignored. But gauge invariance for S_{total} also requires that the second integral must vanish. Since Λ is arbitrary, we require

$$\frac{\partial \rho}{\partial t} + \nabla \cdot \mathbf{J} = 0. \tag{2-21}$$

This formula is simply conservation of electrical charge!

Let us return to our starting point, Equations (2-1)–(2-4). Maxwell's equations are assymmetrical with respect to the source terms. There is no magnetic charge density, $-\rho_m$, on the right-hand side of Equation (2-3). Similarly there is no magnetic current density, $\mathbf{J}_m$, on the right-hand side of Equation (2-4). How would such terms affect gauge invariance? The answer is somewhat surprising at first sight; we can introduce *more* gauge invariance. However, this fact should not be surprising. If $-\rho_m$ and $\mathbf{J}_m$ appear in a fashion analogous to ρ and $\mathbf{J}$, we can derive a conservation of magnetic charge, and as we have just learned, conserved charges are connected to gauge invariance. To see this extra gauge invariance explicitly, we introduce new "magnetic" scalar and vector potentials $(V_m, \mathbf{A}_m)$, define $\mathbf{E}$ and $\mathbf{B}$ by modified equations,

$$\mathbf{E} = -\nabla V - \frac{1}{c}\frac{\partial}{\partial t}\mathbf{A} + \nabla \times \mathbf{A}_m,$$

$$\mathbf{B} = \nabla \times \mathbf{A} + \nabla V_m + \frac{1}{c}\frac{\partial}{\partial t}\mathbf{A}_m,$$

and substitute these equations into (2-1)–(2-4) (modified by our hypothetical magnetic current and charge densities). When this substitution is done we again obtain (2-7) and (2-8), but we also find that (2-3) and (2-4) become

$$\nabla^2 V_m + \frac{1}{c}\frac{\partial}{\partial t}\nabla \cdot \mathbf{A}_m = -4\pi\rho_m,$$

$$\left(\nabla^2 - \frac{1}{c^2}\frac{\partial^2}{\partial t^2}\right)\mathbf{A}_m - \nabla\left(\nabla \cdot \mathbf{A}_m + \frac{1}{c}\frac{\partial}{\partial t}V_m\right) = -\frac{4\pi}{c}\mathbf{J}_m.$$

The extra gauge invariance we mentioned is simply the freedom to subject V_m and $\mathbf{A}_m$ to a transformation like that in Equations (2-10) and (2-11). The point of this theoretical exercise is simply that conserved charges are closely tied to gauge invariance.

Before ending this section, let us look at charge conservation from the point of view of the sources that give rise to the electromagnetic field. The sources in matter, which are described by ρ and $\mathbf{J}$, are principally electrons. (In the following sections we will use the word "matter" to describe the source of other gauge fields also.) How is it that the charge conservation equation gets satisfied? If we consider a single electron moving along a path described by $\mathbf{b}(t)$, the charge density and current densities are given by ($e_0 \equiv$ 1 esu $= 1.6 \times 10^{-19}$ Coulomb):

$$\rho(\mathbf{x}, t) = -e_0\, \delta^{(3)}(\mathbf{R}), \qquad \mathbf{R} = \mathbf{x} - \mathbf{b}(t), \qquad (2\text{-}22)$$

$$\mathbf{J}(\mathbf{x}, t) = -e_0 \left(\frac{d}{dt}\, \mathbf{b}\right) \delta^{(3)}(R), \qquad (2\text{-}23)$$

where $\delta^{(3)}(R)$ is a three-dimensional Dirac delta function. This expression is a mathematical way of saying the charge and current have the same location as the electron. To see if the charge is conserved, we must verify (2-21). We find, by use of the chain rule,

$$\frac{\partial \rho}{\partial t} = -e_0\, [\nabla_R\, \delta^{(3)}(\mathbf{R})] \cdot \frac{\partial \mathbf{R}}{\partial t}$$

$$= e_0 \left(\frac{d\mathbf{b}}{dt}\right) \cdot [\nabla_R\, \delta^{(3)}(\mathbf{R})],$$

$$\nabla \cdot \mathbf{J} = -e_0\, [\nabla_R\, \delta^{(3)}(\mathbf{R})] \cdot \left[\left(\frac{d\mathbf{b}}{dt} \cdot \nabla_R\right)\mathbf{R}\right]$$

$$= -e_0 \left(\frac{d\mathbf{b}}{dt}\right) \cdot [\nabla_R\, \delta^{(3)}(\mathbf{R})]. \qquad (2\text{-}24)$$

So conservation is valid because the charge and the current have the same location.

The reader, having seen the utility of actions, might wonder if an action exists for (2-12). We leave it as an exercise to show that (2-12) can be obtained by variation, with respect to $\mathbf{b}(t)$, of the expression

$$S_{\text{EM-matter}} = \int dt \left[\frac{1}{2}\, m \left|\frac{d\mathbf{b}}{dt}\right|^2 + \frac{q}{c}\left(\frac{d\mathbf{b}}{dt}\right) \cdot \mathbf{A}(\mathbf{b}, t) - qV(\mathbf{b}, t)\right]. \qquad (2\text{-}25)$$

Earlier, we saw in (2-19) how to obtain the charge and current densities by varying the respect to V and $\mathbf{A}$. Performing these variations on (2-25) produces the expressions in (2-22) and (2-23). Finally, we observe that under the gauge transformation of (2-10) and (2-11), the integrand in (2-25) will change by a total time derivative if $\mathbf{b}$ does *not* transform. [That $\mathbf{b}(t)$ should be invariant under a gauge transformation follows from the gauge invariance of ρ and $\mathbf{J}$ and the results of (2-22) and (2-23).]

The sum of (2-18) and (2-25), when suitably generalized to account for quantum mechanics and special relativity, forms the basis for one of the best studied and most successful mathematical descriptions of nature: quantum electrodynamics, or QED. One of the most remarkable features of the generalization is that (2-18) remains unchanged in form.

2-3 QUANTUM MECHANICS AND GAUGE INVARIANCE

At the beginning of the twentieth century, classical Newtonian mechanics was supplanted by a new, radically different view of how nature works. Previously the notion of a point-particle was a fundamental precept in the logical underpinnings of the subject. The heavenly bodies moved along paths that would be traveled by a point-particle with all of the planet's mass concentrated there (at least as a first approximation). The paths of cannonade projectiles were calculable by analyzing the motion of a point-particle moving under the influence of a uniform gravitational field. Even when a problem required that the extent of a body could not be ignored, such as calculating moments of inertia, the answers were found by adding up all of the contributions of the point-particles that made up the extended body.

When the realm of the atom was first probed, a startling realization was gleaned; objects such as electrons, which most of the time appears pointlike, could sometimes demonstrate wavelike behavior! This idea was the catalyst for the heretical new subject of quantum mechanics.

Before we begin the discussion of our particular concern, perhaps a general discussion of a more philosophical nature is in order. Physicists are sometimes asked by questioning people, "How can an electron be both a particle and a wave?" A typical answer to this question is, "You can perform experiments that show that the electron can manifest both particlelike and wavelike behavior." But the question itself reveals that the questioner possesses a particular philosophical view. Let us construct an analogy to attempt to illustrate this argument.

Imagine that you live in a land where for generation upon generation *all* cars built were black and *all* buses built were red. An inhabitant of such a land might say, "I'm going to ride in a black today." or "I saw a red carrying a load of tactical policemen." The fellow inhabitants would clearly understand what is meant by these statements. Now suppose that a particularly adventurous inhabitant traveled to another land where observation revealed that some cars are black and some are red and the same is true for buses. The traveler returns home and when asked about the vehicles seen in his travel says a strange thing: "A car can be red." The fellow inhabitants, owing to their experience, will understand the adventurer to say, "A car can be a bus." They wonder how this is possible and ask, "How can a vehicle be both a car and a bus?"

The point is that when questions are raised about the wave-particle duality of electrons, the questioner has implicitly assumed, owing to experience, that a particle and a wave must be two distinct concepts. Nature, however, is not bound by our assumptions. (We will return to the particle-wave question at the end of this section.)

The fundamental distinction between Newtonian mechanics and quantum mechanics is that the latter replaces the notion of a point-particle with a well-defined position, $\mathbf{b}(t)$, (and possibly other properties) by a density (wave) function, $\psi(\mathbf{x}, t)$. All of the measurable physical characteristics of a particle-wave are properties of the wavefunction. Specifically, the probability of finding a "particle" in a volume d^3x about a point with coordinate $\mathbf{x}$ at a time t is given by $P(\mathbf{x}, t, d^3x)$, where

$$P(\mathbf{x}, t, d^3x) = d^3x \, \bar{\psi}(\mathbf{x}, t)\psi(\mathbf{x}, t).$$

Notice that we have given up all hope of knowing exactly where the "particle" is located (at least for all "reasonable" functions). If we add up the probability of finding the particle at all possible locations, we must obtain a numerical value of one. This argument leads to the statement that

$$1 = \int d^3x \, \bar{\psi}\psi, \tag{2-26}$$

where again we have suppressed the limits of integration.

The dynamics of the wavefunction are determined by Schrödinger's equation, which is analogous to the Newtonian equation of energy conservation

$$E = \frac{1}{2m} |\mathbf{p}|^2 + W \tag{2-27}$$

and follows from Newton's second law in the presence of a conservative force $\mathbf{F} = -\nabla W$. We obtain Schrödinger's equation by making the replacements $E \rightarrow i\hbar \, \partial\psi/\partial t$, $p = -i\hbar \, \nabla \, \psi$, and $W \rightarrow W\psi$, where $\hbar$ is Planck's constant divided by 2π.

$$i\hbar \frac{\partial \psi}{\partial t} = -\frac{\hbar^2}{2m} \nabla^2\psi + W\psi. \tag{2-28}$$

In this construction physically measurable quantities, like energy and momentum, are identified with the average (expectation) value associated with some operator. Thus, for example, the average values of momentum $\langle \mathbf{p} \rangle$ and energy $\langle E \rangle$ are calculated by

$$\langle \mathbf{p} \rangle = -i\tfrac{1}{2}\hbar \int d^3x \, [\bar{\psi}\boldsymbol{\nabla}\psi - \psi\boldsymbol{\nabla}\bar{\psi}], \qquad (2\text{-}29)$$

$$\langle E \rangle = i\frac{1}{2}\hbar \int d^3x \left[\bar{\psi}\,\frac{\partial\psi}{\partial t} - \psi\,\frac{\partial\bar{\psi}}{\partial t} \right]. \qquad (2\text{-}30)$$

The average value of the momentum obeys Newton's second law:

$$\frac{d}{dt}\,\langle \mathbf{p} \rangle = -\langle \boldsymbol{\nabla}W \rangle .$$

This value is verified by differentiating (2-29) appropriately and substituting into this result equation (2-28) and its conjugate. (Keep in mind that we *always* ignore surface terms.) The average value of the total energy obeys (2-27) for the average values of the momentum and potential energy [just multiply (2-28) by $\bar{\psi}$ and integrate].

Finally, having learned in the previous section how useful actions are, we can attempt to construct an action for (2-28). It is not too difficult to convince oneself that

$$S^{QM}_{matter}(\psi, \bar{\psi}) = \int dt \, d^3x \left[i\frac{1}{2}\hbar c \left(\bar{\psi}\,\frac{1}{c}\frac{\partial\psi}{\partial t} - \psi\,\frac{1}{c}\frac{\partial\bar{\psi}}{\partial t} \right) \right.$$
$$\left. - \frac{\hbar^2}{2m}\,(\boldsymbol{\nabla}\bar{\psi}) \cdot (\boldsymbol{\nabla}\psi) - W\bar{\psi}\psi \right] \qquad (2\text{-}31)$$

does the job. [In calculating the variation of this expression to verify that it yields (2-28), ψ and $\bar{\psi}$ must be considered as independent variables. We introduced c for later convenience.] Notice that the values of the action and all physical observables are unchanged if we replace ψ by ψ', where

$$\psi' = e^{i\theta}\psi, \qquad (2\text{-}32)$$

and θ is a constant. This expression too may be regarded as a change of variable or a transformation, like (2-10) and (2-11). But unlike the gauge transformation, the parameter here, θ, is constant, not an arbitrary function. Transformations like (2-32) are called "global" transformations to distinguish them from gauge transformations.

Next we want to obtain a quantum-mechanical description of an electron in the presence of a nontrivial electromagnetic field. Our eventual goal is to obtain an action analogous to (2-24). Let us begin by observing that even our free quantum-mechanical electron has a conserved charge. If we multiply the probability density $\bar{\psi}\psi$ by the fundamental unit of electrical charge $-e_0$, we must obtain a charge density

$$\rho(\mathbf{x}, t) = -e_0\bar{\psi}\psi. \tag{2-33}$$

Differentiating ρ with respect to time and using the free ($W = 0$) Schrödinger equation, we find

$$\frac{\partial\rho}{\partial t} = -\nabla \cdot \mathbf{J}, \qquad \mathbf{j} = i\frac{e_0\hbar}{2m}(\bar{\psi}\nabla\psi - \psi\nabla\bar{\psi}). \tag{2-34}$$

So the quantum-mechanical electron has a conserved charge and an associated current. In obtaining (2-34) we assumed our electron was free. In the presence of some background electromagnetic field this assumption is clearly incorrect. But the charge and current densities obtained in (2-33) and (2-34) are the lowest-order terms, which should act as sources in Maxwell's equations if we neglect these background fields.

Now we require an expression that approaches (2-31) as we "turn off" the electron charge. But we also know that the variation of this unknown expression with respect to V and $\mathbf{A}$ must yield ρ and $\mathbf{J}$. Therefore, using the result of (2-19), we conclude that a term of the form

$$\int dt\, d^3x\left[e_0 V\bar{\psi}\psi + i\frac{e_0\hbar}{2mc}\mathbf{A}\cdot(\bar{\psi}\nabla\psi - \psi\nabla\bar{\psi})\right]$$

must be added to (2-31). Performing this addition leads to

$$S_{\text{EM-matter}}^{\text{QM}} = \int dt\, d^3x\left\{i\frac{1}{2}\hbar c\left[\bar{\psi}\left(\frac{1}{c}\frac{\partial}{\partial t} - i\frac{e_0}{\hbar c}V\right)\psi\right.\right.$$
$$\left.- \bar{\psi}\left(\frac{1}{c}\frac{\partial}{\partial t} + i\frac{e_0}{\hbar c}V\right)\psi\right] - \frac{\hbar^2}{2m}\left[\left(\nabla - i\frac{e_0}{\hbar c}\mathbf{A}\right)\bar{\psi}\right]$$
$$\left.\cdot\left[\left(\nabla + i\frac{e_0}{\hbar c}\mathbf{A}\right)\psi\right] - W\bar{\psi}\psi\right\} \tag{2-35}$$

plus one additional term that we will drop. Strictly speaking, we should use the Hamiltonian formalism to pass from the classical expression in (2-25) to its quantum-mechanical equivalent. Equation (2-35) can then be shown to be correct.

This expression is fairly complicated, so to write it in a more concise manner we can introduce some definitions. (After all, ∇ is a shorthand notation for a complicated operator.) We therefore introduce the notational devices $\mathfrak{D}_0$ and $\mathfrak{D}$, where these operators are defined by

$$\mathfrak{D}_0\psi = \left(\frac{1}{c}\frac{\partial}{\partial t} - i\frac{e_0}{hc}V\right)\psi,$$

$$\mathfrak{D}\psi = \left(\nabla + i\frac{e_0}{\hbar c}\mathbf{A}\right)\psi.$$

The definitions of these operators acting on $\bar{\psi}$ are obtained by taking the complex conjugates of these equations. Using these definitions, (2-35) can be written in the form

$$S_{\text{EM-matter}} = \int dt\, d^3x \left[i\frac{1}{2}\hbar c(\bar{\psi}\mathfrak{D}_0\psi - \psi\mathfrak{D}_0\bar{\psi})\right.$$
$$\left. - \frac{\hbar^2}{2m}(\mathscr{D}\bar{\psi})\cdot(\mathscr{D}\psi) - W\bar{\psi}\psi\right]. \tag{2-36}$$

which looks exactly like (2-31) with the replacements

$$\frac{1}{c}\frac{\partial}{\partial t} \to \mathfrak{D}_0 \quad \text{and} \quad \nabla \to \mathscr{D}.$$

Since this feature is so striking, we call it the "minimal" coupling of the wavefunction ψ to the electromagnetic potentials. The operators $\mathfrak{D}_0$ and $\mathfrak{D}$ also get a special name—*covariant derivatives*. Covariant derivatives have some interesting properties. We can use them to "reconstruct" the $\mathbf{E}$ and $\mathbf{B}$ fields. We see

$$(\mathscr{D} \times \mathscr{D})\psi = i\frac{e_0}{\hbar c}\mathbf{B}\psi, \tag{2-37}$$

$$(\mathscr{D}\mathfrak{D}_0 - \mathfrak{D}_0\mathscr{D})\psi = i\frac{e_0}{\hbar c}\mathbf{E}\psi. \tag{2-38}$$

Now we can ask questions about the gauge invariance of Equation (2-36). Given one set of covariant derivatives $(\mathfrak{D}_0, \mathscr{D})$ we can obtain another set $(\mathfrak{D}_0', \mathscr{D}')$ by performing a gauge transformation. Therefore, using (2-10) and (2-11) yields

$$\mathfrak{D}_0\psi = \mathfrak{D}_0'\psi - i\frac{e_0}{\hbar c}\left(\frac{1}{c}\frac{\partial\Lambda}{\partial t}\right)\psi, \tag{2-39}$$

$$\mathfrak{D}\psi = \mathfrak{D}'\psi - i\frac{e_0}{\hbar c}\,(\boldsymbol{\nabla}\Lambda)\,\psi,\qquad(2\text{-}40)$$

and apparently (2-36) drastically changes its form, which implies a loss of gauge invariance. But, as we take the change of variable form $(V, \mathbf{A})$ to $(V', \mathbf{A}')$, let us also change the wavefunction according to the rule

$$\psi = \exp\left[+i\frac{e_0}{\hbar c}\,\Lambda(\mathbf{x},\,t)\right]\psi'.\qquad(2\text{-}41)$$

Substitution of this result into (2-39) and (2-40) then yields

$$\mathfrak{D}_0\psi = \exp\left[+i\frac{e_0}{\hbar c}\,\Lambda\right](\mathfrak{D}_0'\psi'),\qquad(2\text{-}42)$$

$$\mathscr{D}\psi = \exp\left[+i\frac{e_0}{\hbar c}\,\Lambda\right](\mathscr{D}'\psi'),\qquad(2\text{-}43)$$

and as a consequence we find

$$S_{\text{EM-matter}}^{\text{QM}}\,(\psi,\,\bar{\psi},\,V,\,\mathbf{A}) = S_{\text{EM-matter}}^{\text{QM}}\,(\psi',\,\bar{\psi}',\,V',\,\mathbf{A}').$$

Gauge invariance is, indeed, present but only at the expense of requiring that the matter field ψ also is affected by a gauge transformation. Notice that the transformation in (2-41) is a *local* change in the phase of the wavefunction; the function Λ may assume different values at various points in time and space. This transformation is to be distinguished from the global transformation in (2-32). But it can be regarded as an extension of the global transformation. In fact, we can turn the argument around. Given the action in (2-31) with its global phase invariance under the transformation of (2-32), if we demand that the theory have a local invariance under (2-41) then we are forced to introduce the electromagnetic potential as gauge fields!

The fact that gauge transformations in the quantum-mechanical view affect both matter and gauge fields should not be cause for concern but rather regarded as a step toward unification. After all, if gauge invariance is a fundamental principle, it should have implications for all physical systems. For instance, when we first considered the measurable quantities $\langle\mathbf{p}\rangle$ and $\langle E\rangle$, we defined them using ordinary derivatives. Should this definition imply that these quantities depend on a choice of gauge? Not at all. In the presence of electromagnetic fields, (2-29) and (2-30) are also changed by the minimal substitution $\nabla \to \mathscr{D}$ and $\partial/\partial t \to c\mathfrak{D}_0$ which guarantees that $\langle\mathbf{p}\rangle$ and $\langle E\rangle$ are gauge invariant. In fact, whenever an ordinary derivative is used in the absence of electromagnetism, it is replaced using minimal substitution. This replacement implies gauge invariance.

Minimal substitution and covariant derivatives are one way that quantum-mechanical systems "conspire" to achieve gauge invariance. There is another way that gauge invariance can be maintained by such systems. We can imagine that some measurement in a quantum-mechanical system depends only on the difference of phase angles of a wavefunction measured at two different positions, say x_1 and x_2. If we define phase angle $\Theta(x, t) = -i\frac{1}{2} \ln [\psi/\bar{\psi}]$, clearly the difference $\Theta(x_2, t) - \Theta(x_1, t)$ is a gauge-dependent quantity, and it seems that there is no way to maintain gauge invariance. But note that the expression

$$\Theta(x_2, t) - \Theta(x_1, t) + \frac{e_0}{\hbar c} \int_{x_1}^{x_2} ds \cdot A(s, t) \qquad (2\text{-}44)$$

is gauge invariant. Thus on the basis of gauge invariance alone we can predict that any measurable quantity that depends on such a difference of phase angles *must* also depend on the vector potential as in (2-44). In fact, there are two well-known effects in physics that do depend on the expression in (2-44): the Bohm-Aharanov effect and the Josephson effect.

In the first of these, imagine a beam of electrons that is split into two identical beams that travel from point 1 to point 2 by different paths. (See Fig. 2.1.) In the interior of the region bounded by the two paths, there is a subregion (indicated by the circle and cross) in which there is a magnetic induction field pointed into the paper. But this field exists only in the subregion. Now, if we set (2-44) to zero, it can be used analogously to an expression for the difference of the phase angle for a light ray traveling from points 1 and 2 through a medium with a position-dependent index of refraction. The phase angle at point 2 reached by path I is equal to the phase angle at point 1 plus the contribution obtained by taking the integral along path I. A similar result is obtained by going along path II. If we

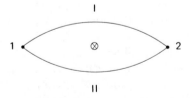

Figure 2-1
Two paths of an electron beam.

difference for the two paths, $\Delta\Theta$, we find

$$\Delta\Theta = \frac{e_0}{\hbar c}\left(\left.\int_{x_1}^{x_2} ds\cdot\mathbf{A}\right|_{\text{I}} - \left.\int_{x_1}^{x_2} ds\cdot\mathbf{A}\right|_{\text{II}}\right)$$

$$= \frac{e_0}{\hbar c}\oint ds\cdot\mathbf{A},$$

where we have used the minus sign in the second integral to change the direction in which it is taken. This last result can be written as $\Delta\Theta = ie_0\Phi_m/\hbar c$ by Stoke's theorem, if Φ_m is the magnetic flux through the region bounded by the electron beams. Thus we expect interference to depend on the size of the magnetic flux. (This argument is by no means rigorous, but it can be made so and yields the same result.) Notice that $\Delta\Theta$ is gauge invariant since performing a gauge transformation yields two further terms that exactly cancel one another.

The Josephson effect depends on the fact that the physics of a superconductor can be described by a single wavefunction, sometimes referred to as a macroscopic wavefunction. When a superconductor is shaped as in Figure 2-2 with a small piece of semiconductor inserted between positions x_1 and x_2, a Superconducting Quantum Interference Device, or SQUID, is formed. Once again a magnetic field can be set up so that there is a magnetic flux through the device. It can be shown that owing to tunneling in the semiconductor, it is possible to set up persistent currents that depend exactly on the factor in (2-44). There is no contraction with gauge invariance because under a gauge transformation the first two terms change by amounts that cancel the terms coming from the integral.

In a similar fashion, if the results of some quantum-mechanical experiment depends on the value of the difference of phase angles at the same position but at different times, say t_2 and t_1, then such a result must depend on the scalar potential also. So the result can only depend on the combination,

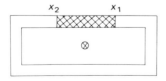

Figure 2-2
A rectangular squid.

$$\Theta(\mathbf{x}, t_2) - \Theta(\mathbf{x}, t_1) + \frac{e_0}{\hbar c} \int_{t_1}^{t_2} d\xi \, V(\mathbf{x}, \xi).$$

In closing this section, we hope the reader has been convinced that quantum mechanics implies a much deeper significance for gauge invariance than does Newtonian mechanics. Quantum mechanics itself may be regarded as a unification. No longer are we constrained to view the world as being composed of two very different quantities: point-particles and fields. Particles are replaced by "probability" fields, $\psi(\mathbf{x}, t)$, which are conceptually similar to the electromagnetic fields $V(\mathbf{x}, t)$ and $\mathbf{A}(\mathbf{x}, t)$. The seemingly localized properties associated with point-particles appear as constructive interference effects of the probability waves. At the macroscopic level, we can ignore the differences between a point-particle and the focus of constructive interference of the ψ-field. Thus we recover, as an approximation, Newtonian mechanics. At the atomic level this approximation breaks down. As the wavelength associated with electrons, given by the de Broglie formula $\lambda = h/|p|$, becomes comparable in size to the atom, wavelike effects become important, and we become aware of the ψ-field, which is affected by gauge transformations. A final implication of quantum mechanics is that wavelike phenomena such as electromagnetism may be capable of exhibiting particlelike behavior. This behavior does, indeed, occur in nature. Two very familiar cases are the Compton effect and the photoelectric effect. In these situations electromagnetic radiation behaves as if it is associated with a particle that is called the photon.

2-4 RELATIVISTIC INTERMEZZO

Before proceeding in our discussion of gauge invariance, we must digress into the other modern heresy in physics, relativity. (We concentrate on the theory of special relativity.) Special relativity has a notorious reputation among nonphysicists, much like quantum mechanics. So before we discuss some mathematics, it may be fruitful to discuss a stick and some shadows.

Consider a stick of fixed length, L_0. Also consider two light bulbs (s_1) and (s_2) arranged near the stick as in Figure 2-3. The stick extends down into the paper as well as above. The circle with a cross merely represents a cross section of the stick. Furthermore, imagine that the line AB is the floor of a room and also extends both above and below the paper. Finally, BC represents a wall that extends into the third dimension. Let us call the floor "space" and the wall "time." Since the light bulbs are very bright, the stick casts shadows in both "space" and "time." We require two rulers, one labeled in centimeters (or inches if preferred by the reader) and the other labeled in seconds. The first ruler will be used to measure the shadow in

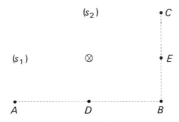

Figure 2-3
A space-time room.

"space" and the second, called a "clock," is used to measure the shadow in "time."

If the stick is parallel to both the floor and the wall, the first ruler laid parallel to the shadow in "space" gives a measurement of L_0 centimeters and the second ruler measuring the "temporal" shadow in the same way reads L_0 seconds. Now, if we rotate the stick so that it remains parallel with the wall but such that one end of the stick is closer to (s_2) than the other end, we find a different set of measurements. The second ruler still tells us that the stick "lasts" L_0 seconds. But the first ruler tells us that the length of the shadow in "space" is ($L_0 \cos \theta$) centimeters, where θ is the angle of rotation of the stick relative to its initial position. Finally, a third situation can occur, where the stick is kept parallel to floor but one end of the stick is closer to (s_1). This situation will yield a "length" measurement of L_0 centimeters and a "temporal" measurement of ($L_0 \cos \theta$) seconds. Obviously we can cause both the "spatial length" and the "temporal duration" to change by combining the latter two situations.

This analogy provides an insight into how the theory of special relativity forces physicists to think about nature. In the first three situations described we produced three sets of "space"-"time" measurements. We can now use some mathematics. Let $x^{\underline{m}}$ denote the ordered pair (x^0, x^1), where x^1 is the result of the spatial measurement and x^0 the result of the temporal measurement. For the first three situations described above [labeled (I), (II), and (III)] we have the results

$$x^{\underline{m}}_{\mathrm{I}} = (L_0 \text{ sec}, L_0 \text{ cm}),$$

$$x^{\underline{m}}_{\mathrm{II}} = (L_0 \text{ sec}, (L_0 \cos \theta) \text{ cm}),$$

$$x^{\underline{m}}_{\mathrm{III}} = ((L_0 \cos \theta) \text{ sec}, L_0 \text{ cm}).$$

Confronted with only these three sets of data *and nothing more,* a physicist might initially conclude that the sets of data refer to three different sticks.

But if told that these sets refer to the same object, the physicist would then conclude that for some reason the clock used in case III runs faster by a factor of (sec θ) *relative* to the clocks used in situations I and II. Similarly, the conclusion is reached that the ruler in situation II, for some reason, was stretched by a factor of (sec θ) *relative* to the other two rulers. (In nature, of course, things work out in the opposite way.) The point is that the measurements that we make with real clocks and rulers only give us the "shadows" of what is really occurring in nature.

Special relativity forces physicists to visualize the world as existing in an essentially four-dimensional way. The notions of space and time are but the "shadows" of what is known as spacetime. The physical measurements of time and position correspond to the "shadow" measurements in our analogy. Time and position do not have a separate existence apart from each other. They should be viewed as the parts of a four-vector. This concept can be stated mathematically. An event is any occurrence that has a specific time t and location $\mathbf{x}$. The speed of light, c, is a constant with the units of velocity. Therefore, the quantity $x^0 \equiv ct$ has the dimension of length. Similarly, the x, y, and z components of the position vector $\mathbf{x}$ have dimensions of length. We will find it more convenient to make the definitions $x^1 \equiv x$, $x^2 \equiv y$, and $x^3 \equiv z$. We now assemble all of these definitions into a four-vector denoted by $x^{\underline{m}}$. We have

$$x^{\underline{m}} = (x^0, x^1, x^2, x^3),$$

where $\underline{m}$ is simply an index that assumes the values 0, 1, 2, or 3 depending on which component of the four-vector we want to specify. The utility of four-vector is seen by recalling the utility of ordinary vectors for Newtonian mechanics. When we write Newton's second law in vector notation,

$$\frac{d}{dt}(m\dot{\mathbf{b}}) = \mathbf{F},$$

we are guaranteed that it describes the same physics if we choose one coordinate system or a second coordinate system that is rotated relative to the first. What is a rotation? Well, one way to think of a rotation is that it is a matrix $R(\boldsymbol{\theta})$ that is a function of three parameters $\boldsymbol{\theta}$. (These parameters θ^{12}, θ^{23}, and θ^{31} are angles that tell us whether the rotation is performed about the 3-axis, 1-axis, 2-axis, or some combination, and how large these rotations are.) Additionally, if the matrix $R(\boldsymbol{\theta})$ is multiplied by $\mathbf{x}$ we get a new vector $\mathbf{w} = R(\boldsymbol{\theta})\mathbf{x}$ such that

$$\mathbf{w} \cdot \mathbf{w} = \mathbf{x} \cdot \mathbf{x}. \qquad\qquad (2\text{-}45)$$

We want to use this type of argument in reverse for four-vectors. We begin by defining the analog of the "dot" product for four-vectors. Let $x^{\underline{m}}$ and $w^{\underline{m}}$ denote two arbitrary four-vectors. The "dot" product for these two quantities is *defined* by

$$x^{\underline{m}}\eta_{\underline{mn}}w^{\underline{n}} \equiv x^0 w^0 - x^1 w^1 - x^2 w^2 - x^3 w^3. \qquad (2\text{-}46)$$

In writing the left-hand side in (2-46) we have introduced a 4×4 matrix $\eta_{\underline{mn}}$ (the Minkowski metric) and made use of the Einstein summation convention of indices. [We sum over an index whenever it appears once up and once down. So we sum over both $\underline{m}$ and $\underline{n}$ in (2-46).] Next we can define the four-vector analog of a rotation of a 4×4 matrix $[M\,(\lambda^{\underline{ab}})]^{\underline{n}}_{\underline{m}}$ that is a function of six parameters, $\lambda^{\underline{ab}} = -\lambda^{\underline{ba}}$. (The fact that $\lambda^{\underline{ab}}$ is antisymmetric in its two four-vector indices implies that it contains six independent quantities. Three of these are equal to θ^{12}, θ^{23}, and θ^{31}. The other three, λ^{01}, λ^{02}, and λ^{03}, tell us whether the temporal axis is being rotated into the 1-, 2-, or 3-axis!) If $[M(\lambda)]^{\underline{n}}_{\underline{m}}$ is a four-vector rotation and if $w^{\underline{m}}$ is related to $x^{\underline{m}}$ by

$$w^{\underline{n}} = [M(\lambda)]^{\underline{n}}_{\underline{m}}\, x^{\underline{m}}, \qquad (2\text{-}47)$$

then the four-vectors must satisfy an equation analogous to (2-45).

$$w^{\underline{m}}\eta_{\underline{mn}}w^{\underline{n}} = x^{\underline{m}}\eta_{\underline{mn}}x^{\underline{n}}.$$

Now the validity of special relativity in physics simply reduces to the statement that all equations in physics must have the property that they can be written in terms of four-vector notation.

Before we start applying this idea, we can return to the six parameters, λ^{ab}, on which our four-dimensional rotation depends. We have already identified λ^{12}, λ^{13}, and λ^{31} as our ordinary three-space rotation angles. We also want to understand the physical significance of λ^{01}, λ^{02}, and λ^{03}. This explanation is done in the following way. Imagine two observers, one moving with uniform velocity $\mathbf{V}$ with respect to the other. We define $\lambda^{01} \equiv V^1/c$, $\lambda^{02} \equiv V^2/c$, and $\lambda^{03} \equiv V^3/c$. The two observers might be watching some events and recording where and when they take place. Afterward, the observers agree to meet and discuss their data. So, observer I forms a set of four-vectors $(x^{\underline{m}}_{\mathrm{I}}, y^{\underline{m}}_{\mathrm{II}} \cdot \cdot \cdot)$ and compares them against an analogous set $(x^{\underline{m}}_{\mathrm{II}}$ $y^{\underline{m}}_{\mathrm{II}} \cdot \cdot \cdot)$ formed from the data of observer II. For values of $|\mathbf{V}| \ll c$ the two sets of data will seem to agree. But as soon as $|\mathbf{V}| = O(c)$, the two sets begin to disagree, and this becomes worse as $|\mathbf{V}| \to c$. Both observers were watching the same events and know this fact. Therefore, one observer concludes that the clocks and rulers of the other were somehow "changed" while they were moving. (Remember our poor physicists in the analogy at the beginning of this section.) Here our two observers are simply victims of a four-vector rotation! The two sets of data are related to each other. Upon careful checking it can be shown that

$$x_{\mathrm{I}}^m = [M(\lambda)]_{\underline{n}}^{\underline{m}}\, x_{\mathrm{II}}^m, \quad y_{\mathrm{I}}^m = [M(\lambda)]_{\underline{n}}^{\underline{m}}\, y_{\mathrm{II}}^n \cdots ,$$

where $[M(\lambda)]_{\underline{n}}^{\underline{m}}$ is the four-vector rotation. For small relative velocities, $|\mathbf{V}|$, the rotation $M(\lambda)$ is very close to the identity matrix. The two sets of data are almost identical. But when $|\mathbf{V}|$ is comparable in size to c, the rotation of the temporal axis into spatial directions is noticeable. This rotation causes clocks to run more slowly and rulers to contract relative to a moving observer.

Now we return to the search for four-vector physics equations. Since x^m is a four-vector, so too must be the derivatives with respect to x^m. (This result is the same as for $\mathbf{x}$ and $\mathbf{\nabla}$.) We write the four-vector gradient as ∂_m, where

$$\partial_{\underline{m}} = \frac{\partial}{\partial x^m} \equiv \left(\frac{1}{c}\frac{\partial}{\partial t}, \mathbf{\nabla} \right).$$

It should now be clear why we introduced extraneous factors of c in previous sections. Equations (2-10) and (2-11) can be written as a single four-vector equation. Defining $A_m \equiv (V, -\mathbf{A})$, these equations can be written as

$$A'_{\underline{m}} = A_{\underline{m}} - \partial_{\underline{m}}\Lambda.$$

Here it is convenient to introduce another 4×4 matrix, $\eta^{\underline{mn}}$. This matrix is by definition the inverse of the matrix $\eta_{\underline{mn}}$ that appears in (2-46). In particular, this definition imples that $\eta^{\underline{mn}}\, \eta_{\underline{nr}} = \delta_{\underline{r}}^{\underline{m}}$. Using this new matrix we can "raise" the index on A_m to find that

$$A^{\underline{m}} \equiv \eta^{\underline{mn}}\, A_{\underline{n}} = (V, \mathbf{A}). \tag{2-48}$$

Clearly we have $\partial_m\, x^m = 4$. Four, of course, is a number—a scalar. A four-vector rotation should not affect a scalar. In (2-47) we saw how a four-vector like x^m can be rotated. From (2-47) it follows that

$$\frac{\partial}{\partial x^m} = \frac{\partial w^n}{\partial x^m}\frac{\partial}{\partial w^n} = [M(\lambda)]_{\underline{m}}^{\underline{n}}\frac{\partial}{\partial w^n}. \tag{2-49}$$

In order to solve for $\partial/\partial w^n$, the four-vector rotation above must be moved to other side of the equation. Recall that $M(\lambda)$ is like a rotation in the counterclockwise direction. To remove the unwanted factor multiplying $\partial/\partial w^n$, we can perform another rotation on (2-49) by the same amount $\lambda^{\underline{ab}}$, but in the clockwise direction. This rotation can be denoted by $[M(-\lambda)]_r^m$, and it will undo the rotation on the left-hand side of (2-49). It implies

$$\frac{\partial}{\partial w^{\underline{m}}} = [M(-\lambda)]_{\underline{m}}^{\underline{n}} \frac{\partial}{\partial x^{\underline{n}}}.$$ (2-50)

Comparing (2-50) to (2-47), we see that the four-gradient rotates in the opposite sense as the coordinates. All upper indices of four-vector quantities rotate as in (2-47) and all lower indices rotate as in (2-50). This fact is important. To see why, (2-50) can be multiplied from the right by $w^{\underline{m}}$ to show

$$\frac{\partial}{\partial w^{\underline{m}}} w^{\underline{m}} = [M(-\lambda)]_{\underline{m}}^{\underline{n}} \frac{\partial}{\partial x^{\underline{n}}} w^{\underline{m}}$$

$$= [M(-\lambda)]_{\underline{m}}^{\underline{n}} \frac{\partial}{\partial x^{\underline{n}}} [M(\lambda)]_{\underline{r}}^{\underline{n}} x^{\underline{r}}$$

$$= [M(\lambda)]_{\underline{r}}^{\underline{m}} [M(-\lambda)]_{\underline{m}}^{\underline{n}} \frac{\partial}{\partial x^{\underline{n}}} x^{\underline{r}}$$

$$= \delta_{\underline{r}}^{\underline{n}} \frac{\partial x^{\underline{r}}}{\partial x^{\underline{n}}} = 4.$$

So the number calculated from $\partial_m x^m$ is the same as the number obtained using rotated coordinates $(\partial/\partial w^{\underline{m}}) w^{\underline{m}}$. We call the upper indices, which rotate counterclockwise, contravariant indices, and the lower indices are called covariant indices.

Now we can consider the four-gradient acting on $A_{\underline{m}}$ or $A^{\underline{m}}$. For instance, we see

$$\partial_{\underline{m}} A^{\underline{m}} = \frac{1}{c} \frac{\partial}{\partial t} V + \nabla \cdot \mathbf{A}.$$

Setting this equation equal to zero, we obtain the Lorentz gauge condition, which imples that this gauge condition is consistent with special relativity. We can also compute the four-vector curl of $A_{\underline{m}}$. The symbol $F_{\underline{mn}}$ may used to denote this four-vector curl, which is defined by

$$F_{\underline{mn}} = \partial_{\underline{m}} A_{\underline{n}} - \partial_{\underline{n}} A_{\underline{m}}.$$ (2-51)

This quantity is actually a 4×4 antisymmetric matrix (or second-rank tensor.) Writing out all of its components yields

$$F_{\underline{mn}} = \begin{bmatrix} 0 & -E_1 & -E_2 & -E_3 \\ E_1 & 0 & B_3 & -B_2 \\ E_2 & -B_3 & 0 & B_1 \\ E_3 & B_2 & -B_1 & 0 \end{bmatrix}. \tag{2-52}$$

The entries of this matrix are the electric and magnetic fields. Because of this, $F_{\underline{mn}}$ is referred to as the field strength tensor. Using $F_{\underline{mn}}$ allows Maxwell's equations to be written in the concise forms

$$\eta^{\underline{mn}} \partial_{\underline{m}} F_{\underline{nr}} = \frac{4\pi}{c} J_{\underline{r}}, \qquad J_{\underline{r}} \equiv (c\rho, -\mathbf{J}), \tag{2-53}$$

$$\partial_{\underline{m}} F_{\underline{nr}} + \partial_{\underline{n}} F_{\underline{rm}} + \partial_{\underline{r}} F_{\underline{mn}} = 0. \tag{2-54}$$

Equations (2-1) and (2-2) are contained in (2-53), and Equations (2-3) and (2-4) are contained in (2-54). Finally, the action for the free electromagnetic field given in (2-18) also has a four-vector form,

$$S_{EM}(A_{\underline{m}}) = \frac{1}{4\pi c} \int d^4x \left[-\frac{1}{4} \eta^{\underline{mn}} \eta^{\underline{rs}} F_{\underline{mr}} F_{\underline{ns}} \right]. \tag{2-55}$$

In other words, classical electromagnetism as described by Maxwell's equations is perfectly consistent with four-vector notation and is relativistic (i.e., consistent with special relativity).

In the previous section, quantum mechanics was shown to be consistent with gauge invariance. It was helpful to introduce covariant derivatives that we now realize can be assembled into a four-vector $\mathcal{D}_{\underline{m}} = (\mathcal{D}_0, \mathcal{D})$ with the property

$$[\mathcal{D}_{\underline{m}}, \mathcal{D}_{\underline{n}}] = -i \frac{e_0}{\hbar c} F_{\underline{mn}}. \tag{2-56}$$

This result follows from (2-37), (2-38), and (2-52), but does not imply that the previous discussion of quantum mechanics was relativistic. Recall that Schrödinger's equation is an extension of Newton's second law. The time coordinate plays a special role there. From the point of view of special relativity a special role cannot be permitted. Time is simply one component of a four-vector. Reconsidering (2-31), we see that temporal derivatives and spatial derivatives play very different roles. Temporal derivatives appear linearly and spatial derivatives appear quadratically. Two solutions can be found after some consideration: Replace (2-31) by another equation that is quadratic in temporal derivative, such as

$$S_{KG} = \frac{\hbar^2}{2mc} \int d^4x \left[\eta^{mn} (\partial_m \bar{\psi})(\partial_n \psi) - \frac{2mW}{\hbar^2} \bar{\psi}\psi \right]. \quad \text{(2-57)}$$

or use another equation that is linear in spatial derivatives, such as

$$S_{\text{Dirac}} = \hbar \int d^4x \left[i\bar{\psi} (\gamma^m \partial_m \psi) - i(\partial_m \bar{\psi}) \gamma_m \psi - \frac{W}{\hbar c} \bar{\psi}\psi \right],$$

where γ^m is a four-vector. (We are not permitted to let $\gamma^m = x^m$ since this assumption implies that $x^m = 0$ is somehow special.) Both solutions are relativistic. By variation and choice of W the first solution yields the Klein-Gordon equation and the second solution yields the Dirac equation. Both can be coupled to the electromagnetic field by minimal substitution. But only the second solution with $W = 2 m_e c^2$ (where m_e is the mass of the electron) and some other mathematical tricks (see Appendix A),

$$S_{\text{Dirac}} = \hbar \int d^4x \left[i\bar{\psi} (\gamma^m \mathcal{D}_m \psi) - i (\mathcal{D}_m \bar{\psi}) \gamma^m \psi - \frac{2 m_e c}{\hbar} \bar{\psi}\psi \right], \quad \text{(2-58)}$$

describes an electron. Here we have entered the regime of relativistic quantum mechanics. Equation (2-58) is quite remarkable. A thorough analysis of it results in two predictions: (1) the electron has a property that makes it appear as though the electron is "spinning," and (2) there must be a particle in nature that has all of the properties of an electron but the opposite electrical charge. The prediction of spin contained in (2-58) can be regarded as a real triumph for the mathematical description of nature. We now know that most elementary particles are "spinning" about an axis. This spin is quantized, however. The angular momentum associated with the spin can be measured in units of $\hbar$. All particles have spins that are either one-half an odd integer or integer units (including zero) of $\hbar$. Particles like the electron, proton, and neutron have one-half a unit of spin. The photon, on the other hand, has one unit of spin. All half-odd integer spinning particles are called fermions. All integer spinning particles are called bosons. Fermins obey the Pauli exclusion principle, but bosons do not.

Having introduced the concept of spin, we are now able to define handedness for a particle. From Newtonian mechanics, we can associate a direction with a rotation by using a right-hand rule. This association gives a three-vector that we can call the spin vector. When the three-space direction of motion is in the same direction as the spin vector, we may say the particle is right-handed. Left-handedness then follows when the spin vector points 180° away from the direction of motion.

The second prediction has also been verified. There does occur in nature a particle that is exactly like the electron but with the opposite charge: the positron. The positron was the first discovery of the existence of antimatter. Antimatter is just like ordinary matter only with all charges reversed. So there also exist antiprotons, for instance. When matter and antimatter are brought together they annihilate one another. All of the mass is converted in energy by the famous equation $E = mc^2$.

To give a completely rigorous treatment of relativistic quantum systems demands that the possibility of particle creation and annihilation be considered. The main effect of this possibility is that ψ can no longer be interpreted as a wavefunction satisfying (2-26). Instead ψ becomes an operator that can be expressed as a linear combination of particle creation and annihilation operators that constitute a complete basis. This so-called second quantization procedure is not strictly a realtivistic phenomenon. It is necessary whenever creation and annihilation become relevant in physical systems.

2-5   CORPUSCULAR INTERMEZZO

When physicists began to probe the nucleus, they noted that some force must keep protons and neutrons bound together in the nucleus. This force must be stronger than the electromagnetic force. Although neutrons have no electric charge, protons do, and therefore some kind of force is required to overcome their electrical repulsion. This force is simply called the strong force. It must have a very short range. When protons at rest are at distances greater than 10^{-13} meters, their mutual electrical repulsion easily drives them apart. So the strong force is unlike the electromagnetic (or gravitational) force, which is long ranged, falling off as the squared reciprocal of the distance (i.e., $1/r^2$).

There is another force in nature, one that is responsible for the natural radioactivity of substances like radium and uranium. When an atom of these substances transmutates, one (or more) of the neutrons in the nucleus is suddenly changed into a proton, an electron (to conserve electrical charge), and a third particle, a neutrino, that is similar to an electron but has no mass and no electrical charge. Since it is massless like a photon, the neutrino also moves at the speed of light. When the transmutation occurs, the proton remains in the nucleus, and the electron and neutrino are ejected. In this way an atom of one material, say uranium, changes into an atom of another material such as thorium. The process just described is called β-decay. We know that the force responsible for it is different from the strong force because individual neutrons outside of the nucleus undergo β-decay with a mean life of 1013 seconds. The force that causes β-decay is called the weak force. It derives its name from the fact that it is much weaker than the electromagnetic force. This force, like the strong force, only acts over a very small distance (10^{-6} cm).

The other two known forces are electromagnetism and gravitation. Gravity is by far the weakest. If we work at a scale where the strong force has strength unity, then the respective sizes of the strong, electromagnetic, weak, and gravitational forces are given by the ratios $1:10^{-4}$, $1:10^{-14}$, and $1:10^{-40}$. Electromagnetism is described by Maxwell's equations and gravitation is described by the theory of general relativity.

In preparation for the discussion of how the strong and weak forces seem to arise from gauge invariance, we now describe the basic building blocks of nature as we presently understand them.

When the study of the strong force commenced, it became clear that the properties of this force were the same independent of whether the particles that it acted upon were protons or neutrons. One way to explain the independence is to assume that *with regard to this force* protons and neutrons are not different particles but simply components of the same particle, called the nucleon. The electrical-charge independence of the strong interaction can then be interpreted as saying that any equation that describes the strong force depends only on the nucleon and not on the proton and neutron separately. This idea has a mathematical formulation. The wavefunctions for the proton ψ_p, and neutron, ψ_n, can be "assembled" into a two-component vector ψ_N,

$$\mathbf{\Psi}_N = (\psi_p, \psi_n). \tag{2-59}$$

The quantity $\mathbf{\Psi}_N$ is the nucleon wavefunction, and if it is used to describe the strong force we automatically obtain charge independence. There is nothing mysterious about this use. It is the same as the use of vectors in Newtonian mechanics. Using vector notation ensures that physics is being described in a way that is independent of the coordinate system chosen. Similarly, using the nucleon wavefunction, which is a vector in an imaginary space we call isotopic spin space, guarantees that the strong force is being described in a way that is independent of the "direction" chosen (i.e., whether one considers the proton or the neutron). Equation (2-59) can also be regarded in another way. Under most circumstances the proton and neutron can be considered as very different objects. One has an electrical charge of 1 esu. The other is neutral. The neutron is even a bit more massive than the proton ($m_n = 1.0014\ m_p$). Yet an equation like (2-59) implies that these differences are inconsequential at some level, which again shows the unification, the underlying sameness, of quantities that at first sight seem vastly different. The collection of seemingly different particles into structures like the nucleon in (2-59) is a very important step. These collections are also referred to as *multiplets* and understanding what multiplets exist in nature often leads to insights into the structure of the basic forces in nature. These insights will be demonstrated amply in the next section.

In the present view, the proton and neutron are no longer regarded as being fundamental. (But the concepts that were introduced in the previous paragraph still survive in a modified form.) First, the proton and neutron have been found to be parts of an even larger multiplet of strongly interacting particles. This multiplet itself has been shown to be only one of a number of such multiplets. So that the number of hadrons, or strongly interacting particles, of which the proton and neutron are the most familiar examples, is now known to be over 100. It is hard to regard such a large number of objects as all being fundamental. In order to give some order to this morass of particles, M. Gell-Mann and Y. Ne'eman (and independently G. Zwieg) proposed the existence of quarks (or *aces* as called by Zwieg). These could be used to explain the number and of variety of hadrons in the same way that the tremendous number of molecules are explained in terms of much smaller numbers of atoms. The second reason for abandoning the concept of the hadron as fundamental is experimental (and therefore may be considered more concrete). It has been found that when very high energy electrons are scattered off protons, the results indicate that inside the proton are pointlike structures that interact with the electrons. These structures are presumably the quarks, whose existence explains the large number and characteristics of hadronic multiplets.

Not all elementary particles in nature "feel" the strong force. The most familiar example is the electron. The electron is not a strongly interacting particle, but it does feel the weak force. Recall the elctron, e^-, participates in β-decay. So does the neutrino ν_e (or more accurately its antiparticle $\bar{\nu}_e$), which must therefore also feel the weak force. Particles that interact through only the weak and electromagnetic forces (we are presently ignoring gravity!) are called leptons. In contrast, quarks (or hadrons) interact via the strong, electromagnetic, and weak forces.

We have now "found" the matter building blocks with which, apparently, it is possible to build a universe. We give a complete list of *one family* of these blocks in Figure 2-4. The quark members appear as both left-handed ($-$) and right-handed ($+$) types. The electron also appears in both left-

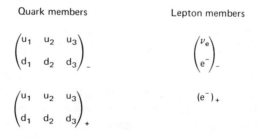

Figure 2-4
First family of "fundamental" particles.

handed and right-handed varieties. But the neutrino, associated with the electron, only exists in the left-handed variety. This is the distinction that nature seems to make with respect to parity, which was mentioned in the Introduction.

In fact, all neutrinos are believed to be only left-handed. The property that distinguishes the upper member of a column from the lower member is called *flavor*. So if the flavor of a neutrino is changed, a left-handed electron results. Similarly, changing the flavor of an up quark (i.e., u_1) gives a down quark (i.e., d_1). Quarks also have another property, called *color*. This is the property that distinguishes the different quarks that lie on the same row. Thus changing the color of the up-one quark (i.e., u_1) gives either an up-two quark (u_2) or an up-three quark (u_3).

Although the quark members possess color, this property has not been observed in any physical particle. It appears as though nature effectively "screens" color to prevent its appearance. A rough analogy of this type of screening occurs when a free positive ion is placed in a solution with free negative ions. The negative ions in the vicinity of the positive ion gather around it and screen it from the rest of the solution. Thus all the hadrons we have observed are believed to be collections of three quarks (baryons) or a quark and antiquark (mesons) assembled in such a way that the composite hadron has no net color. One other fascinating feature of quarks is that they may have fractional units of electrical charge. The up quark must have a charge of two-thirds of that of the proton, while the down quark must have a charge that is one-third that of the electron. Thus a proton is believed to consist of two up quarks and one down quark. In this way the proton has a net charge of one. Similarly, a neutron is assembled from two down quarks and one up quark, yielding a net charge of zero. By choosing one quark of each color (1, 2, 3), we can build a baryon such as the proton or neutron, which has no net color. Another way to construct an object with no net color is to take a quark and antiquark. An example of this method is provided by an up color-1 quark and a down color-1 antiquark. This composite has electric charge +1 and is identified with a particle known as the positively charged pion, which has been observed in many experiments. By forming three quark or quark antiquark composites, hundreds of hadrons can be built up. The quarks thus explain why hadrons are so numerous.

Nature does not seem to stop, however, with just one such family as illustrated in Figure 2-4. At present, there appear to be two more families with precisely the same structure as the first family. The second family is obtained by replacing the up and down quarks by a "strange" (s) and "charmed" quark (c), respectively, and simultaneously replacing the electron, e^-, and its neutrino, ν_e, by the muon, μ^-, and its neutrino ν_μ. The muon is a particle that has all the properties of an electron except that its mass is approximately 200 times larger and it usually decays with a half-life of 10^{-6} seconds into an electron and neutrinos. The third family is again obtained by replacement of the members of the first family in Figure 2-4. This time the up and down quarks are replaced by "truthful" and "beautiful" quarks, and the electron and its neutrino is replaced by the τ-lepton and its neutrino, ν_τ. The τ-lepton has only recently been discovered with a mass that is approximately 4000 times that of the electron. The associated neutrino, ν_τ, has not been seen, but from the structure of the previous families is expected to exist. Presently, there is no real understanding why three such families occur. In fact, if this replication continues it seems more likely that quarks and leptons can no longer be regarded as being fundamental, and perhaps there exist some other objects, "preons" and "haplons," that are fundamental. (This subject is already being discussed among theoretical physicists.)

One striking feature of the family structure, other than the asymmetry between right-handed and left-handed particles, is that the lepton members appear to "know" about the quark members. The electron and its neutrino mirror the second set. The τ-lepton and its neutrino mirror the third set. This symmetry between pairs of leptons and pairs of quarks (when ignoring color) is known as quark-lepton symmetry. One proposed explanation for this symmetry will lead to some surprising implications in the next section.

2-6 YANG-MILLS THEORY: EXACT AND BROKEN

If one is willing to accept the model of elementary particles discussed in the previous section, a very intriguing solution presents itself for the origin of the strong, weak, and electromagnetic forces. We can begin by considering only the quarks. Quarks are fermions, like the electron. Therefore, the quantum mechanics of quarks is determined by the Dirac action (2-58), which takes the form

$$
\begin{aligned}
S_{\text{quark}} = \tfrac{1}{2} \int d^4x \, \text{Tr} \, \{ & i\overline{Q}_-^1 \, (\gamma_{\underline{m}} \partial_{\underline{m}} Q_-^1) - i \, (\partial_{\underline{m}} \overline{Q}_-^1) \, \gamma_{\underline{m}} Q_-^1 \\
& + i\overline{Q}_+^1 (\gamma_{\underline{m}} \partial_{\underline{m}} Q_+^1) - i \, (\partial_{\underline{m}} \overline{Q}_+^1) \, \gamma_{\underline{m}} Q_+^1 \\
& - \overline{Q}_-^1 M_1 Q_-^1 - \overline{Q}_+^1 M_1 Q_+^1 \}.
\end{aligned} \tag{2-60}
$$

In writing (2-60), several conventions have been introduced. First $\hbar$ and c have been set equal to one for convenience. (The reader should be aware that this does not imply that we assume that these physical constants actually have value one. It is only a convenience that we will use throughout the subsequent discussion.) Next the quantities Q^1 and $\overline{Q}^1$ are literally the matrices of quarks introduced in Figure 2-4, where the entries in the matrices are quark wavefunction operators. The barred quantities $\overline{Q}{}^1_-$ and $\overline{Q}{}^1_+$ are obtained by taking the complex matrix transpose (see Appendix B) of Q^1_L and Q^1_R. Finally, since multiplication of matrices yields another matrix, the trace is taken (see also Appendix B) to produce a scalar. The quantity M_1 is a 2×2 "mass matrix" that depends only on the mass of the up and down quarks. Equation (2-60) only describes the free quark members of the first family! A similar expression must be written for the quarks of the second and third families.

Equation (2-60) is complicated. So complicated that we believe using it will obscure the essential simplicity of the concepts we are about to discuss. So instead we will go to a simpler setting to explain these notions. We can imagine a world where quarks are not distinguished by handedness, have no flavor, exist in two not three colors, and only one family. The basic quark multiplet for such a world takes the form $Q = (u_1, u_2)$. This equation, however, is exactly like the nucleon wavefunction in (2-59). So we can use Ψ_N but now with the understanding that its components are quark-fields. The Dirac action for Ψ_N is much simpler than (2-60). Explicitly, for a free nucleon we find

$$S_N = \tfrac{1}{2} \int d^4x \, [i\overline{\Psi}_N \cdot (\gamma^m \partial_{\underline{m}} \Psi_N) - i \, (\partial_{\underline{m}} \overline{\Psi}_N) \, \gamma^m \cdot \Psi_N],$$

$$\Psi_N = (\psi_p, \psi_n), \qquad \overline{\Psi}_N = (\overline{\psi}_p, \overline{\psi}_n), \tag{2-61}$$

where for simplicity the mass matrix has been set to zero. The action clearly possesses a global invariance under the transformation of variable: $\Psi_N = \exp(-i\theta) \, \Psi'_N$. We have already seen how making this into a gauge transformation leads to "electromagnetism."

Remember we are really talking about quark color charge. So the use of the words electromagnetism, photon, and so on should *not* be taken literally. We will use quotes to emphasize the fact that in this toy model we are *not* considering the actual electromagnetic force to which quarks and leptons are subject in the real world. We will return to a discussion of real electromagnetism only *after* we complete the introduction of Yang-Mills theories.

Equation (2-61) however, has more such global invariances! Before discussing these, the formulation of "electromagnetic" gauge invariance can be recast in a manner that will lend it to the simplest discussion of these additional invariances.

Under an *infinitesimal* gauge transformation, from (2-41) it follows that Ψ_N is related to Ψ'_N by $\Psi'_N = \Psi_N - ie_0 \Lambda \Psi_N$. The infinitesimal gauge variation of Ψ_N, denoted by $\delta_G \Psi_N$, is defined to be the difference between Ψ'_N and Ψ_N for infinitesimal Λ. Therefore we find

$$\delta_G \Psi_N \equiv \Psi'_N - \Psi_N = -ie_0 \Lambda \Psi_N. \tag{2-62}$$

Similarly, the infinitesimal gauge variation for A_m, denoted by $\delta_G A_m$, is defined as the difference between A'_m and A_m for infinitesimal Λ. Thus $\delta_G A_m$ is given by

$$\delta_G A_{\underline{m}} \equiv A'_{\underline{m}} - A_{\underline{m}} = -\partial_{\underline{m}} \Lambda. \tag{2-63}$$

The gauge invariance of all of the actions previously discussed can be shown to be equivalent to the statement that

$$S(A_{\underline{m}} + \delta_G A_{\underline{m}}, \psi + \delta_G \psi) = S(A_{\underline{m}}, \psi). \tag{2-64}$$

(We encourage the interested reader to verify this equation in any of the gauge invariant actions in Sections 2-2 or 2-3.) The actions have a global invariance when Λ is chosen to be a constant.

We can now demonstrate the existence of three additional global invariances of the action in (2-61) besides the one in (2-62), which is related to "electromagnetism." Equation (2-61) is left invariant under the variations

$$\delta\psi_p = -i\tfrac{1}{2}g\Lambda^1\psi_n, \qquad \delta\psi_n = -i\tfrac{1}{2}g\Lambda^1\psi_p, \tag{2-65}$$

$$\delta\psi_p = -\tfrac{1}{2}g\Lambda^2\psi_p, \qquad \delta\psi_n = \tfrac{1}{2}g\Lambda^2\psi_p, \tag{2-66}$$

$$\delta\psi_p = -i\tfrac{1}{2}g\Lambda^3\psi_p, \qquad \delta\psi_n = i\tfrac{1}{2}g\Lambda^3\psi_n, \tag{2-67}$$

where g, Λ^1, Λ^2, and Λ^3 are some constants. Since ψ_p and ψ_n are components of Ψ_N, Equations (2-65)–(2-67) can be more compactly written as

$$\delta\Psi_N = -i\tfrac{1}{2}g\,(\Lambda^i\sigma_i)\,\Psi_N, \tag{2-68}$$

where $(\Lambda^i\sigma_i)$ is a 2×2 matrix. Explicitly, this is given by

$$\Lambda^i\sigma_i = \Lambda^1\sigma_1 + \Lambda^2\sigma_2 + \Lambda^3\sigma_3 = \begin{pmatrix} \Lambda^3 & \Lambda^1 - i\Lambda^2 \\ \Lambda^1 + i\Lambda^2 & -\Lambda^3 \end{pmatrix}.$$

The matrices $(\sigma_1, \sigma_2, \sigma_3)$ usually denoted by $\boldsymbol{\sigma}$ are known as the Pauli spin matrices. These matrices arise in many contexts and have many interesting properties. Some of these properties are discussed in Appendix B. Among them is the fact that

$$[\sigma_i, \sigma_j] = \sigma_i\sigma_j - \sigma_j\sigma_i = i2\epsilon_{ij}^k\sigma_k, \tag{2-69}$$

where i and j can take any of the values 1, 2, or 3 and ϵ_{ijk} is the totally antisymmetric Levi-Civita tensor. We will make use of this property shortly.

In classical electromagnetism, it was found that a global invariance could be made into a gauge or local invariance by the introduction of a gauge field. Comparing the global transformation in (2-68) to the gauge transformation of (2-62), the only major differences are seen to be: (1) Λ is a function while Λ^1, Λ^2, and Λ^3 are constants; and (2) the 2×2 identity matrix, **1**, implicitly contained in (2-62) (since $\mathbf{1}\,\Psi_N = \Psi_N$) is replaced by the 2×2 Pauli spin matrices. Now we can demand that the action in (2-61) still possesses the invariance property of (2-64) even if the Λ^i are functions. This necessitates the introduction of more gauge fields, one for each parameter Λ^i ($i = 1, 2,$ or 3). We have already seen how to covariantize (2-61) with respect "electromagnetic" gauge invariance by making the minimal substitution $\partial_m \Psi_N \rightarrow (\partial_m - ie_0 A_m)\,\Psi_N$. So if we want to extend the covariant derivative to allow for gauge transformations of the form of (2-68), the simplest modification is to again extend the definition of the covariant derivative to

$$\mathcal{D}_{m\ N} = [\partial_m - ie_0 A_m \mathbf{1} - i\tfrac{1}{2}g_0\,(G^i_m \sigma_i)]\,\Psi_N, \qquad (2\text{-}70)$$

where the G^i_m ($i = 1, 2,$ or 3) are three new gauge fields. There are two questions that must now be answered. What are the gauge transformations, analogous to (2-63), for the new gauge fields? It is possible to write an action analogous (2-55) that describes the dynamics of these gauge fields in the absence of matter sources?

The answer to the first question is suggested by a careful investigation of Equations (2-41)–(2-43). The latter two equations written in relativistic form imply

$$\mathcal{D}'_m \psi' = \exp\,[-ie_0 \Lambda]\,\mathcal{D}_m \exp\,[ie_0 \Lambda]\,\psi' \qquad (2\text{-}71)$$

where (2-41) has been used on the right-hand side to express ψ in terms of ψ'. This equation must be valid independent of the configuration of ψ'. So it is actually a statement about the relation of $\mathcal{D}'_m$ to $\mathcal{D}_m$. For infinitesimal values of Λ, (2-71) implies

$$\delta_G \mathcal{D}_m = ie_0\,[\mathcal{D}_m, \Lambda], \qquad (2\text{-}72)$$

where we introduced the notation of the commutator in Section 2-3. For the "photon," (2-72) is just another way of rewriting (2-63). However, for the new gauge fields (2-72) contains some new information. This transformation of Ψ_N in (2-68) only differs from that in (2-62) by the replacement $e_0 \Lambda \mathbf{1} \rightarrow \tfrac{1}{2}\,g\,(\Lambda^i \sigma_i)$. Thus one is led to suspect that the same replacement in (2-72) will yield the gauge variations of the new fields G_m^i. On making this substitution, we have for *arbitrary* functions Λ^i

$$\delta_G \Psi_N = -i\tfrac{1}{2}g\,(\Lambda^i \sigma_i)\,\Psi_N,$$

$$\delta_G \mathfrak{D}_m = i\tfrac{1}{2}g\,\mathfrak{D}_m,\,\Lambda^i \sigma_i], \qquad (2\text{-}73)$$

as the variations of the matter and gauge fields, respectively. Explicitly for the gauge fields, by using (2-69) and (2-70), Equation (2-73) implies

$$\delta_G A_m = 0, \qquad (2\text{-}74)$$

$$\delta_G G_m^k = -\partial_m \Lambda^k + g\epsilon_{ij}^k G_m^i \Lambda^j. \qquad (2\text{-}75)$$

Thus, the "photon" does not transform under the Λ^i gauge transformation, which is reasonable in view of the fact that new gauge fields do not transform under the "electromagnetic" gauge invariance. But the new gauge fields do transform with respect to the new gauge transformations. However, comparing (2-75) to (2-72) shows that an additional term appears that does not appear for the photon. The origin of this term lies in (2-69). Since the order of multiplication of the Pauli spin matrices is important, their commutator is nonvanishing. This is not true of the multiplication of the 2×2 identity matrix; that is, $[\mathbf{1}, \mathbf{1}] = 0$. To distinguish the types of gauge theories in which the order of multiplication of matrices is important from those in which it is not, the former are called non-Abelian gauge theories and the latter are called Abelian gauge theories. Thus the "photon" is said to be an Abelian gauge field and can be distinguished from the G_m^i gauge fields, which are referred to as non-Abelian gauge fields. Utilizing (2-74) and (2-75), it is possible to prove that (2-61), with the substitution of (2-70), satisfies Equation (2-64). In other words, the minimally coupled "nucleon" action now has *four* gauge invariances.

The answer to the second question is also straightforward. The action for the free electromagnetic gauge field was given in (2-55) in terms of the field strength tensor F_{mn}, which in turn satisfied a relationship to the covariant derivative in (2-56). The same strategy can be pursued again. Using the covariant derivative given in (2-70), fields strength, F_{mn}, for the "photon" *and* field strengths, G_{mn}^i, for the non-Abelian gauge fields can be defined:

$$[\mathfrak{D}_m, \mathfrak{D}_n] = -ie_0 F_{mn} - i\tfrac{1}{2}g G_{mn}^i \sigma_i.$$

The quantity F_{mn} defined by this equation is exactly the same as in (2-51). However, G_{mn}^i takes the form

$$G_{\underline{mn}}^k = \partial_{\underline{m}} G_{\underline{n}}^k - \partial_{\underline{n}} G_{\underline{m}}^k + \epsilon_{ij}^k G_{\underline{m}}^k G_{\underline{n}}^j.$$

Once again the noncommutativity of the matrix multiplication of the Pauli spin matrices causes new terms to appear. Having computed the non-Abelian field strength, we are able to write the action for the free non-Abelian gauge field by analogy,

$$S_{\text{YM}} = \frac{1}{4\pi} \int d^4x \left[-\frac{1}{4} \eta^{\underline{mn}} \eta^{\underline{rs}} G_{\underline{mr}}^i G_{\underline{ns}}^j \delta_{ij} \right]. \qquad (2\text{-}76)$$

We have just completed the construction of non-Abelian gauge theories, which was discovered by Yang and Mills and independently by Shaw. The class of theories that can be constructed by the generalization of these arguments lies at the center of the current understanding of nature. We are now ready to "turn on" the strong, electromagnetic, and weak forces for real-world quarks.

Returning to (2-60), we note that the Q^1_- and Q^1_+ are 2×3 matrices. This fact implies that they can be multiplied from the left by 3×3 matrices and from the right by 2×2 matrices. In fact, (2-60) has some *global* invariance for a transformation that utilizes these matrix multiplications. With some calculations, we can show that (2-60) is invariant under the variation

$$\delta Q^1_- = [-ig_r \Lambda_r \mathbf{1} - i\tfrac{1}{2} g_w (\Lambda_w^i \sigma_i)] \, Q^1_- - i\tfrac{1}{2} g_c Q^1_- (\Lambda_c^\alpha \lambda_\alpha), \qquad (2\text{-}77)$$

where g_r, g_w, g_c, Λ_r, Λ_w^i, and Λ_c^α ($\alpha = 1, \dots, 8$) are constants. The quantities Λ_α, with α taking values from 1 to 8, are a set of 3×3 matrices that are traceless and Hermitian (see Appendix B). These matrices are a 3×3 generalization of the Pauli spin matrices. The variation of Q^1_+ is the same as in (2-77). By now the reader has probably guessed where (2-77) will lead. Making the global invariance of (2-77) into a gauge invariance requires the introduction of gauge fields! So by the same line of reasoning as in our toy model, we find a covariant derivative that permits (2-77) to become a gauge invariance. This covariant derivative takes the form (and is the same for all other Q fields)

$$\mathcal{D}_{\underline{m}} Q^1_- = [\partial_{\underline{m}} - ig_r Y_{\underline{m}} \mathbf{1} - i\tfrac{1}{2} g_w (W_{\underline{m}}^i \sigma_i)] \, Q^1_- - i\tfrac{1}{2} g_c Q^1_- (G_{\underline{m}}^\alpha \lambda_\alpha). \qquad (2\text{-}78)$$

Thus a total of 12 gauge bosons $G_{\underline{m}}^\alpha$, $W_{\underline{m}}^i$, and $Y_{\underline{m}}$ are introduced. (The astute reader may notice that the photon of Maxwell's equations has yet to appear. We will solve this mystery after the discussion of the strong force.)

Eight of the gauge fields, $G_{\underline{m}}^\alpha$, are associated with the 3×3 matrices λ_α. In general, if we start with a quark in a particular column in Q^1_- and multiply by one of the λ_α matrices, that quark will end in a different column (but same row). But since the column position indicates the color of the quark, this change in position implies that the color of the quark has been changed. Therefore the gauge fields G_m^α are associated with changes of quark color. As changes in the state of electrons (i.e., accelerations) produce electrodynamic effects, changes in the color states of quarks produce chromodynamic effects. When the quarks and the non-Abelian gauge fields G_m^α called *gluons* are treated rigorously as relativistic second quantized fields, one obtains the theory of quantum chromodynamics (QCD). As the same QCD suggests, gluons also possess dynamics that are described by an action of the form of (2-75). The real gluonic field strength takes the form

$$G_{mn}^\gamma = \partial_m G_n^\gamma - \partial_n G_m^\gamma + f_{\alpha\beta}^\gamma G_m^\alpha G_n^\beta,$$

where $f_{\alpha\beta}^\gamma$ are a set of numbers defined by

$$[\lambda_\alpha, \lambda_\beta] = i f_{\alpha\beta}^\gamma \lambda_\gamma.$$

These numbers are called structure constants, and they determine how the gluons interact among themselves. Gluons are believed to play a very important role. No quark has ever been seen directly! Yet they are believed to exist inside of hadrons. Gluons, as their name suggests, are thought to be the "glue" that keep quarks trapped in the interior of hadrons. Thus gluons are considered to be the primary reason why physically observable hadrons consist only of composites of three quark states (baryons, such as the proton) or quark-antiquark states (mesons, such as the π-meson) with no net color. The process by which gluons trap quarks is called infrared slavery, and is one of the outstanding problems in this field. No mathematically rigorous proof exists to date that verifies that infrared slavery really occurs in QCD. The actual strong force required to keep protons and neutrons in the nucleus is regarded as having the same relation to chromodynamics as the van der Waals force of molecular physics has to electrodynamics. That is, it arises as a higher-order effect that is the "residue" of the infinitely stronger color force.

We now turn our attention to the relation of the four remaining gauge bosons Y_m and W_m^i of Equation (2-78) to the electromagnetic and weak forces. As a prelude, however, we can introduce the leptons into our mathematical model. This is done by introducing the lepton wavefunction operators L_-^1 and e_+^1, where L_-^1 and e_+^1 are the two-component "isovector" and "isoscalar" introduced to describe the lepton members of the first family in Figure 2-4. Once again the symbols in Figure 2-4 are to be regarded as mathematical wavefunction operators. The analogous procedures are carried out for the leptons of the second and third families. Being fermions, the free leptons are also described by the Dirac action

$$S^1_{\text{lepton}} = \tfrac{1}{2} \int d^4x \, [i\bar{L}^1_- \, (\gamma^m \partial_m L^1_-) - i \, (\partial_m \bar{L}^1_-) \, \gamma^m L^1_-]$$

$$+ \tfrac{1}{2} \int d^4x \, [i\bar{e}^1_+ \, (\gamma^m \partial_m e^1_+) - i \, (\partial_m \bar{e}^1_+) \, \gamma^m e^1_+]. \qquad (2\text{-}79)$$

(Note the lack of a term proportional to the mass of the electron. This lack will be remedied shortly). As with the quarks, the first step is to identify a set of global invariances. For the leptons these take the forms

$$\delta L^1_- = (-i\tfrac{1}{2}g_r \Lambda_r - i\tfrac{1}{2}g_w \Lambda^i_w \sigma_i) \, L^1_-, \qquad (2\text{-}80)$$

$$\delta e^1_+ = -ig_r \Lambda_r e^1_+. \qquad (2\text{-}81)$$

It can be seen that it is not possible to multiply L^1_- by λ_α since the latter are 3×3 matrices. Similarly, the scalar e^1_+ cannot be multiplied by λ_α or by σ_i in such a way so as to produce another scalar. The fact that the leptons cannot be multiplied by λ_α (and therefore cannot be made to interact with the gluons) is the mathematical expression that leptons are "colorless." Upon reflection it is seen that the flavor of leptons and quarks consists of two parts. The part associated with $g_r \Lambda_r$ is called the weak hypercharge and the part associated with $g_w \Lambda^i_w$ is called the weak isospin. From (2-81) it follows that e^1_+ only has a weak isospin. The interactions are again turned on by requiring that the invariances in (2-80) and (2-81) be gauge invariances. Thus, the ordinary derivatives in (2-79) are replaced by covariant derivatives that take the forms

$$\mathcal{D}_m L^1_- = \partial_m L^1_- - i(\tfrac{1}{2}g_r Y_m + \tfrac{1}{2}g_w W^i_m \sigma_i) \, L^1_-,$$

$$\mathcal{D}_m e^1_+ = \partial_m e^1_+ - ig_r Y_m e^1_+.$$

The actions for the gauge fields Y_m and W^i_m are given by (2-55) with the replacement $A_m \rightarrow Y_m$ and (2-76) with the replacement $G^i_m \rightarrow W^i_m$. The stage has now been set for the introduction of "spontaneously" broken Yang-Mills theories. Notice that unless a new feature intercedes, the weak isospin gauge fields W^i_m will introduce the phenomenon of infrared slavery (just like the gluons) and render the electron and other leptons unobservable!

Infrared slavery is associated with the presence of massless non-Abelian gauge fields. So clearly to avoid this problem the field W^i_m must be given a mass. In addition it has been known since the work of Yukawa that if a particle of mass M is associated with a force then the strength of that force is given by $|\mathbf{F}| = \exp(-Mr)/r^2$. From experimental observation it is known that the weak force is indeed short-ranged. But introducing a mass for a gauge field is not as simple as it might seem. For particles like the quarks a mass is introduced by writing the last two terms in Equation (2-60). If a similar approach is attempted for gauge fields, a peculiar situation develops. A careful study shows that adding a mass term

$$S_{\text{YM mass}} = -\frac{1}{8\pi} \int d^4x \, \eta^{mn} \, (W^i_{\underline{m}} M^2_{ij} W^j_{\underline{n}} + M^2 Y^i_{\underline{m}} \delta_{ij} Y^j_{\underline{n}}) \qquad (2\text{-}82)$$

to the pure gauge action of (2-76) describes a theory that contains particles that have negative mass! There is no evidence that such particles exist in nature. Another interesting point is that $S_{\text{YM mass}}$ is *not* gauge invariant. To add further problems, when quantum properties are properly taken into account, the addition of (2-82) to (2-76) with $G^i_{\underline{m}} \rightarrow W^i_{\underline{m}}$ yields a theory that cannot be used in a perturbative approach to make calculably unique predictions. Such a theory is said to be unrenormalizable. For many years this was the situation in which particle physicists found themselves.

The ingenious solution to this problem was provided by the Glashow-Salam-Weinberg unified theory of the weak and electromagnetic interaction. As its name implies, this theory provides a deeper, richer understanding of physics that permits the weak force and the electromagnetic force to be viewed as different manifestations of a single force. Nothing of this nature has occurred in physics since Maxwell's equations provided the means of understanding why electric and magnetic phenomena are interrelated. Instead of adding the mass term in (2-82) to (2-75), the G-S-W theory proposes that there should exist in nature a fundamental particle that is described by the Klein-Gordon type action of Equation (2-57). This particle, sometimes called the Higgson, is the only physical remnant of a most remarkable phenomenon: the spontaneous breaking of Yang-Mills theories. This phenomenon is achieved by introducing a complex scalar field ϕ that is actually a complex double like L^1_-. Thus ϕ can be written as

$$\phi = \begin{pmatrix} \phi_+ \\ \phi_0 \end{pmatrix},$$

and since it has the same form as L^1_- its covariant derivative takes the form

$$\mathcal{D}_{\underline{m}}\phi = \partial_{\underline{m}}\phi - i \, (\tfrac{1}{2}g_Y Y_{\underline{m}}\mathbf{1} + \tfrac{1}{2}g_w W^i_{\underline{m}}\sigma_i) \, \phi.$$

As we have learned to associate particles with each real field operator, the complex doublet ϕ is associated with four bosonic states (i.e., the real and imaginary parts of ϕ_+ and ϕ_0). This quantity can now be substituted into the action formula in (2-57) to determine the dynamics of the Higgs or ϕ-field:

$$S_{\text{Higgs}} = \int d^4x \, [\eta^{\underline{mn}} \, (\mathcal{D}_{\underline{m}}\phi)(\mathcal{D}_{\underline{n}}\phi) - W\bar{\phi}\phi]. \tag{2-83}$$

In order for the whole theory, including the Higgs field, to be renormalizable, the most general form of W can only be

$$W = \tfrac{1}{4}\lambda_0\bar{\phi}\phi \pm m_0^2, \qquad \lambda_0 > 0. \tag{2-84}$$

[The action described by (2-83) and (2-84) is known as the Landau-Ginsberg action. It was first discovered in the study of the spontaneous magnetization of some materials below the Curie temperature.] The constant m_0 in (2-84) has the interpretation of the mass of the ϕ-field, and the sign in the equation is of paramount importance. If the plus sign is chosen, then (2-83) and (2-84) describe a spin-zero particle doublet of mass m_0 interacting with the gauge fields W_m^i and Y_m. This situation certainly does nothing to solve the infrared slavery problem owing to the masslessness of the non-Abelian gauge fields. On the other hand, if the minus sign is chosen, the Higgs fields have negative mass! At first this result seems to simply yield further difficulties. But, at this point the remarkable Higgs phenomenon intervenes in such a way that all the problems are solved at once.

To show this solution, we begin by varying (2-83) to obtain the equation of motion for ϕ,

$$-\eta^{\underline{mn}}\mathcal{D}_{\underline{n}}\mathcal{D}_{\underline{n}}\phi + m_0^2\phi - \tfrac{1}{2}\lambda_0|\phi|^2\phi = 0, \tag{2-85}$$

where the minus sign was chosen in (2-84). Clearly $\phi = 0$ is a solution of this equation, and if a small fluctuation about this solution is studied, then (2-85) implies $\langle E^2 \rangle = \langle \mathbf{p}^2 \rangle - m_0^2$. So there do appear to be negative-mass particles present. These are called Goldstone bosons. However, Equation (2-85) has another spatial isotropic solution given by $\bar{\phi}_s = m_0\sqrt{2/\lambda_0}\,(1, 0)$. Studying small fluctuations about this solution implies that only positive mass particles are being described. This implication is most simply seen by making the change of variable

$$\phi = m_0\sqrt{\frac{2}{\lambda_0}}\begin{pmatrix}1\\0\end{pmatrix} + \tilde{\phi} = \bar{\phi}_s + \tilde{\phi}$$

in (2-85) and noting that the terms purely linear in $\tilde{\phi}$ obtained all have mass terms opposite in sign to that in (2-85), that is, positive mass.

Next, we find that some of the gauge fields gain masses necessary to avoid the infrared slavery problem. To see the gain, it is only necessary to substitute ϕ_s into (2-83) and keep the terms proportional to the gauge fields:

$$S'_{mass} = \int d^4x \, m_0^2 \left(\frac{2}{\lambda_0}\right)\eta^{mn}\left[\frac{1}{2}\, g_w^2 W_{\underline{m}}^+ W_{\underline{n}}^- \right.$$

$$\left. + \left(\frac{1}{2}\, g_Y Y_{\underline{m}} + \frac{1}{2}\, g_w W_{\underline{m}}^3\right)\left(\frac{1}{2}\, g_Y Y_{\underline{n}} + \frac{1}{2}\, g_w W_{\underline{n}}^3\right)\right] \quad \text{(2-86)}$$

$$W_{\underline{m}}^\pm \equiv \frac{1}{\sqrt{2}}\,(W_{\underline{m}}^1 \pm iW_{\underline{m}}^2).$$

The complex field $W_{\underline{m}}^+$ is immediately seen to have gained a mass of $g_w^2 m_0^2/\lambda_0$. Of course, its complex conjugate $W_{\underline{m}}^-$ has the same mass. These fields or the associated particles have physical significance. They are called charged, intermediate vector bosons and are the agents responsible for carrying the weak force. Very recently these particles have been produced in the elegant experiment at CERN and have been found to have a mass of approximately 80 BeV, or 85 times the mass of the proton. The remaining two gauge fields in (2-86) are "mixed up" and require that we "diagonalize" their mass matrix, done by defining new fields $A_{\underline{m}}$ and $Z_{\underline{m}}^0$ by

$$A_{\underline{m}} = \cos\theta\, Y_{\underline{m}} - \sin\theta\, W_{\underline{m}}^3,$$

$$Z_{\underline{m}}^0 = \sin\theta\, Y_{\underline{m}} + \cos\theta\, W_{\underline{m}}^3,$$

where θ is called the weak mixing angle. Now, picking $\tan\theta = g_Y/g_w$, we find that (2-86) takes the form

$$S'_{mass} = \int d^4x \, m_0^2 \left(\frac{1}{\lambda_0}\right)g_w^2\left[W_{\underline{m}}^+ W_{\underline{n}}^- + \frac{1}{2}\sec^2\theta Z_{\underline{m}}^0 Z_{\underline{n}}^0\right]\eta^{mn}. \quad \text{(2-87)}$$

Therefore the mass of the $Z_{\underline{m}}^0$ or neutral intermediate vector boson is related to the mass of the $W_{\underline{m}}^+$ by $M_z^2 = M_W^2 \cos^2\theta$. The $Z_{\underline{m}}^0$ has also been seen experimentally and has a mass around 95 BeV, which implies that the weak mixing angle has a value of approximately 0.5 radians.

That $A_{\underline{m}}$ does not appear in (2-87) is very fortunate for the theory because $A_{\underline{m}}$ is precisely the photon of Maxwell's equations! Since we still require electromagnetic gauge invariance, a mass term for $A_{\underline{m}}$ in (2-87) would be inconsistent with nature.

The insightful reader at this point may wonder what happened. We started with massless gauge fields interacting with scalars that "spontaneously" obtained a nontrivial solution to their equations of motion and we ended with three massive nongauge fields and one massless gauge field. Although this result may seem highly unreasonable, it is nevertheless completely and rigorously correct. When the scalar equation of motion admits the solution ϕ_s, called a nonvanishing vacuum expectation value, the gauge symmetry was spontaneously broken by this solution because ϕ_s is *not* left-invariant by a general gauge transformation. However, the equation

$$\phi_s' = \exp\left[-i\tfrac{1}{2}(g_Y\Lambda_Y + g_w\Lambda_w^i\sigma_i)\right]\phi_s$$

is satisfied if $\Lambda_Y = \Lambda\cos\theta$, $\Lambda_w^1 = \Lambda_w^2 = 0$, and $\Lambda_w^3 = -\Lambda\sin\theta$, so that only the usual electromagnetic gauge invariance associated with Λ (x, t) remains. Three of the negative mass or Goldstone bosons are "eaten" by the would-be massless gauge fields, which "grow" a mass as a result. The remaining spin-zero field remains as the positive-mass, physically relevant particle we called the Higgson, h_0. The most important reward of this roundabout way of giving mass to the gauge fields is that the resulting theory is renormalizable, that is, yields unique calculable answers in a perturbative approach.

Finally we can give the electron (also the muon and tauon) mass by introducing a Yukawa interaction

$$S_{\text{Yukawa}} = \int d^4x\, h_0\, [\bar{e}_+\hat{\phi}^+L_-^1 + L_-^1\hat{\phi}e_+],\qquad (2\text{-}88)$$
$$\hat{\phi} \equiv -i\sigma^2\phi^*.$$

When ϕ acquires a nonvanish vacuum value, Equation (2-88) must be expressed in terms of $\tilde{\phi}$ and $\hat{\phi}$. This "shift" of variables then yields the mass term for the electron.

Just as Maxwell's equations unify electric and magnetic phenomena, the G-S-W theory unifies electromagnetic phenomena with weak interaction phenomena. This unification can be seen by calculating the modifications to Maxwell's equations (2-53) and (2-54), which arise from the contributions of W_m^+, W_m^- and Z_m^0. To the first set of equations (2-53), these modifications are what should be expected; W_m^+ and W_m^- have respective electric charges of $+1$ and -1 and couple appropriately to the photon by contributing to J_m. The particle Z_m^0 is neutral and does not contribute to J_m. The second set of equations (2-54) is unmodified.

The model we have described so far is what has become to be known as the standard model. It explains a great deal of the structure we observe in nature. The strong, electromagnetic, and weak forces are all manifestations of a single principle: gauge invariance. There are some puzzling features, however. Among these is the quark-lepton symmetry noted in the previous section. Why should a symmetry exist? One possible explanation provides a most far-reaching and surprising implication: Protons may *not* be stable. (We shall return to the importance of this implication shortly.) A simple explanation of quark-lepton symmetry was first suggested by Pati and Salam and leads to the concept of Grand Unified Theory (GUT) models. The essential point of these models is the assumption that quarks and leptons are *not* disjoint fields but instead parts of even larger multiplets. There are many ways to realize this idea, but one of the simplest can be seen by replacing the separate left-handed quark and lepton multiplets in Figure 2-4 by a single left-handed Fermion multiplet F_-, where

$$F_- = \begin{pmatrix} u_1 & u_2 & u_3 & \nu_e \\ d_1 & d_2 & d_3 & e^- \end{pmatrix}_-.$$

Since F_- is a 2×4 matrix it can be post-multiplied by 4×4 matrices. These matrices will, in general, exchange quark fields with lepton fields. Multiplication by such matrices will be a global symmetry of the action constructed from F_- and gauging this symmetry allows quarks to change into leptons by emitting "lepto-quark" gauge bosons $X_m(\alpha)$. This change is exactly analogous to β-decay, where a down quark emits a W_m^- and changes into an up quark). Therefore, it would be possible for a proton to decay into leptons plus other hadrons! Proton decay is a general feature of most GUT models.

Presently, there are a number of experiments being conducted in which physicists are attempting to detect proton decay. If such decays are detected, then we will be in exactly the same position as Becquerel, who discovered natural radioactivity. The discovery of proton decay would be of monumental importance. Looking back on history, we can see that there was direct path from the discovery of natural radioactivity to the study, understanding, and manipulation of nuclear processes. Should proton decay occur in nature, then it is conceivable that such processes could, eventually, be induced artificially. High-luminosity beams of lepto-quark bosons would literally constitute the disintegrator rays of science fiction.

ACKNOWLEDGMENTS

I would like to thank Abdus Salam, Jogesh Pati, Bryan Lynn, Martin Rocek, Anders Karlhede, Ulf Linstrom, and Ray Gelinas for conversations that contributed to this work. Additional thanks are extended to Dianna Abney, Martha Adams, and Geoffrey Baker-Roberts for preparation of the manuscript. I further wish to acknowledge for their generous hospitality ICTP (Trieste, Italy), ITP University of Stockholm (Stockholm, Sweden) and A. Lindstrom, CERN (Geneva, Switzerland), and LAPP (Annecy-Lex Vieus, France). Finally I wish to thank the editor, Ronald Mickens, for his assistance, support, and constant encouragement without which this work would never have been undertaken or completed. This work was supported in part by the National Science Foundation under grant PHY81-07394.

BIBLIOGRAPHY

Abers, E. S., and B. W. Lee, "Gauge Theories," *Physics Reports* 9(1) (1973).

Bransden, B. H., D. Evans, and J. V. Major, *The Fundamental Particles* (New York: Van Nostrand Reinhold, 1973).

Cheng, D. C., and G. K. O'Neill, *Elementary Particle Physics* (Reading, Mass: Addison-Wesley, 1979).

Eisele, J. A., *Modern Quantum Mechanics with Applications to Elementary Particle Physics* (New York: John Wiley & Sons, 1969).

Feld, B. T., *Models of Elementary Particles* (Boston: Blaisdell Publishing Company, 1969).

Ford, K. W., *The World of Elementary Particles* (Boston: Blaisdell Publishing Company, 1963).

French, A. P., *Special Relativity* (New York: W. W. Norton & Company, 1968).

Lopes, J. L., *Gauge Field Theories* (New York: Pergamon Press, 1981).

Purcell, E. M., *Electricity and Magnetism,* vol. 2 (New York: McGraw-Hill Book Company, 1965).

Sakurai, J. J., *Advanced Quantum Mechanics* (Reading, Mass.: Addison-Wesley, 1973).

Soper, D. E., *Classical Field Theory* (New York: John Wiley & Sons, 1976).

Tassie, L. J., *The Physics of Elementary Particles* (New York: A. Halsted Press Book, John Wiley & Sons, 1973).

Tipler, P. A., *Foundations of Modern Physics* (New York: Worth Publishers, 1969).

APPENDIX A: THE DIRAC EQUATION

In this appendix we discuss the mathematical background of the Dirac equation in some detail. First, the Dirac wavefunction is not a single complex wavefunction but instead corresponds to four complex functions, ψ_1, ψ_2, ψ_3, and ψ_4. These may be regarded as components arranged in the form of a column vector

$$\Psi = \begin{pmatrix} \psi_4 \\ \psi_2 \\ \psi_3 \\ \psi_4 \end{pmatrix}.$$

However, under a Lorentz tranformation ψ_1, ψ_2, ψ_3, and ψ_4 *do not* transform like the components of a four-vector. Since Ψ is a column vector it can be multiplied by 4×4 matrices. In fact, the quantities γ^m that appear in the Dirac equation are such matrices and can be chosen as

$$\gamma^0 = \begin{pmatrix} 0 & 1 \\ 1 & 0 \end{pmatrix}, \qquad \gamma^1 = \begin{pmatrix} 0 & \sigma^1 \\ -\sigma^1 & 0 \end{pmatrix}.$$

$$\gamma^2 = \begin{pmatrix} 0 & \sigma^2 \\ -\sigma^2 & 0 \end{pmatrix}, \qquad \gamma^3 = \begin{pmatrix} 0 & \sigma^3 \\ -\sigma^3 & 0 \end{pmatrix},$$

where 1, σ^1, σ^2, and σ^3 are 2×2 matrices (see Appendix B). The most important property of these matrices is

$$\gamma^m \gamma^n + \gamma^n \gamma^m = 2 \eta^{mn} 1_4, \tag{A2-1}$$

where η^{mn} is the inverse Minkowski metric [see (2-46) and (2-48)] and 1_4 is the 4×4 identity matrix. The reason for imposing (A2-1) is so that the Dirac equation of motion arising from variation of the action (2-58),

$$[i\gamma^n \partial_n - m 1_4] \Psi = 0. \tag{A2-2}$$

can be combined with the quantum-mechanical identifications $E = i\partial_0$ and $\mathbf{p} = i\nabla$ to yield the relativistic equation $E^2 = p^2 + m^2$. To show this combination, we multiply (A2-2) by $[i\gamma^m \partial_m + m 1_4]$,

$$[i\gamma^m \partial_m + m 1_4]\, [i\gamma^n \partial_n - m 1_4] \Psi = 0,$$

$$[-\gamma^m \gamma^n \partial_m \partial_n - m^2 1_4] \Psi = 0,$$

$$[-\tfrac{1}{2}(\gamma^m \gamma^n \partial_m \partial_n + \gamma^n \gamma^m \partial_n \partial_m) - m^2 1_4] \Psi = 0,$$

$$[-\tfrac{1}{2}(\gamma^m \gamma^n + \gamma^n \gamma^m)\, \partial_m \partial_n - m^2 1_4] \Psi = 0,$$

$$[-\eta^{mn} \partial_m \partial_n - m^2]\, 1_4 \Psi = 0,$$

$$[E^2 - |\mathbf{p}|^2 - m^2] = 0,$$

where we have used the relation $\partial_m \partial_n = \partial_n \partial_m$ and the result in (A2-1). As the m index on γ^m suggests, the four gamma matrices together constitute a Lorentz four-vector. On the other hand, the suppressed index on the Dirac wavefunction Ψ and those on the gamma matrices are called spinor indices.

A distinguishing feature of a spinor is its transformation law under a Lorentz transformation. Whereas a Lorentz covector transforms as in (2-50), $(\partial_m)' = [M(-\lambda)]_m^n \, \partial_n$, a spinor instead transforms as

$$\Psi' = \exp\{\tfrac{1}{4}[M(+\lambda)]^{mn}\sigma_{mn}\} \, \Psi, \qquad (A2\text{-}3)$$

where $\sigma_{mn} \equiv \gamma_m\gamma_n - \gamma_n\gamma_m$. This transformation leads to some peculiar results. For example, if $M(\lambda)$ in (A2-3) describes an ordinary rotation by $2\pi N$ radians, then $\Psi' = (-)^N\Psi$. So the rotation of a spinor through any odd multiple of $360°$ yields minus the spinor. The adjoint of a Dirac spinor denoted by $\overline{\Psi}$ is defined by

$$\overline{\Psi} = (\psi_3^*, \, \psi_4^*, \, \psi_1^*, \, \psi_2^*)$$

$$= (\Psi')^* \, (\gamma^0) = \Psi^+\psi^0,$$

and thus the Dirac action is of the form of an inner product of a row spinor $\overline{\Psi}$ with a column spinor Ψ. (Of course, the equation can always be rewritten in terms of the functions $\psi_1, \ldots, \psi_4$.)

Since the Dirac wavefunction "lives" in a four-dimensional space, the set of all possible linear transformations acting on this space must be expressable as linear combinations of sixteen 4×4 matrices. It is convenient to identify these matrices as

$$1_4, \, \gamma^{\underline{m}}, \, \sigma^{\underline{mn}}, \, \gamma^5\gamma^{\underline{m}}, \, \gamma^5,$$

where $\gamma^5 \equiv i\gamma^0\gamma^1\gamma^2\gamma^3$ and $\sigma^{\underline{mn}}$ was defined in Equation (A2-3). These matrices permit the construction of bilinear covariants via the definitions

$$S = \overline{\Psi}1_4\Psi, \qquad\qquad P = \overline{\Psi}\gamma^5\Psi,$$

$$V^{\underline{m}} = \overline{\Psi}\gamma^{\underline{m}}\Psi, \qquad A^{\underline{m}} = \overline{\Psi}\gamma^5\gamma^{\underline{m}}\Psi,$$

$$T^{\underline{mn}} = \overline{\Psi}\sigma^{\underline{mn}}\Psi.$$

These quantities transform simply under Lorentz transformation. Using (A2-3) and properties of the complete set of 4×4 matrices, we see

$$S' = S, \qquad\qquad P' = P,$$

$$(V^{\underline{m}})' = [M(\lambda)]_{\underline{n}}^{\underline{m}}V^{\underline{n}}, \qquad (A^{\underline{m}})' = [M(\lambda)]_{\underline{n}}^{\underline{m}}A^{\underline{n}},$$

$$(T^{\underline{mn}})' = [M(\lambda)]_{\underline{r}}^{\underline{m}}[M(\lambda)]_{\underline{s}}^{\underline{n}}T^{\underline{rs}}.$$

Therefore, S and P transform as Lorentz scalars. The quantities $V^{\underline{m}}$ and $A^{\underline{m}}$ transform as four-vectors. Finally, $T^{\underline{mn}}$ transforms (like $\eta^{\underline{mn}}\eta^{\underline{ns}}F_{\underline{rs}}$) as a second-rank antisymmetric tensor.

APPENDIX B: MATRIX MANIPULATION

In this appendix we review manipulations of matrices and the properties of some specific matrices. For us it is sufficient to regard a matrix, A, as an array of quantities, a_{ij}, $i = 1, \ldots , M, j = 1, \ldots , N$,

$$A = \begin{pmatrix} a_{11} & \cdots & a_{1N} \\ \vdots & & \vdots \\ a_{M1} & \cdots & a_{MN} \end{pmatrix},$$

where the entries a_{ij} may be real or complex numbers or functions depending on spatial and temporal variables. The multiplication of matrices is given by the standard row-by-column rule

$$AB = \begin{pmatrix} \sum_{j=1}^{N} a_{1j} b_{j1} & \cdots & \sum_{j=1}^{N} a_{1j} b_{jN} \\ \vdots & & \vdots \\ \sum_{j=1}^{N} a_{Mj} b_{j1} & \cdots & \sum_{j=1}^{N} a_{mj} b_{jN} \end{pmatrix}$$

The complex conjugate, transpose, and Hermitian conjugate of A denoted respectively by A^*, A^t, and A^+ are defined by

$$A^* \equiv \begin{pmatrix} a_{11}^* & & a_{1N}^* \\ & & \\ a_{M11}^* & \cdots & a_{MN}^* \end{pmatrix}, \quad A^t = \begin{pmatrix} a_{11} & \cdots & a_{M1} \\ \vdots & & \\ a_{1N} & \cdots & a_{MN} \end{pmatrix},$$

$$A^+ = (A^t)^*$$

The trace of A [denoted $\mathrm{Tr}(A)$] is defined by

$$\mathrm{Tr}(A) = \sum_{j=1}^{N} a_{jj}$$

and only exists for square (i.e., $M = N$) matrices.

A special set of 2×2 matrices, the Pauli matrices, σ^1, σ^2, and σ^3, are defined by

$$\sigma^1 \equiv \begin{pmatrix} 0 & 1 \\ 1 & 0 \end{pmatrix}, \qquad \sigma^2 \equiv \begin{pmatrix} 0 & -i \\ i & 0 \end{pmatrix}, \qquad \sigma^3 \equiv \begin{pmatrix} 1 & 0 \\ 0 & -1 \end{pmatrix}.$$

These matrices possess the properties

$$\text{Tr}(\sigma^i) = 0, \qquad \sigma^i = (\sigma^i)^+,$$

$$\sigma^i \sigma^j \equiv \sigma^{ij} 1 + i\epsilon^{ijk}\sigma^k,$$

where δ^{ij} is the Kronecker delta ($\delta^{ij} = 1$ if $i = j$ and $\delta^{ij} = 0$ if $i \neq j$) and ϵ^{ijk} denotes the three-dimensional Levi-Civita tensor. The components of ϵ^{ijk} satisfy

$$\epsilon^{123} = \epsilon^{231} = \epsilon^{312} = -\epsilon^{132} = -\epsilon^{213} = -\epsilon^{321} = 1$$

and $\epsilon^{ijk} = 0$ for all other values.

The set of 3×3 matrices denoted by λ^α in the text are defined by

$$\lambda^1 \equiv \begin{pmatrix} 0 & 1 & 0 \\ 1 & 0 & 0 \\ 0 & 0 & 0 \end{pmatrix}, \qquad \lambda^2 \equiv \begin{pmatrix} 0 & -i & 0 \\ i & 0 & 0 \\ 0 & 0 & 0 \end{pmatrix}, \qquad \lambda^3 \equiv \begin{pmatrix} 1 & 0 & 0 \\ 0 & -1 & 0 \\ 0 & 0 & 0 \end{pmatrix},$$

$$\lambda^4 \equiv \begin{pmatrix} 0 & 0 & 1 \\ 0 & 0 & 0 \\ 1 & 0 & 0 \end{pmatrix}, \qquad \lambda^5 \equiv \begin{pmatrix} 0 & 0 & -i \\ 0 & 0 & 0 \\ i & 0 & 0 \end{pmatrix}, \qquad \lambda^6 \equiv \begin{pmatrix} 0 & 0 & 0 \\ 0 & 0 & 1 \\ 0 & 1 & 0 \end{pmatrix},$$

$$\lambda^7 \equiv \begin{pmatrix} 0 & 0 & 0 \\ 0 & 0 & -i \\ 0 & i & 0 \end{pmatrix}, \qquad \lambda^8 \equiv \frac{1}{\sqrt{3}} \begin{pmatrix} 1 & 0 & 0 \\ 0 & 1 & 0 \\ 0 & 0 & -2 \end{pmatrix}.$$

Analogous to the Pauli matrices, the λ-matrices satisfy

$$\text{Tr}(\lambda^\alpha) = 0, \qquad \lambda^\alpha = (\lambda^\alpha)^+,$$

$$\lambda^\alpha \lambda^\beta = \sum_{\gamma=1}^{8} (d^{\alpha\beta\gamma}\lambda^\gamma + if^{\alpha\beta\gamma}\lambda^\gamma) + \tfrac{2}{3}\delta^{\alpha\beta},$$

where λ^0 is simply the 3×3 identity matrix and the coefficients $d^{\alpha\beta\gamma}$ and $f^{\alpha\beta\gamma}$ are given in Table B2-1.

Table B2-1
Nonvanishing Values of $d^{\alpha\beta\gamma}$ and $f^{\alpha\beta\gamma}$

$\alpha\beta\gamma$	$f^{\alpha\beta\gamma}$	$\alpha\beta\gamma$	$d^{\alpha\beta\gamma}$
123	1	118	$1/\sqrt{3}$
147	$\frac{1}{2}$	146	$\frac{1}{2}$
156	$-\frac{1}{2}$	157	$\frac{1}{2}$
246	$\frac{1}{2}$	228	$1/\sqrt{3}$
257	$\frac{1}{2}$	247	$-\frac{1}{2}$
345	$\frac{1}{2}$	256	$\frac{1}{2}$
367	$-\frac{1}{2}$	338	$1/\sqrt{3}$
458	$\sqrt{3}/2$	344	$\frac{1}{2}$
678	$\sqrt{3}/2$	355	$\frac{1}{2}$
		366	$-\frac{1}{2}$
		377	$-\frac{1}{2}$
		448	$-\frac{1}{2}\sqrt{3}$
		558	$-\frac{1}{2}\sqrt{3}$
		668	$-\frac{1}{2}\sqrt{3}$
		778	$-\frac{1}{2}\sqrt{3}$
		888	$-1/\sqrt{3}$

3

The Radon Transform

Stanley R. Deans

University of South Florida

ABSTRACT: The Radon transform is an integral transform named in honor of the mathematician Johann Radon, who pioneered work in this area earlier in this century. While its close relative, the Fourier transform, has been widely used throughout the scientific world for many years, it is only during the past decade that the Radon transform has begun to receive recognition as a useful tool in a variety of practical applications. Its properties and applications still are not widely known and appreciated, but this situation is rapidly changing.

In this chapter, some of the more fundamental properties of the Radon transform are presented and, when appropriate, compared with corresponding properties of the Fourier transform. The applications are very diverse; they emerge in such fields as molecular biology, medical imaging, optics, geophysics, and astrophysics.

In a discussion confined to a chapter of this size, it is not possible to do justice to either the theory or the applications. Thus the approach is to emphasize essential mathematical concepts and present a few simple illustrative examples directed toward those scientists and engineers not yet familiar with the transform. These concepts are covered is Sections 3-2 through 3-8. Topics covered include useful definitions, connection between Radon and Fourier transforms, elementary properties (such as similarity, linearity, shifting, derivatives, and convolution), finding transforms of specific functions, inversion in two and three dimensions, symmetry and related transforms, and series.

Section 3-1 contains a brief introduction and some historical material. This is still incomplete, since uncovering original work and giving proper credit is understandably a slow process in an area encompassing such a wide range of applications.

Finally, in Section 3-9, some of the more highly developed applications are discussed briefly. A few recent sources are provided for each area. These, in turn, contain additional references to prior work in the respective area.

3-1 BACKGROUND

Johann Radon's 1917 paper,[1] "Über die Bestimmung von Funktionen durch ihre Integralwerte längs gewisser Mannigfaltigkeiten" (On the determination of functions from their integrals along certain manifolds), provides the foundation for the theory of the Radon transform. An English translation, by R. Lohner, of this important paper appears in Appendix A of the monograph by Deans.[2] There have been several "rediscoveries" of major results presented in Radon's pioneering work. This is not surprising since the problem of determining a function from knowledge of its line integrals (the two-dimensional problem) or determining a function from knowledge of its integrals over planes (the three-dimensional problem) emerges in widely diverse fields, and individuals working in such different areas may have little natural contact with each other.

The Dutch physicist H. A. Lorentz may have been the first person to find a solution to the three-dimensional problem. In any case, it is clear from an early paper by Bockwinkel[3] that Lorentz knew of a solution. Also, it is clear from Radon's 1917 paper that he was building on work by Minkowski[4] and Funk[5,6] and private communication with G. Herglotz . Many individuals have extended and/or rediscovered Radon's work. An incomplete but certainly important list of contributions prior to 1965 includes Uhlenbeck;[7] Mader;[8] John;[9] Cramér and Wold;[10] Ambartsumian;[11,12] Szarski and Wazewski;[13] Rényi;[14] John;[15] Bracewell;[16] Smerd and Wild;[17] Tetel'baum;[18] Korenblyum, Tetel'baum, and Tyutin;[19] Kalos, Davis, Mittelman, Mastras, and Hutton;[20] Cormack;[21,22] and Pincus.[23] (In connection with references 18 and 19, see Barrett, Hawkins, and Joy.[24]) For those interested in the history of the development of the Radon transform, Allan Cormack's remarks are highly recommended.[25]

By 1966, the transform was widely known among mathematicians;[26-28] however, it was the advent of computerized tomography in the 1970s that really drew attention to applications of the Radon transform.[29-31] Many of the applications are discussed by Deans[2] and Barrett.[32] These sources also provide considerable reference to orignal work.

Theory of the Radon transform plays a fundamental role in providing certain foundations for what has come to be known as *reconstruction from projections*. In this context, the importance has been succinctly expressed by Richard Gordon.[33] He states, "Reconstruction from projections is as fundamental to science as calculus. It is a battery of methods for unfolding a structure from observations" (p. 321).

3-2 DEFINITIONS

Three spaces are fundamental to a discussion of the radon transform and the connection between the Radon transform and the Fourier transform. These three spaces are conveniently called *feature space*, *Radon space*, and *Fourier space*.

Feature space is just n-dimensional Euclidean space R^n. For most applications, $n = 2$ or $n = 3$. Vectors in R^n are designated by $\mathbf{x} = (x_1, x_2, \ldots, x_n)$. For R^2 and R^3, it is convenient to use the notation $\mathbf{x} = (x, y)$ and $\mathbf{x} = (x, y, z)$, respectively. Functions defined on feature space are designated in the conventional notation $f(\mathbf{x}) = f(x_1, x_2, \ldots, x_n)$. For convenience, these functions are usually selected from some nice class, such as $\mathcal{D}$ [infinitely differentiable (C^∞) functions with compact support] or S (rapidly decreasing C^∞ functions).[34] This assumption serves well for most of the current discussion; however, it can be relaxed to include distributions[28] in more general treatments.

As the names imply, Radon space is the space in which the Radon transform of f lives and Fourier space is the space in which the Fourier transform of f lives. In an obvious notation, functions defined on these spaces are designated by

$$\check{f} = \mathcal{R}f$$

for Radon space, and

$$\tilde{f} = \mathcal{F}f$$

for Fourier space. Appropriate designation of points in these spaces follows naturally from the definitions.

Two Dimensions

Radon transform

The Radon transform of the feature space function $f(x, y)$ is defined by

$$\check{f}(p, \phi) = \int_{-\infty}^{\infty} \int_{-\infty}^{\infty} f(x, y)\, \delta(p - x \cos \phi - y \sin \phi)\, dx\, dy. \quad \textbf{(3-1)}$$

The coordinates are indicated in Figure 3-1. Observe that the delta function in (3-1) is one-dimensional and thus reduces the integral over the xy plane to a line integral along the line L defined by

$$p = x \cos \phi + y \sin \phi$$

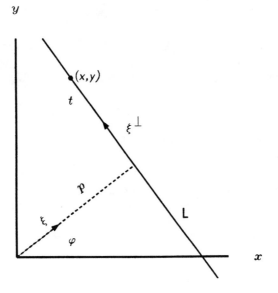

Figure 3-1
Coordinates used to define the Radon transform.

in normal form such that p is the (perpendicular) distance from the origin to the line

To completely define $\Re f$, the function $\check{f}(p, \phi)$ must be known for all p and ϕ. If $\check{f}$ is known for all p and a single fixed angle Φ, we write $\check{f}_\Phi(p)$, which indicates a single *sample* of the Radon transform. For the two-dimensional case, it is just a *projection* or *profile* of $f(x, y)$ at constant angle Φ.

Remark.   In many applications, several samples of $\check{f}$ may be deduced from raw data. These samples enable one to approximate $\check{f}$. In this situation, the interesting problem is the inversion of (3-1); that is, given $\check{f}$, find an approximation to f.

For purposes of generalization, it is useful to rewrite (3-1) using vector notation. The equation of the line **L** can be written as

$$p = \xi \cdot \mathbf{x},$$

where

$$\xi = (\cos \phi, \sin \phi) \tag{3-2}$$

is a unit vector, and

$$\mathbf{x} = (x, y)$$

is an arbitrary point (or vector) on the line **L**. In terms of these variables, (3-1) becomes

$$\check{f}(p, \xi) = \Re f = \int f(\mathbf{x}) \, \delta(p - \xi \cdot \mathbf{x}) \, d\mathbf{x}. \qquad (3\text{-}3)$$

Here the shorthand notation

$$\int d\mathbf{x} \equiv \int_{-\infty}^{\infty} \int_{-\infty}^{\infty} d x \, dy \qquad (3\text{-}4)$$

has been used for integration over R^2.

Another form for $\Re f$ follows immediately from the geometry in Figure 3-1. The vector $\mathbf{x}$ can be written as

$$\mathbf{x} = p\xi + t\xi^{\perp}$$

in terms of scalars p and t. The unit vector ξ is defined in (3-2) and the unit vector perpendicular to ξ is given by

$$\xi^{\perp} = (-\sin \phi, \cos \phi).$$

Hence,

$$f(\mathbf{x}) = f(p\xi + t\xi^{\perp})$$

and

$$\check{f}(p, \xi) = \int_{-\infty}^{\infty} f(p\xi + t\xi^{\perp}) \, dt.$$

Radon space

It is important to relize that $\check{f}(p, \phi)$, or equivalently $\check{f}(p, \xi)$, is not defined on a circular polar coordinate system. If such a system is assumed, serious problems are created as p approaches zero for different values of ϕ.[2] The appropriate space for $\check{f}$ is on the surface of a unit cylinder, as indicated in Figure 3-2.

A line on the cylinder is defined if the angle ϕ is held constant as p varies. The real numbers associated with each point on this line define a sample or profile $\check{f}_\phi(p)$. For purposes of illustration, it is convenient to cut the cylinder

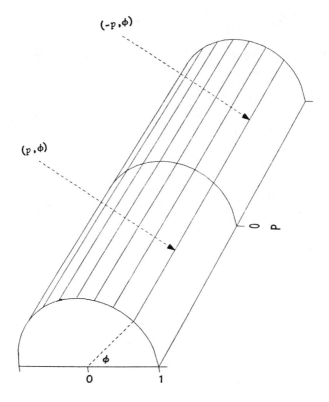

Figure 3-2
Two-dimensional Radon space.

along the $\phi = 0$ line and unroll the cylinder. Then the variables p and ϕ define a plane, and $\check{f}(p, \phi)$ can be represented as a surface in the third dimension.

Fourier transform

The two-dimensional Fourier transform of $f(x, y)$ is given by[35]

$$\tilde{f}(u, v) = \mathcal{F}_2 f = \int_{-\infty}^{\infty} \int_{-\infty}^{\infty} f(x, y) \exp\left[-i2\pi(xu + yv)\right] dx \, dy. \quad \textbf{(3-5)}$$

In vector notation, this becomes

$$\tilde{f}(\mathbf{u}) = \int f(\mathbf{x}) \exp\left(-i2\pi\mathbf{x} \cdot \mathbf{u}\right) d\mathbf{x}, \quad \textbf{(3-6)}$$

where $\mathbf{u} = (u, v)$, $\mathbf{x} = (x, y)$, and convention (3-4) is used. Note that both feature space and Fourier space are rectangular and no cylindrical surfaces are involved, as with Radon space.

Three Dimensions

Radon transform

The extension to three dimensions follows easily from (3-3) with an obvious generalization of the variables from two to three dimensions.

$$\check{f}(p, \xi) = \Re f = \int f(\mathbf{x}) \, \delta(p - \xi \cdot \mathbf{x}) \, d\mathbf{x}. \qquad (3\text{-}7)$$

Here the integration is over planes defined by

$$p = \xi \cdot \mathbf{x}. \qquad (3\text{-}8)$$

If $\mathbf{x} = (x, y, z)$ is a general point in the plane and ξ is a unit vector directed along the perpendicular from the origin to the plane, then p is the (perpendicular) distance from the origin to the plane. The convention (3-4) is extended in the obvious way:

$$\int d\mathbf{x} \equiv \int_{-\infty}^{\infty} \int_{-\infty}^{\infty} \int_{-\infty}^{\infty} dx \, dy \, dz. \qquad (3\text{-}9)$$

Fourier transform

The three-dimensional Fourier transform of $f(x, y, z)$ is given by (3-6) if $\mathbf{x} = (x, y, z)$, $\mathbf{u} = (u, v, w)$, and convention (3-9) is used.

Higher Dimensions

The extension from R^3 to R^n is accomplished in the same way as the extension from R^2 to R^3. Equation (3-6) still defines the Fourier transform and (3-7) still defines the Radon transform. Equation (3-8) now represents a hyperplane at a distance p from the origin, ξ is a unit vector directed toward the hyperplane, and $\mathbf{x} = (x_1, x_2, \dots, x_n)$ is a general point on the hyperplane. Convention (3-9) is extended to an integration over R^n.

A question of generality

No generality is lost by selecting ξ as a unit vector. To see this, consider

$$\check{f}(sp, s\xi) = \int f(\mathbf{x})\, \delta(sp - s\xi \cdot \mathbf{x})\, d\mathbf{x}$$

$$= |s|^{-1} \int f(\mathbf{x})\, \delta(p - \xi \cdot \mathbf{x})\, d\mathbf{x}$$

$$= |s|^{-1} \check{f}(p, \xi) \tag{3-10}$$

for $s \neq 0$. An important *symmetry property* follows from (3-10) with $s = -1$,

$$\check{f}(-p, -\xi) = \check{f}(p, \xi).$$

Thus $\check{f}$ is an even homogenous function of degree -1.

It is useful to consider $\check{f}(p, \zeta)$, where ζ is not a unit vector. Clearly, ζ can be written in terms of the unit vector ξ,

$$\zeta = s\xi,$$

where $s = |\zeta|$. It follows from (3-10) that[2]

$$\check{f}(p, \zeta) = |s|^{-1} \check{f}\left(\frac{p}{s}, \xi\right)$$

$$= |\zeta|^{-1} \check{f}\left(\frac{p}{|\zeta|}, \frac{\zeta}{|\zeta|}\right).$$

3-3   CONNECTION BETWEEN SPACES

The connection between spaces can be conveniently represented by the following schematic diagram:

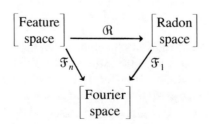

The *n*-dimensional Fourier transform of $f(\mathbf{x})$ is equivalent to the Radon transform of $f(\mathbf{x})$ followed by a one-dimensional Fourier transform of $\check{f}(p, \xi)$ on the variable p.

To see this for $n = 2$, rewrite (3-5) as

$$\tilde{f}(u, v) = \int_{-\infty}^{\infty} dx \int_{-\infty}^{\infty} dy \int_{-\infty}^{\infty} dt \, f(x, y) e^{-i2\pi t} \, \delta(t - ux - vy). \quad (3\text{-}11)$$

Now interchange the order of integration and let $t = qp$ with $q \geq 0$. Then (3-11) becomes

$$\tilde{f}(u, v) = q \int_{-\infty}^{\infty} dp \int_{-\infty}^{\infty} dx \int_{-\infty}^{\infty} dy \, f(x, y) e^{-i2\pi qp}$$

$$\cdot \, \delta(qp - qx \cos \phi - qy \sin \phi),$$

where $u = q \cos \phi$ and $v = q \sin \phi$ in Fourier space. By use of the property

$$\delta(ax) = \frac{1}{|a|} \delta(x),$$

it follows that

$$\tilde{f}(u, v) = \int_{-\infty}^{\infty} dp \int_{-\infty}^{\infty} dx \int_{-\infty}^{\infty} dy \, f(x, y) e^{-i2\pi qp}$$

$$\cdot \, \delta(p - x \cos \phi - y \sin \phi)$$

$$= \int_{-\infty}^{\infty} \check{f}(p, \xi) e^{-i2\pi qp} \, dp.$$

This equation demonstrates that

$$\tilde{f} = F_1 (\check{f}) \quad (3\text{-}12)$$

for $n = 2$. The generalization to higher dimensions is straightforward.[2]

The fundamental result (3-12), sometimes referred to as the *projection-slice theorem,* can be used to compute Fourier transforms when the Radon transforms are known, or it can be used to compute Radon transforms when the Fourier transforms are known.

Example.   Given that

$$f(x, y) = \exp(-x^2 - y^2),$$

compute $\tilde{f}$ and $\check{f}$. Let $x = r \cos\theta$ and $y = r \sin\theta$. Then the polar form of (3-5) is

$$\tilde{f}(u, v) = \int_0^{\infty} dr\, r \exp(-r^2) \int_0^{2\pi} d\theta \exp[-i2\pi qr \cos(\theta - \phi)],$$

where $u = q \cos\phi$ and $v = q \sin\phi$. The θ integral yields $2\pi J_0(2\pi qr)$, where J_0 is the zeroth-order Bessel function of the first kind.[35, 36] The remaining integral is a Hankel transform of order zero,[35, 36]

$$\tilde{f} = 2\pi \int_0^{\infty} r \exp(-r^2) J_0(2\pi qr)\, dr$$

$$= \pi \exp(-\pi^2 q^2).$$

Now (3-12) can be used in the form

$$\check{f} = \mathfrak{F}_1^{-1} \tilde{f}$$

to find the Radon transform of f. From Bracewell,[35]

$$\check{f} = \pi \int_{-\infty}^{\infty} \exp(-\pi^2 q^2) \exp(i2\pi qp)\, dq$$

$$= \sqrt{\pi} \exp(-p^2).$$

The important result in the above example can be expressed as

$$\mathfrak{R}\,\{\exp(-x^2 - y^2)\} = \sqrt{\pi} \exp(-p^2). \qquad \textbf{(3-13)}$$

In this example $\check{f}$ was computed by going through Fourier space. It is actually simpler to compute $\check{f}$ directly.[2] Also, it is simpler to compute $\tilde{f}$ by going through Radon space because no Bessel functions arise. This observation suggests a fast numerical procedure for computing Hankel transforms: first compute the Radon transform, and then use a fast Fourier transform (FFT) algorithm. Since the route from Radon space to Fourier space is always by a one-dimensional Fourier transform, the FFT will always be one-dimensional. This topic is currently under investigation.[37]

3-4 PROPERTIES

Several basic properties of the Radon transform follow directly from the definition (3-7)

$$\check{f}(p, \xi) = \Re f(\mathbf{x}).$$

It is interesting to compare these properties with the corresponding properties of the Fourier transform (3-6),

$$\tilde{f}(\mathbf{u}) = \Im_n f(\mathbf{x}).$$

Very little generality is lost by giving results for R^2 rather than R^n. The slight loss is more than compensated for by the advantage of showing details of computation for a specific case. Consequently, most of the properties are developed from the definitions for two dimensions,

$$\check{f}(p, \xi_1, \xi_2) = \Re f \tag{3-14}$$

$$= \int_{-\infty}^{\infty} \int_{-\infty}^{\infty} f(x, y) \, \delta(p - x\xi_1 - y\xi_2) \, dx \, dy,$$

and

$$\tilde{f}(u, v) = \Im_2 f = \int_{-\infty}^{\infty} \int_{-\infty}^{\infty} f(x, y) e^{-i2\pi(ux + vy)} \, dx \, dy. \tag{3-15}$$

In (3-14) the components of the unit vector ξ are shown explicitly.

Similarity

If the Radon transform of $f(x, y)$ is given by $\check{f}(p, \xi_1, \xi_2)$, then the Radon transform of $f(ax, by)$ is given by

$$\frac{1}{|ab|} \check{f}\left(p, \frac{\xi_1}{a}, \frac{\xi_2}{b}\right).$$

This result follows immediately from

$$\Re f(ax, by) = \int_{-\infty}^{\infty} \int_{-\infty}^{\infty} f(ax, by) \, \delta(p - x\xi_1 - y\xi_2) \, dx \, dy$$

after a change of variables: $X = ax$, $Y = by$.

$$\Re f(ax, by) = \frac{1}{|ab|} \int_{-\infty}^{\infty} \int_{-\infty}^{\infty} f(X, Y)\, \delta\left(p - X\frac{\xi_1}{a} - Y\frac{\xi_2}{b}\right) dX\, dY$$

$$= \frac{1}{|ab|} \check{f}\left(p, \frac{\xi_1}{a}, \frac{\xi_2}{b}\right).$$

The corresponding result for the Fourier transform also follows directly from the definition (3-15). If the Fourier transform of $f(x, y)$ is given by $\tilde{f}(u, v)$, then

$$\mathcal{F}_2 f(ax, by) = \frac{1}{|ab|} \tilde{f}\left(\frac{u}{a}, \frac{v}{b}\right).$$

Linearity

Both transforms are linear. Given two functions $f(x, y)$ and $g(x, y)$, and constants a and b,

$$\Re\{af + bg\} = a\check{f} + b\check{g},$$

and

$$\mathcal{F}_2\{af + bg\} = a\tilde{f} + b\tilde{g}.$$

Shift

The Radon transform of $f(x - a, y - b)$ is given by

$$\Re f(x - a, y - b)$$

$$= \int_{-\infty}^{\infty} \int_{-\infty}^{\infty} f(x - a, y - b)\, \delta(p - x\xi_1 - y\xi_2)\, dx\, dy$$

$$= \int_{-\infty}^{\infty} \int_{-\infty}^{\infty} f(X, Y)\, \delta(p - a\xi_1 - b\xi_2 - X\xi_1 - Y\xi_2)\, dX\, dY$$

$$= \check{f}(p - a\xi_1 - b\xi_2, \xi).$$

$$(3\text{-}16)$$

Note the change of variables $X = x - a$, $Y = y - b$ in the second step. The corresponding result for the Fourier transform is

$$\mathcal{F}_2 f(x - a, y - b) = e^{-i2\pi(au + bv)} \tilde{f}(u, v).$$

Example. Given (3-13), find the Radon transform of $\exp[-(x - a)^2 - (y - b)^2]$. Assume the unit vector ξ is given by $(\cos \phi, \sin \phi)$. The prescription in (3-16) is to replace p in (3-13) by $p - a \cos \phi - b \sin \phi$. Thus

$$\mathcal{R} \{\exp[-(x - a)^2 - (y - b)^2]\} = \sqrt{\pi} \exp[-(p - a \cos \phi - b \sin \phi)^2].$$

This example shows that a Gaussian surface in feature space transforms to a sinusoidal Gaussian surface in Radon space. This transformation is illustrated for $a = b = 4$ in Figure 3-3. Note that the cylinder in Radon space has been cut along the $\phi = 0$ line and unrolled for purposes of illustration.

Differentiation

The partial derivative of $f(x, y)$ with respect to x can be written in the form

$$\frac{\partial f}{\partial x} = \lim_{\epsilon \to 0} \frac{f(x + (\epsilon/\xi_1), y) - f(x, y)}{\epsilon/\xi_1}.$$

The Radon transform of this expression yields

$$\mathcal{R} \left\{ \frac{\partial f}{\partial x} \right\} = \xi_1 \lim_{\epsilon \to 0} \frac{\breve{f}(p + \epsilon, \xi) - \breve{f}(p, \xi)}{\epsilon},$$

where (3-16) has been used with $a = -\epsilon/\xi_1$. It follows that

$$\mathcal{R} \left\{ \frac{\partial f(x, y)}{\partial x} \right\} = \xi_1 \frac{\partial \breve{f}(p, \xi)}{\partial p}.$$

Likewise,

$$\mathcal{R} \left\{ \frac{\partial f(x, y)}{\partial y} \right\} = \xi_2 \frac{\partial \breve{f}(p, \xi)}{\partial p}.$$

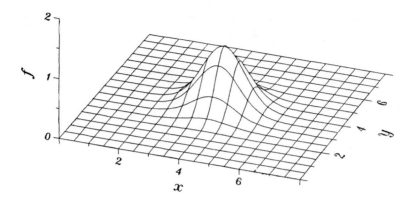

Figure 3-3(a)
Gaussian surface in feature space.

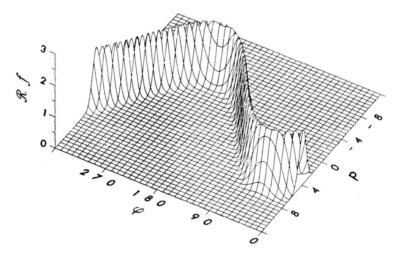

Figure 3-3(b)
Sinusoidal Gaussian surface in Radon space.

The extension of these results to higher derivatives is straightforward. For second order,

$$\Re\left\{\frac{\partial^2 f}{\partial x^2}\right\} = \xi_1^2 \frac{\partial^2 \tilde{f}}{\partial p^2}\,,$$

$$\Re\left\{\frac{\partial^2 f}{\partial x\,\partial y}\right\} = \xi_1 \xi_2 \frac{\partial^2 \tilde{f}}{\partial p^2}\,,$$

$$\mathcal{R}\left\{\frac{\partial^2 f}{\partial y^2}\right\} = \xi_2^2 \frac{\partial^2 \tilde{f}}{\partial p^2}.$$

The corresponding derivative theorems for Fourier transforms are

$$\mathcal{F}_2\left\{\frac{\partial f}{\partial x}\right\} = 2\pi i u \tilde{f}(u, v),$$

$$\mathcal{F}_2\left\{\frac{\partial f}{\partial y}\right\} = 2\pi i v \tilde{f}(u, v),$$

$$\mathcal{F}_2\left\{\frac{\partial^2 f}{\partial x^2}\right\} = -4\pi^2 u^2 \tilde{f}(u, v),$$

$$\mathcal{F}_2\left\{\frac{\partial^2 f}{\partial x \, \partial y}\right\} = -4\pi^2 u v \tilde{f}(u, v),$$

$$\mathcal{F}_2\left\{\frac{\partial^2 f}{\partial y^2}\right\} = -4\pi^2 v^2 \tilde{f}(u, v).$$

Convolution

Thus far, the effort has been directed toward the most basic properties similar for the two transforms. Other properties, such as transforms of convolution

$$f ** g = \int_{-\infty}^{\infty} \int_{-\infty}^{\infty} f(x', y') g(x - x', y - y') \, dx' \, dy',$$

are considerably different. The Fourier convolution theorem[35]

$$\mathcal{F}_2 \{f ** g\} = \tilde{f}(u, v)\tilde{g}(u, v)$$

yields a simple product in Fourier space, whereas the Radon transform gives the more complicated one-dimensional convolution in Radon space,[2]

$$\mathcal{R} \{f ** g\} = \check{f} * \check{g}$$

$$= \int_{-\infty}^{\infty} \check{f}(x, \xi)\check{g}(p - s, \xi) \, ds.$$

Note that ** is used for two-dimensional convolution and * is used for one-dimensional convolution.

Further properties emerge in sections that follow. More detail and examples can be found in Bracewell[35] and Sneddon[38] for the Fourier transform and in Helgason, [39] Deans,[2] and Barrett[32] for the Radon transform. Many other mathematical aspects are covered in the proceedings of the 1980 Oberwolfach conference edited by Herman and Natterer.[40]

3-5 FINDING TRANSFORMS

The properties developed in the previous section can be used to find transforms of various functions. A very nice illustration is provided by use of the Rodrigues formula for Hermite polynomials $H_k(t)$,[41,42]

$$\exp(-t^2)\, H_k(t) = (-1)^k \left(\frac{\partial}{\partial t}\right)^k \exp(-t^2). \qquad \textbf{(3-17)}$$

This formula is used in the section on Hermite polynomials to find transforms of functions of the form

$$x^k y^l \exp(-x^2 - y^2).$$

Transforms of impulse functions and impulse lines are considered in the section on impulse functions. Although these are distributions,[34] finding their transforms causes no difficulty.

Hermite Polynomials

It follows easily from (3-17) that

$$H_k(x)H_l(y)\exp(-x^2 - y^2) = (-1)^{k+l}\left(\frac{\partial}{\partial x}\right)^k \left(\frac{\partial}{\partial y}\right)^l \exp(-x^2 - y^2),$$

where k and l are nonnegative integers. By extending the results of the section on differentiation,

$$\mathcal{R}\left\{\left(\frac{\partial}{\partial x}\right)^k \left(\frac{\partial}{\partial y}\right)^l f(x, y)\right\} = \xi_1^k \xi_2^l \left(\frac{\partial}{\partial p}\right)^{k+l} \check{f}(p, \xi).$$

It follows, after a brief calculation,[2] that

$\Re\{H_k(x)H_l(y)\exp(-x^2-y^2)\}$

$$= \sqrt{\pi}(\cos\phi)^k(\sin\phi)^l\exp(-p^2)H_{k+l}(p),\quad \textbf{(3-18)}$$

where $\xi=(\xi_1,\xi_2)=(\cos\phi,\sin\phi)$.

Example. Verify (3-12) for

$$f(x,y)=x^2\exp(-x^2-y^2).$$

First, observe that x^2 can be expanded in terms of Hermite polynomials. Since

$$H_0(x)=1,\qquad H_1(x)=2x,\qquad H_2(x)=4x^2-2,$$

it follows that

$$x^2=\tfrac{1}{4}H_2(x)+\tfrac{1}{2}H_0(x).$$

Use of (3-18) yields[2]

$$\Re\{x^2\exp(-x^2-y^2)\}=\frac{\sqrt{\pi}}{4}\cos^2\phi\exp(-p^2)H_2(p)+\frac{\sqrt{\pi}}{2}H_0(p)\exp(-p^2)$$

$$=\sqrt{\pi}\left(p^2\cos^2\phi+\frac{1}{2}\sin^2\phi\right)\exp(-p^2).$$

Now, the Fourier transform of this expression gives

$$\mathcal{F}_1\Re\{x^2\exp(-x^2-y^2)\}=\sqrt{\pi}\cos^2\phi\,\mathcal{F}_1[p^2\exp(-p^2)]$$

$$+\frac{\sqrt{\pi}}{2}\sin^2\phi\,\mathcal{F}_1[\exp(-p^2)]$$

$$=\sqrt{\pi}\cos^2\phi\left[2\sqrt{\pi}\exp(-\pi^2 q^2)\left(\frac{1}{4}-\frac{1}{2}\pi^2 q^2\right)\right]$$

$$+\frac{\sqrt{\pi}}{2}\sin^2\phi[\sqrt{\pi}\exp(-\pi^2 q^2)]$$

$$=\pi\left(\frac{1}{2}-\pi^2 q^2\cos^2\phi\right)\exp(-\pi^2 q^2).$$

Next, compute $\mathcal{F}_2 f$ directly,

$$\mathcal{F}_2\{x^2 \exp(-x^2 - y^2)\} = \int_{-\infty}^{\infty}\int_{-\infty}^{\infty} x^2 \exp(-x^2 - y^2)$$

$$\cdot \exp[-i2\pi(ux + vy)] \, dx \, dy$$

$$= 2\sqrt{\pi} \exp(-\pi^2 v^2) \int_0^{\infty} x^2 \cos(2\pi ux) \exp(-x^2) \, dx$$

$$= \pi(\tfrac{1}{2} - \pi^2 u^2) \exp[-\pi^2(u^2 + v^2)]$$

$$= \pi(\tfrac{1}{2} - \pi^2 q^2 \cos^2 \phi) \exp(-\pi^2 q^2),$$

where $u = q \cos \phi$ and $v = q \sin \phi$.

In this example note that for $\phi_1 \neq \phi_2$ in Radon space

$$\lim_{p \to 0} \check{f}(p, \phi_1) \neq \lim_{p \to 0} \check{f}(p, \phi_2).$$

However, in Fourier space

$$\lim_{q \to 0} \mathcal{F}_2 f = \lim_{q \to 0} \mathcal{F}_1 \mathcal{R} f.$$

Impulse Functions

It is clear from the example in Section 3-4 that as the Gaussian surface approaches an impulse function in feature space, the transformed surface approaches a sinusoidal curve in Radon space. To see this directly, consider the Radon transform of $\delta(x - x_0) \, \delta(y - y_0)$,

$$\mathcal{R}\{\delta(x - x_0) \, \delta(y - y_0)\} = \int_{-\infty}^{\infty}\int_{-\infty}^{\infty} \delta(x - x_0) \, \delta(y - y_0)$$

$$\cdot \delta(p - x \cos \phi - y \sin \phi) \, dx \, dy$$

$$= \delta(p - x_0 \cos \phi - y_0 \sin \phi).$$

Thus the "surface" in Radon space is along the sinusoidal curve

$$p = x_0 \cos \phi + y_0 \sin \phi.$$

Before considering how line segments in feature space transform, consider the following example with a Gaussian surface along a line.

Example. Find the Radon transform of

$$f(x, y) = \begin{cases} \exp[-(p_0 - x \cos \phi_0 - y \sin \phi_0)^2] \\ \qquad \text{for} \quad -L < x < L, \quad -L < y < L \\ 0 \qquad \text{otherwise.} \end{cases}$$

A function of this form is shown in Fig. 3-4.
First, consider the case $p \neq p_0$, $\phi \neq \phi_0$. From the definition

$$\check{f}(p, \phi) = \int_{-L}^{L} \int_{-L}^{L} \exp[-(p_0 - x \cos \phi_0 - y \sin \phi_0)^2]$$

$$\cdot \, \delta(p - x \cos \phi - y \sin \phi) \, dx \, dy$$

$$= \frac{1}{|\cos \phi|} \int_{-L}^{L} \int_{-L}^{L} \exp[-(p_0 - x \cos \phi_0 - y \sin \phi_0)^2]$$

$$\cdot \, \delta(p \sec \phi - x - y \tan \phi) \, dx \, dy$$

$$= \frac{1}{|\cos \phi|} \int_{-L}^{L} \exp[-(ay - b)^2] \, dy$$

$$= \frac{1}{|\cos \phi|} \int_{-aL - b}^{aL - b} \exp(-t^2) \, dt$$

$$= \frac{\sqrt{\pi}}{2a |\cos \phi|} [\text{erf}(aL - b) + \text{erf}(aL + b)],$$

where

$$a = \sin(\phi_0 - \phi)/\cos \phi,$$

$$b = (p_0 \cos \phi - p \cos \phi_0)/\cos \phi,$$

and the error function is defined by[43]

$$\text{erf}(x) = \frac{2}{\sqrt{\pi}} \int_0^x \exp(-t^2) \, dt.$$

The Radon space function $\check{f}(p, \phi)$ is illustrated in Figure 3-5.

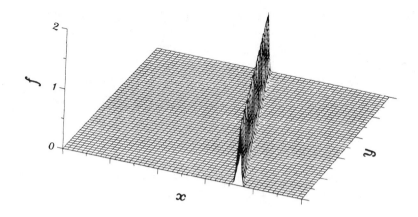

Figure 3-4
Feature space function.

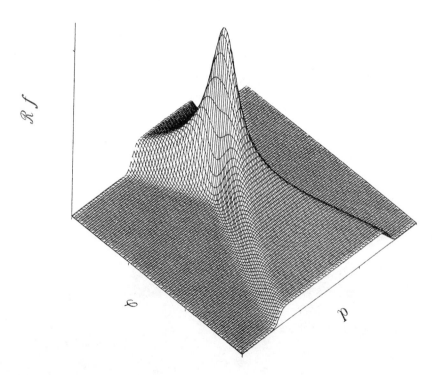

Figure 3-5
Radon transform of function in Figure 3-4.

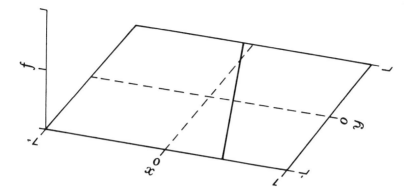

Figure 3-6
Limiting case of Figure 3-4.

Note that as $\phi \to \phi_0$ with $p \ne p_0$, $\check{f} \to 2L \exp[-(p - p_0)^2]/|\cos \phi_0|$. Then as $p \to p_0$, $\check{f} \to 2L/|\cos \phi_0|$. This result is to be expected from Figure 3-6.

It is interesting to consider the direct computation of the Radon transform of the line segment of Figure 3-6. Let $A(x, y)$ represent a density function along the line. Then

$$f(x, y) = A(x, y) \delta(p_0 - x \cos \phi_0 - y \sin \phi_0).$$

For $p \ne p_0$ and $\phi \ne \phi_0$,

$$\check{f}(p, \phi) = \int_{-L}^{L} \int_{-L}^{L} A(x, y) \delta(p_0 - x \cos \phi_0 - y \sin \phi_0)$$

$$\cdot \delta(p - x \cos \phi - y \sin \phi) \, dx \, dy$$

$$= \frac{1}{|\cos \phi_0|} \int_{-L}^{L} A(p_0 \sec \phi_0 - y \tan \phi_0, y)$$

$$\cdot \delta\left(\frac{y \sin(\phi_0 - \phi) + p \cos \phi_0 - p_0 \cos \phi}{\cos \phi_0}\right) dy$$

$$= \frac{1}{|\sin(\phi_0 - \phi)|} A\left(\frac{p \sin \phi_0 - p_0 \sin \phi}{\sin(\phi_0 - \phi)}, \frac{p_0 \cos \phi - p \cos \phi_0}{\sin(\phi_0 - \phi)}\right).$$

The arguments of the density function are easy to interpret by solving for the point (x, y) of intersection of the two lines

$$p = x \cos \phi - y \sin \phi,$$
$$p_0 = x \cos \phi_0 - y \sin \phi_0.$$

Thus the arguments of A represent the x and y values of the point of intersection. The $|\sin(\phi_0 - \phi)|^{-1}$ term ensures the correct limit when the lines coincide. To see this result, consider the case $\phi = \phi_0, p \neq p_0$. Then

$$\check{f}(p, \phi), = \frac{1}{|\cos \phi_0|} \int_{-L}^{L} A(p_0 \sec \phi_0 - y \tan \phi_0, y) \, \delta(p - p_0) \, dy$$

$$= \frac{\delta(p - p_0)}{|\cos \phi_0|} \int_{-L}^{L} A\left(\frac{p_0 - y \sin \phi_0}{\cos \phi_0}, y\right) dy.$$

If A is unity along the line defined by $x = (p_0 - y \sin \phi_0)/\cos \phi_0$, then

$$\check{f}(p, \phi_0) = \frac{2L}{|\cos \phi_0|} \delta(p - p_0).$$

Clearly, $\check{f}(p, \phi)$ is concentrated at the point $(p, \phi) = (p_0, \phi_0)$.

3-6   INVERSION

There are several routines from Radon space to feature space. One obvious possibility is by use of the inverse Radon transform,

$$\begin{bmatrix} \text{Feature} \\ \text{space} \end{bmatrix} \xleftarrow{\mathcal{R}^{-1}} \begin{bmatrix} \text{Radon} \\ \text{space} \end{bmatrix}.$$

Another route is through Fourier space,

These and other approaches to inversion for the two- and three-dimensional cases are discussed in this section. Very little consideration is given to numerical procedures and the implementation of algorithms. These matters are discussed in recent books and review articles.[31,44-47] Those who desire a more rigorous treatment may wish to consult the book by Helgason[39] and articles by Shepp and Kruskal,[48] Smith, Solmon, and Wagner,[49] and Quinto.[50] More detail about many of the methods can be found in the book by Deans.[2] Also, the very nice summary by Barrett[32] is highly recommended.

Two Dimensions

A direct route to the inverse Radon transform is presented by John;[15] however, the path through Fourier space may be easier to follow.

First, observe that the Fourier transform of (3-14) is just $\mathcal{F}_2 f = \tilde{f}(q\xi)$,

$$\tilde{f}(q\xi) = \mathcal{F}_1 \mathcal{R} f = \int_{-\infty}^{\infty} \check{f}(p, \xi) e^{-i2\pi qp} \, dp. \tag{3-19}$$

As usual, the notation $\xi = (\cos \phi, \sin \phi)$, $\mathbf{x} = (x, y)$, and $(u, v) = (q \cos \phi, q \sin \phi)$ is followed. The inverse Fourier transform of $\tilde{f}$ can be written in the polar form,

$$f(\mathbf{x}) = \mathcal{F}_2^{-1} \tilde{f} = \int_{-\infty}^{\infty} |q| \, dq \int_0^{\pi} d\phi \, \tilde{f}(q\xi) e^{i2\pi q\xi \cdot \mathbf{x}}$$

$$= \int_0^{\pi} d\phi \left[\int_{-\infty}^{\infty} dq \, |q| \tilde{f}(q\xi) e^{i2\pi qp} \right]_{p = \xi \cdot \mathbf{x}}. \tag{3-20}$$

The term in square brackets is the inverse Fourier transform of the product $|q| \tilde{f}$ evaluated at $p = \xi \cdot \mathbf{x}$. From the Fourier convolution theorem[35]

$$\mathcal{F}_1^{-1}\{|q| \tilde{f}(q\xi)\} = \mathcal{F}_1^{-1}\{|q|\} * \mathcal{F}_1^{-1}\{\tilde{f}(q\xi)\}.$$

But, from (3-19), $\mathcal{F}_1^{-1}\{\tilde{f}(q\xi)\} = \check{f}(p, \xi)$. Thus (3-20) simplifies to

$$f(\mathbf{x}) = \int_0^{\pi} d\phi \, [\check{f}(p, \xi) * \mathcal{F}_1^{-1}\{|q|\}]_{p = \xi \cdot \mathbf{x}}. \tag{3-21}$$

The inverse Fourier transform of $|q|$ must be interpreted in terms of generalized functions.[35,51] The desired form for the inverse is obtained by writing

$$|q| = q \operatorname{sgn} q = 2\pi i q \, \frac{\operatorname{sgn} q}{2\pi i},$$

where

$$\operatorname{sgn} q = \begin{cases} +1 & q > 0 \\ 0 & q = 0. \\ -1 & q < 0 \end{cases}$$

Then[35,51]

$$\mathscr{F}_1^{-1}\{|q|\} = \mathscr{F}_1^{-1}\left\{(2\pi i q)\left(\frac{\operatorname{sgn} q}{2\pi i}\right)\right\}$$

$$= \mathscr{F}_1^{-1}\{2\pi i q\} * \mathscr{F}_1^{-1}\left\{\frac{\operatorname{sgn} q}{2\pi i}\right\}. \qquad (3\text{-}22)$$

The first transform on the right can be evaluated by use of the derivative theorem.

Derivative theorem.   If

$$\mathscr{F}_1 f(p) = \tilde{f}(q)$$

then

$$\mathscr{F}_1 f'(p) = 2\pi i q \tilde{f}(q),$$

where the prime indicates differentiation with respect to p.

If $f(p)$ in this theorem is identified with the impulse function $\delta(p)$, then

$$\mathscr{F}_1 \{\delta'(p)\} = 2\pi i q \mathscr{F}_1\{\delta(p)\} = 2\pi i q,$$

since the Fourier transform of the delta function is unity. Thus by taking the inverse transform,

$$\mathscr{F}_1^{-1}\{2\pi i q\} = \delta'(p). \qquad (3\text{-}23)$$

Bracewell[35] shows that

$$\mathscr{F}_1\left\{\frac{1}{p}\right\} = -i\pi \operatorname{sgn} q, \qquad (3\text{-}24)$$

where $1/p$ is to be interpreted in terms of a Cauchy principal value. To emphasize this interpretation, we write the inverse of (3-24), after dividing by $2\pi^2$, as

$$\mathcal{F}_1^{-1}\left\{\frac{\operatorname{sgn} q}{2\pi i}\right\} = \frac{1}{2\pi^2}\,\mathcal{P}\left(\frac{1}{p}\right). \tag{3-25}$$

By use of (3-23) and (3-25) in (3-22), it follows that

$$\mathcal{F}_1^{-1}\{|q|\} = \delta'(p) * \frac{1}{2\pi^2}\,\mathcal{P}\left(\frac{1}{p}\right), \tag{3-26}$$

and (3-21) becomes

$$f(x) = \frac{1}{2\pi^2}\int_0^\pi d\phi\left[\check{f}(p, \xi) * \delta'(p) * \mathcal{P}\left(\frac{1}{p}\right)\right]_{p\,=\,\xi\,\cdot\,x}. \tag{3-27}$$

Application of the derivative theorem for consolution $f * g' = f' * g$ yields

$$\check{f}(p, \xi) * \delta'(p) = \check{f}_p(p, \xi) * \delta(p) = \check{f}_p(p, \xi),$$

where $\check{f}_p = \partial\check{f}/\partial p$.

Now (3-27) can be written as

$$\begin{aligned}
f(\mathbf{x}) &= \frac{1}{2\pi^2}\int_0^\pi d\phi\left[\check{f}_p\,(p, \xi) * \mathcal{P}\left(\frac{1}{p}\right)\right]_{p\,=\,\xi\,\cdot\,x}\\
&= \frac{1}{2\pi^2}\int_0^\pi d\phi\left[\mathcal{P}\int_{-\infty}^\infty \frac{\check{f}_t\,(t, \xi)}{p - t}\,dt\right]_{p\,=\,\xi\,\cdot\,x}\\
&= \frac{-1}{2\pi^2}\,\mathcal{P}\int_0^\pi d\phi\int_{-\infty}^\infty \frac{\check{f}_t\,(t, \xi)}{t - \xi \cdot \mathbf{x}}\,dt.
\end{aligned} \tag{3-28}$$

It is understood that in the last step the $\mathcal{P}$ refers to the integral over t and not over ϕ.

Introduction of the Hilbert transform

Equation (3-28) can be rewritten in terms of a Hilbert transform. The Hilbert transform of f is given by[35, 36]

$$f_H(x) = \mathcal{H}[f(t); t \rightarrow x] = \frac{1}{\pi}\int_{-\infty}^\infty \frac{f(t)/dt}{t - x} = -\frac{1}{\pi}\,\mathcal{P}\left(\frac{1}{x}\right) * f(x). \tag{3-29}$$

From (3-28) and (3-29)

$$f(\mathbf{x}) = \frac{-1}{2\pi} \int_0^\pi d\phi \; |\mathcal{H}[\check{f}_p \; (p, \xi); p \to \xi \bullet \mathbf{x}] \; . \qquad (3\text{-}30)$$

Introduction of backprojection

The integral over ϕ can be expressed in terms of the *backprojection* operator $\mathcal{B}$. If $\psi(t, \xi)$ is some function defined on Radon space then the backprojection operation yields a function of x and y,[52]

$$[\mathcal{B}\psi] \; (x, y) = \int_0^\pi \psi(x \cos \phi + y \sin \phi, \xi) \; d\phi$$

$$= \int_0^\pi \psi \; (\xi \bullet \mathbf{x}, \xi) \; d\phi \; .$$

In terms of backprojection, (3-30) becomes

$$f(\mathbf{x}) = \frac{-1}{2\pi} \mathcal{B}\mathcal{H}[\check{f}_p \; (p, \xi); p \to t] \; . \qquad (3\text{-}31)$$

Note that now $p \to t$ in the Hilbert transform since the $\mathcal{B}$ operator affects the $t \to \xi \bullet \mathbf{x}$ change during the backprojection.

The operator form of (3-31) is

$$f = \frac{-1}{2\pi} \mathcal{B}\mathcal{H} \frac{\partial}{\partial p} \mathcal{R}f \; .$$

This equation implies the identity operator equation

$$I = \frac{-1}{2\pi} \mathcal{B}\mathcal{H} \, \partial\mathcal{R} \; , \qquad (3\text{-}32)$$

and a symbolic form for the inverse Radon transform,

$$\mathcal{R}^{-1} = \frac{-1}{2\pi} \mathcal{B}\mathcal{H} \, \partial \; . \qquad (3\text{-}33)$$

It is understood that the operator identity in (3-32) operates on functions defined on feature space. The inverse operator (3-33) operates on functions defined on Radon space. The symbol ∂ means differentiate with respect to the radial variable, usually designated by p.

Other versions

Barrett[32] points out that another form for $\mathcal{R}^{-1}$ follows from (3-28) in the form

$$f(\mathbf{x}) = \frac{1}{2\pi^2} \int_0^\pi d\phi \left[\breve{f}_p(p, \xi) * \mathcal{P}\left(\frac{1}{p}\right) \right]_{p \,=\, \xi \,\cdot\, \mathbf{x}}, \qquad (3\text{-}34)$$

and the observation that

$$\mathcal{P}\left(\frac{1}{p}\right) = \frac{d}{dp} \log|p|. \qquad (3\text{-}35)$$

(Note that natural logarithms are designated by log.) If $\mathcal{P}\,(1/p)$ from (3-35) is substituted into (3-34), an integration by parts in the convolution integral yields

$$f(\mathbf{x}) = \frac{1}{2\pi^2} \int_0^\pi d\phi \left[\breve{f}_{pp}(p, \xi) * \log|p| \right]_{p \,=\, \xi \,\cdot\, \mathbf{x}}.$$

And by the properties of convolution involving derivatives, another form is

$$f(\mathbf{x}) = \frac{1}{2\pi^2} \int_0^\pi d\phi \left\{ \frac{\partial^2}{\partial p^2} \left[\breve{f}(p, \xi) * \log|p| \right] \right\}_{p \,=\, \xi \,\cdot\, \mathbf{x}}. \qquad (3\text{-}36)$$

This equation can be written in yet another form by the observation

$$\nabla^2 \psi(\xi \cdot \mathbf{x}) = (\xi_1^2 + \xi_2^2) \left[\frac{\partial^2 \psi(p)}{\partial p^2} \right]_{p \,=\, \xi \,\cdot\, \mathbf{x}} = \left[\frac{\partial^2 \psi(p)}{\partial p^2} \right]_{p \,=\, \xi \,\cdot\, \mathbf{x}}, \qquad (3\text{-}37)$$

since ξ is a unit vector. Here, ∇^2 is the Laplacian operator

$$\nabla^2 = \frac{\partial^2}{\partial x^2} + \frac{\partial^2}{\partial y^2}.$$

From (3-36) and (3-37)

$$f(\mathbf{x}) = \frac{1}{2\pi^2} \nabla^2 \int_0^\pi d\phi \left[\breve{f}(p, \xi) * \log|p| \right]_{p \,=\, \xi \,\cdot\, \mathbf{x}}.$$

For future reference, it is useful to observe that (3-37) holds for higher dimensions with the appropriate generalization of the ∇^2 operator.

Convolution methods

Although no theoretical difficulties are presented by going from Radon space to feature space by way of Fourier space, there are problems associated with numerical implementation owing to the presence of the $|q|$ factor. One way to eliminate some of the numerical difficulties is to avoid Fourier space. It would be desirable to have some well-behaved function $Q(s)$ such that [48]

$$\mathcal{F}_1 Q = |q|. \tag{3-38}$$

Then the part of (3-20) in brackets could be written as

$$\bar{f}(s, \phi) = \int_{-\infty}^{\infty} |q| \check{f}(q\xi) e^{i2\pi sq} \, dq$$

$$= \mathcal{F}_1^{-1} \{(\mathcal{F}_1 Q)(\mathcal{F}_1 \check{f})\},$$

since $\check{f} = \mathcal{F}_1 \, \check{f}$. From the Fourier convolution theorem,

$$\bar{f}(s, \phi) = Q * \check{f} = \check{f} * Q. \tag{3-39}$$

Finally, $f(x, y)$ could be recovered by backprojection,

$$f(x, y) = \int_0^{\pi} \bar{f}(x \cos \phi + y \sin \phi, \phi) \, d\phi. \tag{3-40}$$

Equations (3-39) and (3-40) form the basis for convolution and backprojection methods that avoid Fourier space. Symbolically,

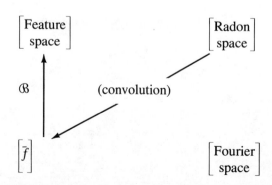

We have already seen in (3-26) that Q in (3-38) is a singular distribution. The way around this is to find a well-behaved function whose Fourier transform approximates $|q|$. These approximation methods have received much attention since they work much better with real (and therefore noisy) input data. Extensive literature exists on this subject, and there are several good entry points.[44, 47, 52, 53]

Three Dimensions

The inversion formula in three dimensions can be found by going through Fourier space as with the two-dimensional case. The details are considerably less complicated since no Hilbert transforms or convolutions with $\log |p|$ emerge.

The starting point is (3-19), with ξ a unit vector in three dimensions,

$$\xi = (\sin \theta \cos \phi, \sin \theta \sin \phi, \cos \theta),$$

in terms of the polar angle θ and azimuthal angle ϕ. The three-dimensional counterpart of (3-20) is

$$f(x) = \mathfrak{F}_3^{-1} \tilde{f} = \int_0^\infty q^2 \, dq \int d\xi \, \tilde{f}(q\xi) e^{i2\pi q\xi \cdot x}$$

$$= \tfrac{1}{2} \int_{-\infty}^\infty q^2 \, dq \int d\xi \, \tilde{f}(q\xi) e^{i2\pi q\xi \cdot x}. \qquad \text{(3-41)}$$

Here $x = (x, y, z)$ and the $\int d\xi$ is over the unit sphere. In terms of θ and ϕ,

$$\int d\xi = \int_0^{2\pi} \int_0^\pi \sin \theta \, d\theta \, d\phi \; .$$

Since $\tilde{f}(q\xi) = \mathfrak{F}_1 f(p, \xi)$ and the Fourier derivative theorem yields

$$\mathfrak{F}_1 \left\{ \frac{\partial^2 f}{\partial p^2} \right\} = -4\pi^2 q^2 \tilde{f}(q\xi),$$

equation (3-41) can be expressed as

$$f(x) = \frac{1}{2} \int d\xi \left[\int_{-\infty}^\infty dq \, q^2 \tilde{f}(q\xi) e^{i2\pi qp} \right]_{p \, = \, \xi \, \cdot \, x}$$

$$= \frac{1}{2} \int d\xi \, \mathfrak{F}_1^{-1} \, [q^2 \tilde{f}(q\xi)]_{p \, = \, \xi \, \cdot \, x}$$

$$= \frac{-1}{8\pi^2} \int d\xi \, [\check{f}_{pp}(p, \xi)]_{p \, = \, \xi \, \cdot \, x}, \qquad \text{(3-42)}$$

where $\check{f}_{pp} = \partial^2 \check{f}/\partial p^2$. The extension of (3-37) to three dimensions is

$$\nabla^2 \psi(\xi \cdot \mathbf{x}) = \left[\frac{\partial^2 \psi(p)}{\partial p^2} \right]_{p = \xi \cdot \mathbf{x}},$$

where ∇^2 is the Laplacian operator

$$\nabla^2 = \frac{\partial^2}{\partial x^2} + \frac{\partial^2}{\partial y^2} + \frac{\partial^2}{\partial z^2}.$$

Consequently, (3-42) can be rewritten as

$$f(\mathbf{x}) = \frac{-1}{8\pi^2} \nabla^2 \int \check{f}(\xi \cdot \mathbf{x}, \xi) \, d\xi. \tag{3-43}$$

Note that the integral in (3-43) is an extension of the concept of backprojection to three dimensions.

For a further general discussion of three-dimensional inversion, see Barrett;[32] for a discussion of numerical considerations, see Louis[54] and references therein.

Higher Dimensions

The development of inversion formulas for higher dimensions ($n > 3$) can be patterned after the R^2 and R^3 cases going through Fourier space and using the general operator relationship

$$f = \mathfrak{F}_n^{-1} \mathfrak{F}_1 \mathfrak{R} f.$$

Details will not be given here; there are several readily available sources.[2, 27, 32, 29] Other methods not depending so heavily on Fourier transforms are also available.[2, 15]

3-7   SYMMETRY AND RELATED TRANSFORMS

When the feature space function exhibits certain symmetry properties, the Radon transform of f reduces to more familiar transforms. Some of these interrelationships are discussed in this section. Other related transforms emerge more naturally in the context of series in Section 3-8.

Abel Transform

If $f(x, y)$ is circularly symmetric in feature space, the Radon transform of f is identical to the Abel transform. Since f depends only on $r = (x^2 + y^2)^{1/2}$ and not on the polar angle, $f = f(r)$ and all profiles (projections) are equivalent. Suppose the angle ϕ in the Radon transform formula (3-1) is taken as zero, then $\check{f}(p, \phi) = \check{f}(p, 0)$. Since $\check{f}$ does not depend on ϕ, this can be written as $\check{f}(p)$. Now (3-1) becomes

$$
\begin{aligned}
\check{f}(p) &= \int_{-\infty}^{\infty} \int_{-\infty}^{\infty} f(\sqrt{x^2 + y^2})\, \delta(p - x)\, dx\, dy \\
&= \int_{-\infty}^{\infty} f(\sqrt{p^2 + y^2})\, dy \\
&= 2 \int_{0}^{\infty} f(\sqrt{p^2 + y^2})\, dy.
\end{aligned}
\tag{3-44}
$$

After changing variables $r^2 = p^2 + y^2$, $r\, dr = y\, dy$, (3-44) becomes

$$
\begin{aligned}
\check{f}(p) &= 2 \int_{|p|}^{\infty} \frac{f(r) r\, dr}{(r^2 + p^2)^{1/2}} \\
&= f_A(p).
\end{aligned}
\tag{3-45}
$$

This is just the Abel transform of $f(r)$, indicated by writing $f_A(p)$.

The Abel transform can be inverted by Fourier methods,[35] by Laplace transform methods,[36] or by Radon transform methods.[2] It is interesting to follow Barrett [32] and derive the inverse directly from

$$
\mathcal{F}_2 f = \mathcal{F}_1 \mathcal{R} f = \mathcal{F}_1 f_A.
$$

If $f = f(r)$, then $\mathcal{F}_2 f$ is just a zero-order Hankel transform of f,[35]

$$
\mathcal{F}_2 f = \mathcal{H}_0 f = 2\pi \int_0^{\infty} f(r) J_0(2\pi q r)\, r\, dr.
$$

J_0 is a zeroth-order Bessel function of the first kind, and q is the radial coordinate in Fourier space.

Notation. Note that the symbol $\mathcal{H}$ refers to the Hilbert transform defined in (61) and $\mathcal{H}_n$ refers to a Hankel transform of order n.

Since $\mathfrak{K}_0^{-1} = \mathfrak{K}_0,^{36}$ it follows that $\mathfrak{K}_0 f = \mathfrak{F}_1 f_A$, and

$$f = \mathfrak{K}_0^{-1} \mathfrak{F}_1 f_A = \mathfrak{K}_0 \mathfrak{F}_1 f_A .$$

Thus the inverse Abel transform operator is just $\mathfrak{K}_0 \mathfrak{F}_1$. The explicit calculation follows the approach used by Barrett,[32]

$$f(r) = 2\pi \int_0^\infty q \, dq \, J_0(2\pi qr) \int_{-\infty}^\infty f_A(p) e^{-2\pi qp} \, dp . \qquad \textbf{(3-46)}$$

Since $f_A(p)$ is even, the last integral can be written as

$$2 \int_0^\infty f_A(p) \cos(2\pi qp) \, dp .$$

An integration by parts yields

$$\frac{-1}{\pi q} \int_0^\infty f_A'(p) \sin(2\pi qp) \, dp .$$

It is assumed that $f_A(p) \to 0$ as $p \to \infty$, and the prime indicates differentiation with respect to p. Now (3-46) can be expressed as

$$f(r) = -2 \int_0^\infty dp \, f_A'(p) \int_0^\infty dq \, J_0(2\pi qr) \sin(2\pi qp) .$$

The last integral vanishes for $0 < p < r$ and yields[55]

$$\left(\frac{1}{(2\pi p)^2 - (2\pi r)^2}\right)^{1/2} \qquad \text{for} \quad 0 < r < p .$$

Thus for $r > 0$,

$$f(r) = -\frac{1}{\pi} \int_r^\infty f_A'(p) \, (p^2 - r^2)^{-1/2} \, dp . \qquad \textbf{(3-47)}$$

Equations (3-45) and (3-47) constitute an Abel transform pair. For a further discussion of Abel transforms and applications, see Bracewell[35] and Vest.[56]

Spherical Symmetry

If $f(x, y, z)$ has spherical symmetry, then f depends on $(x^2 + y^2 + z^2)^{1/2}$. And it is sufficient to consider the Radon transform for both θ and ϕ equal to zero. Thus, the unit vector $\xi = (0, 0, 1)$ and $\xi \cdot \mathbf{x} = z$. The three-dimensional formula (3-7) reduces to

$$\check{f}(p) = \int_{-\infty}^{\infty} \int_{-\infty}^{\infty} \int_{-\infty}^{\infty} f(\sqrt{x^2 + y^2 + z^2})\, \delta(p - z)\, dx\, dy\, dz$$

$$= \int_{-\infty}^{\infty} \int_{-\infty}^{\infty} f(\sqrt{x^2 + y^2 + p^2})\, dx\, dy.$$

In terms of polar coordinates, $x = \rho \cos \phi$, $y = \rho \sin \phi$,

$$\check{f}(p) = \int_{0}^{2\pi} \int_{0}^{\infty} f(\sqrt{\rho^2 + p^2})\, \rho\, d\rho\, d\phi = 2\pi \int_{0}^{\infty} f(\sqrt{\rho^2 + p^2})\rho\, d\rho.$$

Now let $\rho^2 + p^2 = r^2$ so that $\rho\, d\rho = r\, dr$ and $r = (p^2)^{1/2}$ when $\rho = 0$. Then

$$\check{f}(p) = 2\pi \int_{|p|}^{\infty} f(r)r\, dr$$

$$= 2\pi \int_{p}^{\infty} f(r)r\, dr, \qquad p > 0.$$

In this case, $f(r)$ can be recovered from $f(p)$ by differentiation!

$$\frac{d\check{f}(p)}{dp} = -2\pi p f(p).$$

Or, with p replaced by r,

$$f(r) = \frac{-1}{2\pi r}\check{f}'(r). \tag{3-48}$$

The same result follows after a short computation [57] directly from inversion formula (3-42) or (3-43). Barrett[32] points out that (3-48) was given by Vest and Steel[58] in 1978 but was known much earlier in connection with Compton scattering [59] and positron annihilation.[60, 61]

Hankel and Tchebycheff Transforms

Suppose $f(x, y) = f(r, \theta)$ can be expanded in a Fourier series

$$f(r, \theta) = \sum_{n} f_n(r) e^{in\theta}, \tag{3-49}$$

where the Fourier coefficients are given by

$$f_n(r) = \frac{1}{2\pi} \int_0^{2\pi} f(r,\theta) e^{-in\theta} \, d\theta .$$

(The sum on n is from $-\infty$ to ∞ throughout this section.) The polar form of $\mathcal{F}_2 f$ may be expanded in a series of the same form,

$$\tilde{f}(q, \phi) = \sum_n \tilde{f}_n(q) e^{in\phi} , \qquad (3\text{-}50)$$

where

$$\tilde{f}_n(q) = \frac{1}{2\pi} \int_0^{2\pi} \tilde{f}(q, \phi) e^{-in\phi} \, d\phi .$$

The two-dimensional Fourier transform of (3-49) in polar form is

$$\tilde{f}_n(q, \phi) = \sum_n \int_0^{2\pi} \int_0^{\infty} f_n(r) e^{in\theta} e^{-i2\pi qr\cos(\theta - \phi)} \, r \, dr \, d\theta$$

$$= \sum_n e^{in\phi} \int_0^{\infty} dr \, rf_n(r) \int_0^{2\pi} d\beta e^{in\beta - i2\pi qr\cos\beta} , \qquad (3\text{-}51)$$

where the change of variables $\beta = \theta - \phi$ was made in the last step. The integral over β can be expressed in terms of a Bessel function of the first kind,[62]

$$J_n(x) = \frac{e^{in\pi/2}}{2\pi} \int_0^{2\pi} e^{i(n\beta - x\cos\beta)} d\beta .$$

Thus (3-51) becomes

$$\tilde{f}_n(q, \phi) = 2\pi \sum_n e^{in\phi} e^{-in\pi/2} \int_0^{\infty} f_n(r) J_n(2\pi qr) r \, dr .$$

The Hankel transform of order n is given by[35, 36]

$$\mathcal{H}_n[F(r); r \to q] = 2\pi \int_0^{\infty} F(r) J_n(2\pi qr) r \, dr .$$

In terms of the Hankel transform,

$$\tilde{f}_n(q, \phi) = \sum_n e^{in\phi} e^{-in\pi/2} \mathcal{H}_n[f_n(r)] . \qquad (3\text{-}52)$$

Since the functions $e^{in\phi}$ are linearly independent, by comparing (3-50) to (3-52) it follows that

$$\tilde{f}_n(q) = e^{-in\pi/2}\mathcal{H}_n[f_n(r)] = (-i)^n\mathcal{H}_n[f_n(r)]. \tag{3-53}$$

To make the connection with functions in Radon space, expand $\check{f}(p, \phi)$,

$$\check{f}(p, \phi) = \sum_n \check{f}_n(p)e^{in\phi}, \tag{3-54}$$

where

$$\check{f}_n(p) = \frac{1}{2\pi}\int_0^{2\pi} \check{f}(p, \phi)e^{-in\phi}\, d\phi.$$

The Fourier transform of $\check{f}$ on the variable p gives

$$\tilde{f}(q, \phi) = \sum_n e^{in\phi}\int_{-\infty}^{\infty} \check{f}_n(p)e^{-i2\pi qp}\, dp. \tag{3-55}$$

Again, by linear independence, from (3-50) and (3-55),

$$\tilde{f}_n(q) = \mathcal{F}_1\check{f}_n(p). \tag{3-56}$$

By combining (3-53) and (3-56),

$$\mathcal{H}_n[f_n(r)] = i^n\mathcal{F}_1\check{f}_n(p). \tag{3-57}$$

Equation (3-57) provides an alternative method for computing Hankel transforms, a method that does not require the evaluation of Bessel functions.[37] Finally, to make the interconnections complete it is necessary to find the relationship between $f_n(r)$ and $\check{f}_n(p)$. The essential result is presented here; for more details see Cormack[21,22] or Deans.[2] The Radon transform of (3-49) yields

$$\check{f}(p, \phi) = \sum_n \int_0^{2\pi}\int_0^{\infty} e^{in\theta}f_n(r)\, \delta[p - r\cos(\theta - \phi)]r\, dr\, d\theta$$

$$= \sum_n e^{in\phi}\int_0^{\infty} dr\, rf_n(r)\int_0^{2\pi} d\beta e^{in\beta}\, \delta(p - r\cos\beta), \tag{3-58}$$

where the change of variables $\beta = \theta - \phi$ was made in the last step as with (3-51). By comparing (3-54) and (3-58),

$$\check{f}_n(p) = \int_0^\infty dr \, r f_n(r) \int_0^{2\pi} d\beta e^{in\beta} \, \delta(p - r \cos \beta)$$

$$= 2 \int_{|p|}^\infty f_n(r) \, T_n\left(\frac{p}{r}\right) \left(1 - \frac{p^2}{r^2}\right)^{-1/2} dr. \qquad \textbf{(3-59)}$$

The functions T_n are Tchebycheff polynomials of the first kind and (3-59) is a Tchebycheff transform.

Dipole-Sheet Transform

Barrett[32, 63] observed that the inversion formula (3-43) could be written as

$$f(\mathbf{x}) = \frac{-1}{4\pi^2} \int \check{f}(p, \xi) \, \delta''(p - \xi \cdot \mathbf{x}) \frac{d\mathbf{p}}{p^2}, \qquad \textbf{(3-60)}$$

where the integral is understood to mean

$$\int d\mathbf{p} = \int_0^\infty p^2 \, dp \int d\xi.$$

Integration over the unit sphere is as in the discussion on three dimensions in Section 6. Barrett further observed that there is an asymmetry between (3-60) and the formula for $\check{f}(p, \xi)$ in (3-7),

$$\check{f}(p, \xi) = \int f(\mathbf{x}) \, \delta(p - \xi \cdot \mathbf{x}) \, d\mathbf{x}.$$

A more symmetrical pair can be constructed by defining a new transform,

$$f_D(\mathbf{p}) = \int \psi(\mathbf{p}, \mathbf{x}) f(\mathbf{x}) \, d\mathbf{x}, \qquad \textbf{(3-61)}$$

where

$$\psi(\mathbf{p}, \mathbf{x}) = \frac{i}{2\pi p} \delta'(p - \xi \cdot \mathbf{x}), \qquad p = |\mathbf{p}|.$$

By using the formula for the inverse Radon transform, Barrett[32, 63] was able to show that

$$f(\mathbf{x}) = \int \psi^*(\mathbf{p}, \mathbf{x}) f_D(\mathbf{p}) \, d\mathbf{p}, \qquad \textbf{(3-62)}$$

where the $*$ means complex conjugation. The transform pair (3-61) and (3-62) clearly possesses a desirable symmetry since it can be shown that the transform is unitary.[32]

The physical interpretation of $\psi(\mathbf{p}, \mathbf{x})$ as a double layer or dipole sheet led Barrett[63] to call $f_D(\mathbf{p})$ the *dipole-sheet transform*. The functions ψ satisfy orthogonality and completeness conditions[63] and the transform itself has many interesting properties discussed by Barrett.[32,63]

3-8 SERIES

Series methods can be used when $f(\mathbf{x})$ can be expressed as a linear superposition of terms of the form

$$g_l(r)\, S_{lm}(\omega),$$

where $\mathbf{x} \in R_n$ and

$$r = |\mathbf{x}| \geq 0, \qquad \omega = \mathbf{x}/|\mathbf{x}|.$$

The doubly subscripted term $S_{lm}(\omega)$ is a real surface harmonic[64] of degree l.

If

$$f(\mathbf{x}) = g_l(r)\, S_{lm}(\omega), \tag{3-63}$$

then[2]

$$\check{f}(p, \xi) = \check{g}_l(p)\, S_{lm}(\xi), \tag{3-64}$$

where $g_l(r)$ and $\check{g}_l(p)$ are a Gegenbauer transform pair,[2]

$$\check{g}_l(p) = \frac{(4\pi)^\nu \Gamma(l+1)\Gamma(\nu)}{\Gamma(l+2\nu)} \int_p^\infty r^{2\nu} g_l(r)\, C_l^\nu\left(\frac{p}{r}\right) \left[1 - \frac{p^2}{r^2}\right]^{\nu - 1/2} dr,$$

$$g_l(r) = \frac{(-1)^{2\nu+1}\Gamma(l+1)\Gamma(\nu)}{2\pi^{\nu+1}\Gamma(l+2\nu)r} \int_r^\infty \check{g}_l^{(2\nu+1)}(t)\, C_l^\nu\left(\frac{t}{r}\right) \left[\frac{t^2}{r^2} - 1\right]^{\nu - 1/2} dt.$$

Here, $p \geq 0$, $r \geq 0$, and $\nu = (n-2)/2$. C_l^ν is a Gegenbauer polynomial. The superscript $(2\nu+1)$ indicates the order of the derivative with respect to the argument of $\check{g}_l$. If $n = 2$, the transform pair reduces to a Tchebycheff pair, and if $n = 3$, the transform pair reduces to a Legendre pair.[2]

An especially interesting situation arises for $n = 2$ when $f(x, y)$ vanishes outside the unit disk. In this case, if the radial functions in (3-63) are Zernike polynomials, then the radial functions in (3-64) are Tchebycheff polynomials of the second kind. The interesting and important point is that one set of orthogonal functions transforms to another set of orthogonal functions. A further interesting point is that the corresponding radial functions in Fourier space are an orthogonal set of Bessel functions. For details and reference to the early literature, see Deans.[2]

A simple example is illustrated in Figures 3-7 and 3-8. Here

$$f = (x^2 + y^2)x$$

in feature space, and

$$\check{f} = \frac{2p}{3} (2p^2 + 1) \sqrt{1 - p^2} \cos \phi$$

in Radon space.

A significant number of subtle mathematical points emerge in connection with the actual imploementation of inversion through series techniques. Discussions by Verly,[65] Louis,[54, 66, 67] Louis and Natterer,[46] and Hawkins[68] are recommended for $n = 2, 3$ along with general theorems and extensions by Quinto.[50]

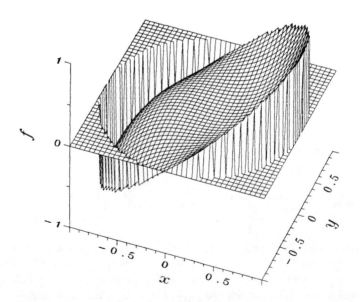

Figure 3-7
Feature space function.

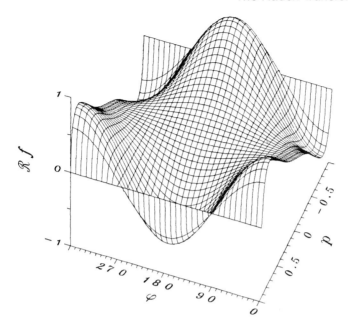

Figure 3-8
Radon transform of function in Figure 3-7.

3-9 APPLICATIONS

Applications of the Radon transform emerge in a wide variety of areas. Partly overlapping subsets of these have been reviewed by Deans[2] and Barrett.[32] The purpose here is to indicate where various applications arise and to cite a few relevant references for each application.

Transmission-Computed Tomography

In transmission-computed tomography, x-rays with initial intensity I_0 are passed through the body along many straight-line paths. As the x-rays pass through the body, the intensity of the beam is reduced from I_0 to I. The natural logarithm of the ratio of I_0 to I is equal to the line integral of the linear attenuation coefficient μ (dependent on density and nuclear composition) along the line from the source to the detector. This integral is a sample of the Radon transform of μ. The entire two-dimensional transform over a plane can be determined by measurement of a sufficient number of samples. The ultimate goal is to invert the transform and thus find the feature space function μ. The three-dimensional structure can be inferred by making measurements over several parallel planes.

The March 1983 issue of *Proceedings of the IEEE* is an excellent source for this important application.[69-74] Other sources include Deans,[2] Herman,[31] and Barrett.[32]

Emission Computed Tomography

In one type of emission computed tomography, a single-photon emitting radioactive pharmaceutical is introduced into the body. The emitted photons are detected by gamma cameras, and the goal is to reconstruct the source density distribution through the body or in a specific organ being studied. Since the detected photon flux depends on both attenuation and source density, a natural extension of the Radon transform emerges in the analysis of projection data.[75-79]

A more direct application of the Radon transform emerges in connection with positron-emission tomography. Positron-emitting isotopes are introduced into the body. Within a few millimeters, the emitted positrons annihilate (essentially at rest) with a nearby electron in the surrounding tissue. This annihilation produces two high-energy gamma rays that travel away from each other at a 180° angle. When these emitted gamma rays are detected in coincidence by external detectors, line-integral data are produced to yield a function proportional to the Radon transform of the isotope distribution. Several discussions of positron-emission tomography are available.[80-84] For a comprehensive review of both types of emission-computed tomography, see Budinger, Gullberg, and Huesman.[53]

Nuclear Magnetic Resonance

Basic principles of nuclear magnetic resonance (NMR) have been familiar to physicists and chemists for many years. Since 1945, NMR spectroscopy has become an indispensable analytical and structural tool in physics and chemistry.[85,86] Recently, NMR techniques have been successfully adapted to imaging internal structure. Some of these techniques provide excellent examples of utilization of both the two-dimensional and the three-dimensional Radon transform.[2,32,87] The prospects for NMR imaging in medicine are especially exciting since no ionizing radiation is involved. Several commercial systems are now available.[74] Excellent reviews are provided by Andrew,[88] Pykett,[89] Barrett,[32] and Hinshaw and Lent.[90]

Other Probes

Probes other than x-rays, gamma rays, and NMR signals can be used to collect projection data. For example, an increasingly important probe in medical imaging involves the use of ultrasound.[91] Other possible probes include heavy ions and heavy particles such as pions, neutrons, and protons.[92-94] Use of weak electrical currents for electrical conductivity imaging may hold promise in medicine[95] and geophysics.[96] All of these probes and others are discussed and analyzed relative to current and future status for computerized tomography by Bates, Garden, and Peters.[74]

Geophysics

The utility and importance of the Radon transform in geophysics was demonstrated by Chapman.[97, 98] The important result is that if the record section or seismogram is identified with the feature space function, then the Radon space function is identified with what geophysicists call the "slant stack seismogram," or simply "slant stack." The Radon transform thus becomes a first step in the interpretation of seismic record sections or its inversion as the last step in the computation of synthetic seismograms.[99, 100] The informative review of the spectral approach to geophysical inversion by Robinson[101] and the review of theoretical seismology by Richards[102] are recommended for background material. Another rapidly expanding field in geophysics is seismic tomography for exploration between boreholes. Details and references are given by McMechan.[103]

Astronomy

A variety of applications of the Radon transform have been discovered by researchers in astronomy. Apparently, the first numerical inversion of the Radon transform was by the Armenian astronomer Ambartsumian,[11, 12] who was unaware of Radon's work until two years after his own independent work was published in 1936.[25] Ambartsumian solved a problem put forward by Eddington:[104] Is it possible to find the distribution function of the components of stellar space velocities from radial velocities alone without making any special assumption about the form of the distribution function? The radial velocities are deduced from the Doppler shifts of their spectra. The actual distribution of velocities in three dimensions is just Radon's problem in a three-dimensional velocity space.

Another pioneering contribution in astronomy was by the radio astronomer Bracewell,[16] who solved Radon's problem in connection with mapping emitted solar microwave radiation. As with Ambartsumian, Bracewell also solved the problem without prior knowledge of Radon's work.

The Radon transform emerges in a number of other astronomical connections: aperture synthesis and radar astronomy,[105, 106] occultation studies,[107, 108] x-ray astronomy,[109] and determination of the magnetic-field and electron-density distributions of the solar corona.[110] For a review of image reconstruction in radio astronomy, see Bracewell.[111]

Electron Microscopy

Reconstruction of the three-dimensional structure of biological macro-molecules from electron micrographs was initiated independently by DeRosier and Klug[112] and by Vainshtein[113] and co-workers in 1968. This work constitutes yet another solution to the Radon problem of reconstruction from projections. Excellent reviews are provided by DeRosier,[114] Crowther and Klug,[115] and Vainshtein.[113] The significance of Aaron Klug's contributions in this area was further recognized in 1982 when he was awarded the Nobel Prize in chemistry, in part for his pioneering work on three-dimensional reconstruction of biological molecules from electron micrographs.[116]

Electron Momentum in Solids

The Radon transform of the momentum space probability density function for a bound electron can be determined from Compton scattering spectra taken at various angles.[32, 117] A similar situation emerges in studying the electronic structure of solids by investigating the electron momentum density with the aid of low-energy positron annihilation radiation. The angular correlations of the annihilation radiation yield Radon space information on the electronic structure of crystalline solids and on the shape of the Fermi surface in metals. The analogy with the corresponding computerized tomography inversion problem in radiology is discussed by Mijnarends,[61] and a more recent review is also given by Mijnarends.[118] The books edited by Williams[119] and Hautojärvi[120] are especially valuable for background and current methods for the determination of electron momentum densities in solids.

Microwave Imaging

The problem of radar target-shape estimation can be formulated as an image-reconstruction problem. Das and Boerner[121] applied Radon transform methods to an important inverse problem of electromagnetics—the identification of a target by a radar system. Further developments of physical optics inverse-scattering techniques using Radon and Fourier transform theory are discussed by Boerner[122] and Bojarski.[123] Devaney's unified discussion of both inverse-source and inverse-scattering problems serves to help bridge the gap between methods of inverse scattering and reconstruction from projections used in computerized tomography.[124] Further extensions using full three-dimensional image reconstruction,[125] and frequency swept-imaging theory suggest the feasibility of tomographic radar.[126, 127]

The discussion of coherent Doppler tomography for microwave imaging by Mensa, Halevy, and Wade[128] is especially interesting since it involves going from Fourier space to feature space by use of a Radon transform. The basic idea is that f can be recovered from $\tilde{f}$ by

$$f = \mathcal{F}_1 \, \Re \, \tilde{f}. \tag{3-65}$$

(In Deans,[2] this is illustrated in Figure 1.3 and proved on page 100.) To obtain a formula equivalent to (9) of Mensa, Halevy, and Wade,[128] let $\tilde{f}$ be given in polar coordinates in Fourier space, $\tilde{f} = \tilde{f}(q, \phi)$. Furthermore, let $\tilde{f}$ be sampled such that the values of $\tilde{f}(2/\lambda, \phi)$ are precisely the observed data,

$$\tilde{f}(q, \phi) \rightarrow \tilde{f}(q, \phi) \, \delta(q - 2/\lambda). \tag{3-66}$$

Now, from (3-65) and (3-66), with $x + iy = re^{i\theta}$ and $u + iv = qe^{i\phi}$,

$$
\begin{aligned}
f(r, \theta) &= \int_{-\infty}^{\infty} dp \, e^{-i2\pi rp} \int_0^{2\pi} \int_0^{\infty} \tilde{f}(q, \phi) \, \delta(q - 2/\lambda) \\
&\qquad \cdot \, \delta[p - q \cos(\theta - \phi)] q \, dq \, d\phi \\
&= \frac{2}{\lambda} \int_0^{2\pi} d\phi \, \tilde{f}(2/\lambda, \phi) \int_{-\infty}^{\infty} dp \, e^{-i2\pi rp} \, \delta\left[p - \frac{2}{\lambda} \cos(\theta - \phi)\right] \\
&= \frac{2}{\lambda} \int_0^{2\pi} \tilde{f}(2/\lambda, \phi) e^{-i(4\pi/\lambda)r \cos(\theta - \phi)} \, d\phi.
\end{aligned}
$$

Mensa, Halevy, and Wade[128] point out that this equation can be interpreted as a circular convolution integral between data $\tilde{f}(2/\lambda, \phi)$ and the exponential.

Another interesting example of the emergence of the Radon transform in microwave scattering is discussed by Barrett[32] in connection with the temporal impulse response of a weak volume scatterer (such as a tenuous vapor) in the first Born approximation. His results are similar to those obtained for metallic reflectors.[121-125]

Other Applications and Problems

There are several other areas where the Radon transform emerges in a natural way. The connection between Radon theory and electrostatics is discussed by Barrett.[32] Uses of the Radon transform in holographic interferometry, aerodynamics, flow visualization, and plasma physics have been reviewed by Vest[56] and Deans.[2] Willis[129-131] used the Radon transform in studying stresses in a composite body. Other work in stress analysis is reviewed by Hildebrand and Hufferd.[132] Its relevance in picture restoration is illustrated by Rosenfeld and Kak.[44] In pattern analysis, it is useful in line and curve detection.[133] For extensions to more general unresolved targets and references to important related image-analysis work, see Cowart, Snyder, and Ruedger.[134] Other applications include air-pollution studies,[135-136] nondestructive testing,[137-140] image transmission,[141] and optical signal processing.[142] Optical computers for Radon transform inversion have been developed,[143] and coded-aperture imaging methods are still under investigation.[144, 145]

Some of the more mathematical applications have been in partial differential equations, wave propagation, radiation, and scattering.[15, 27, 32, 39, 50, 146-148] Other important mathematical applications emerge in geometry[39, 149] and mathematical statistics.[150, 151]

There are a number of important considerations of a technical mathematical nature that relate to the inherent finiteness of the measurement process. Since $\check{f}$ is usually determined by physical measurements, it only represents an approximation to $\check{f}$—and a *discrete* approximation at that. Thus it is to be expected that questions of uniqueness, resolution, accuracy, incomplete sampling, and ill-posedness in presence of errors such as noise and discretization will arise. A wealth of literature has been generated regarding these and other questions. Fortunately, there are several recent discussions that serve as excellent starting points for those interested in numerical implementation of the Radon transform and its inversion.[40, 44-47, 74, 152-154]

ACKNOWLEDGMENTS

It is a pleasure to acknowledge stimulating discussions with Kent Cullers and other scientists at NASA Ames Research Center and Stanford University during the summer of 1983. Use of library facilities at both locations and graphics facilities at Stanford combined with the friendly guidance provided by many support people helped make this an enjoyable project. Discussions with Bruce Suter and communications with Harrison Barrett and Allan Cormack proved especially valuable and are gratefully acknowledged. Special thanks are reserved for Karen Lee Smith, who helped with the final preparation of the manuscript.

REFERENCES

1. J. Radon, "Über die Bestimmung von Funktionen durch ihre Integralwerte längs gewisser Mannigfaltigkeiten," *Berichte Sachsische Akademie der Wissenchaften, Leipzig, Mathematisch-Physikalische Klasse* 69 (1917): 262–267.
2. S. R. Deans, *The Radon Transform and Some of Its Applications* (New York: Wiley-Interscience, 1983).

3. H. B. A. Bockwinkel, "Over de voortplanting van licht in een twee-assig kristal rondom een middelpunt van trilling," *K. Akad. van Wet. Amsterdam Versl. Natuurk.* 14(2) (1906): 636–651.
4. H. Minkowski, "Über die Körper konstanter Breite," in *Gesammelte Abhandlungen von Hermann Minkowski,* vol. 2, ed. D. Hilbert, (New York: Chelsea, 1967), pp. 277–279.
5. P. Funk, "Über Flächen mit Lauter geschlossnen geodätischen Linien," *Mathematische Annalen* 75 (1913): 278–300.
6. P. Funk, "Über eine geometrische Anwendung der Abelschen Integralgleichung," *Mathematische Annalen* 77 (1916): 129–135.
7. G. E. Uhlenbeck, "Over een stelling van Lorentz en haar uitbreiding voor meerdimensionale ruimten," *Physica, Nederlandsch Tijdschrift voor Natuurkunde* 5 (1925): 423–428.
8. Ph. Mader;, "Über die Dartstellung von Punktfunktionen im *n*-dimensionalen euklidischen Raum durch Ebenenintegrale," *Mathematische Zeitschrift.* 26 (1927): 646–652.
9. F. John, "Bestimmung einer Funktion aus ihren Integralen über gewisse Mannigfaltigkeiten," *Mathematische Annalen* 109 (1934): 488–520.
10. H. Cramér and H. Wold, "Some theorems on distribution functions," *Journal of the London Mathematical Society* 11 (1936): 290–294.
11. V. Ambartsumian, "On the Derivation of the Frequency Function of Space Velocities of the Stars from the Observed Radial Velocities," *Monthly Notices of the Royal Astronomical Society* 96 (1936): 172–179.
12. V. Ambartsumian, "On the Distribution of Space Velocities of B and F type Stars," *Publications de L'Observatoire Astronomique Leningrad* 7 (1936): 21–32.
13. J. Szarski and T. Wazewski (1947). See remarks by A. Cormack.[25]
14. A. Rényi, "On Projections of Probability Distributions," *Acta Mathematica Academiae Scientiarum Hungaricae (Budapest)* 3 (1952): 131–141.
15. F. John, *Plane Waves and Spherical Means Applied to Partial Differential Equations* (New York: Interscience, 1955).
16. R. N. Bracewell, "Strip Integration in Radio Astronomy," *Australian Journal of Physics* 9 (1956): 198–217.
17. S. F. Smerd and J. P. Wild, "The Effects of Incomplete Resolution on Surface Distributions Derived from Strip-Scanning Observations, with Particular Reference to an Application in Radio Astronomy," *Philosophical Magazine,* ser. 8, 2 (1957): 119–130.
18. S. I. Tetel'baum, "About a Method of Obtaining Volume Images with the Help of X Rays," *Bulletin of Kiev Polytechnic Institute (Izvestiya Kievskoga Politekhnich. Instituta)* 22 (1957): 154–160.
19. B. I. Korenblyum, S. I. Telel'baum, and A. A. Tyutin, "About One Scheme of Tomography," *Bulletin of the Institutes of Higher Education—Radiophysics (Izvestiya Vyshikh Uchebnykh Zavedenii—Radiofizika)* 1 (1958): 151–157.
20. M. H. Kalos, S. A. Davis, P. S. Mittelman, P. Mastras, and J. H. Hutton, *Conceptual Design of a Vapor Volume Fraction Instrument,* Nuclear Development Corporation of America, NDA 2131-34, Contract AT (30-1)-2303 (IX), U.S. Atomic Energy Commission (White Plains, N.Y.: 1961).
21. A. M. Cormack, "Representation of a Function by Its Line Integrals, with

Some Radiological Applications," *Journal of Applied Physics* 34 (1963): 2722–2727.

22. A. M. Cormack, "Representation of a Function by Its Line Integrals, with Some Radiological Applications, II, *Journal of Applied Physics* 35 (1964): 2908–2913.

23. J. D. Pincus, *A Mathematical Reconstruction of a Radioactive Source Density from Its Induced Radiation Pattern,* Applied Mathematics Department, AMD 359 (Upton, N.Y.: Brookhaven National Laboratory, 1964).

24. H. H. Barrett, W. G. Hawkins, and M. L. G. Joy, "Historical Note on Computed Tomography," *Radiology* 147 (1983): 172.

25. A. M. Cormack, "Computed Tomography: Some History and Recent Developments," *Proceedings of Symposia in Applied Mathematics* 27 (1982): 35–42.

26. S. Helgason, "The Radon Transform on Euclidean Spaces, Compact Two-Point Homogeneous Spaces, and Grassmann Manifolds," *Acta Mathematica* 113 (1965): 153–180.

27. D. Ludwig, "The Radon Transform on Euclidean Space," *Communications on Pure and Mathematics* 19 (1966): 49–81.

28. I. M. Gel'fand, M. I. Graev, and N. Ya. Vilenkin, *Generalized Functions,* vol. 5 (New York: Academic Press, 1966).

29. A. M. Cormack, Nobel Prize address, December 8, 1979, "Early Two-Dimensional Reconstruction and Recent Topics Stemming from It," *Medical Physics* 7 (1980): 277–282. (Also in *Journal of Computer Assisted Tomography* 4 (1980): 658–664 and *Science* 209 (1980): 1482–1486.

30. G. N. Hounsfield, Nobel Prize address, December 8, 1979, "Computed Medical Imaging," *Med. Phys.* 7 (1980): 283–290. (also in *Journal of Computer Assisted Tomography* 4, 665–674 and *Science* 210, 22–28.)

31. G. T. Herman, *Image Reconstruction from Projections: The Fundamentals of Computerized Tomography* (New York: Academic Press, 1980).

32. H. H. Barrett, "The Radon Transform and Its Applications," in *Progress in Optics* vol. 21, ed. E. Wolf (Amsterdam: Elsevier Science Publishers B. V., 1984), pp. 219–286.

33. R. Gordon, "Reconstruction from Projections in Medicine and Biology," in *Image Formation from Coherence Functions in Astronomy,* vol. 76, *Astrophysics and Space Sciences Library,* ed. C. van Schoovenveld (Dordrecht, Holland: Reidel, 1979), pp. 317–325.

34. L. Schwartz, *Mathematics for the Physical Sciences* (Paris: Hermann, 1966).

35. R. N. Bracewell, *The Fourier Transform and Its Applications,* 2nd ed. (New York: McGraw-Hill Book Company, 1978).

36. I. N. Sneddon, *The Use of Integral Transforms* (New York: McGraw-Hill Book Company, 1972).

37. S. R. Deans and B. Suter, "On Computing Hankel Transforms" (in preparation).

38. I. N. Sneddon, *Fourier Transforms* (New York: McGraw-Hill Book Company, 1951).

39. S. Helgason, *The Radon Transform* (Boston, Basel, Stuttgart: Birkhäuser, 1980).

40. G. T. Herman and F. Natterer, eds., *Mathematical Aspects of Computerized Tomography*, in *Lecture Notes in Medical Informatics*, vol. 8 (Berlin, Heidelberg, New York: Springer-Verlag, 1981).

41. G. Szegö, *Orthogonal Polynomials*, in *American Mathematical Society Colloquium Publications*, vol. 23 (Providence, R.I.: American Mathematical Society, 1939).

42. E. D. Rainville, *Special Functions* (New York: Chelsea, 1960).

43. M. Abramowitz and I. A. Stegun, eds., *Handbook of Mathematical Functions with Formulas, Graphs, and Mathematical Tables*, National Bureau of Standards Applied Mathematics Series 55 (Washington, D.C.: U.S. Government Printing Office, 1964).

44. A. Rosenfeld and A. C. Kak, *Digital Picture Processing*, 2nd ed., vols. I and II (New York: Academic Press, 1982).

45. A. G. Lindgren and P. A. Rattey, "The Inverse Discrete Radon Transform with Applications to Tomographic Imaging Using Projection Data," in *Advances in Electronics and Electron Physics*, vol. 56, ed. C. Marton (New York: Academic Press, 1981), pp. 359–410.

46. A. K. Louis and F. Natterer, "Mathematical Problems of Computerized Tomography," *IEEE Proceedings* 71 (1983): 379–389.

47. R. M. Lewitt, "Reconstruction Algorithms: Transform Methods," *IEEE Proceedings* 71 (1983): 390–408.

48. L. A. Shepp and J. B. Kruskal, "Computerized Tomography: The New Medical X-Ray Technology," *American Mathematics Monthly* 85 (1978): 420–439.

49. K. T. Smith, D. C. Solmon, and S. L. Wagner, "Practical and Mathematical Aspects of the Problem of Reconstructing Objects from Radiographs," *Bulletin of the American Mathematical Society* 83 (1977): 1227–1270.

50. E. T. Quinto, "Null Spaces and Ranges for the Classical and Spherical Radon Transforms," *Journal of Mathematical Analysis and Applications* 90 (1982): 408–420.

51. M. J. Lighthill, *Introduction to Fourier Analaysis and Generalized Functions* (Cambridge, England: Cambridge University, 1958).

52. S. W. Rowland, "Computer Implementation of Image Reconstruction Formulas," in *Image Reconstruction from Projections*, vol. 32, *Topics in Applied Physics*, ed. G. T. Herman (New York: Springer-Verlag, 1979), pp. 7–79.

53. T. F. Budinger, G. T. Gullberg, and R. H. Huesman, "Emission Computed Tomography," in *Image Reconstruction from Projections*, vol. 32, *Topics in Applied Physics*, ed. G. T. Herman (New York: Springer-Verlag, 1979), pp. 147–246.

54. A. K. Louis, "Optimal Sampling in Nuclear Magnetic Resonance (NMR) Tomography," *Journal of Computer Assisted Tomography* 6 (1982): 334–340.

55. I. S. Gradshteyn and I. M. Ryzhik, *Table of Integrals, Series, and Products*, 4th ed. (New York: Academic Press, 1965).

56. C. M. Vest, *Holographic Interferometry* (New York: John Wiley & Sons, 1979).

57. M. Y. Chiu, H. H. Barrett, and R. G. Simpson, "Three-Dimensional Reconstruction from Planar Projections," *Journal of the Optical Society of America* 70 (1980): 755–762.

58. C. M. Vest and D. G. Steel, "Reconstruction of Spherically Symmetric Objects

from Slit-Imaged Emission: Application to Spatially Resolved Spectroscopy," *Optics Letters* 3 (1978): 54–56.

59. J. W. M. Du Mond, "Compton Modified Line Structure and Its Relation to the Electron Theory of Solids," *Physical Review* 33 (1929): 643–658.

60. A. T. Stewart, "Momentum Distribution of Metallic Electrons by Positron Annihilation," *Canadian Journal of Physics* 35 (1957): 168–183.

61. P. E. Mijnarends, "Determination of Anisotropic Momentum Distribution in Positron Annihilation," *Physical Review* 160 (1967): 512–519.

62. G. N. Watson, *Theory of Bessel Functions,* 2nd ed. repr. (London: Cambridge University, 1966), p. 20.

63. H. H. Barrett, "Dipole-Sheet Transform," *Journal of the Optical Society of America* 72 (1982): 468–475.

64. H. Hochstadt, *The Functions of Mathematical Physics* (New York: John Wiley & Sons, 1971).

65. J. G. Verly, "Circular and Extended Circular Harmonic Transforms and Their Relevance to Image Reconstruction from Line Integrals," *Journal of the Optical Society of America* 71 (1981): 825–835.

66. A. K. Louis, "Ghosts in Tomography—The Null Space of the Radon Transform," *Mathematical Methods in the Physical Sciences* 3 (1981): 1–10.

67. A. K. Louis, "Orthogonal Function Series Expansions and the Null Space of the Radon Transform," *SIAM Journal of Mathematical Analysis* (in press).

68. W. G. Hawkins, Mathematics of Computed Tomography, Ph.D. thesis, (Tucson: University of Arizona, 1983).

69. G. T. Herman, "The Special Issue on Computerized Tomography," *IEEE Proceedings* 71 (1983): 291–292.

70. L. Axel, P. H. Arger, and R. A. Zimmerman, "Applications of Computerized Tomography to Diagnostic Radiology," *IEEE Proceedings* 71 (1983): 293–297.

71. D. P. Boyd and M. J. Lipton, "Cardiac Computed Tomography," *IEEE Proceedings* 71 (1983): 298–307.

72. R. A. Robb, E. A. Hoffman, L. J. Sinak, L. D. Harris, and E. L. Ritman, "High-Speed Three-Dimensional X-Ray Computed Tomography: The Dynamic Spatial Reconstructor," *IEEE Proceedings* 71 (1983): 308–319.

73. P. Block and J. K. Udupa, "Application of Computerized Tomography to Radiation Therapy and Surgical Planning," *IEEE Proceedings* 71 (1983): 351–355.

74. R. H. T. Bates, K. L. Garden, and T. M. Peters, "Overview of Computerized Tomography with Emphasis on Future Developments," *IEEE Proceedings* 71 (1983): 356–372.

75. G. T. Gullberg, The Attenuated Radon Transform: Theory and Application in Medicine and Biology, Ph.D. thesis, Lawrence Berkeley Laboratory Report LBL-7486 (Berkeley: University of California, 1979).

76. F. Natterer, "On the Inversion of the Attenuated Radon Transform," *Numerische Mathematik* 32 (1979): 431–438.

77. O. J. Tretiak and C. Metz, "The Exponential Radon Transform," *SIAM Journal of Applied Mathematics* 39 (1980): 341–354.

78. G. T. Gullberg, "The Attenuated Radon Transform: Application to Single-Photon Emission Computed Tomography in the Presence of a Variable Atten-

uating Medium," Lawrence Berkeley Laboratory Report LBL-10276 (Berkeley, Calif., 1980).

79. G. F. Knoll, "Single-Photon Emission Computed Tomography," *IEEE Proceedings* 71 (1983): 320–329.

80. G. L. Brownell, G. A. Correia, and R. G. Zamenhof, "Positron Instrumentation," in *Recent Advances in Nuclear Medicine,* vol. 5, eds. J. H. Lawrence and T. F. Budinger (New York: Grune & Stratton, 1978), pp. 1–49.

81. T. F. Budinger, S. E. Derenzo, G. T. Gullberg, and R. H. Huesman, "Trends and Prospects for Circular Ring Positron Cameras," *IEEE Transactions on Nuclear Science* NS-26(2) (1979): 2742–2745.

82. M. E. Phelps, E. J. Hoffman, S. Huang, and D. E. Kuhl, "Design Considerations in Positron Computed Tomography (PCT)," *IEEE Transactions on Nuclear Science* (2) (1979): 2746–2751.

83. M. M. Ter-Pogossian, M. E. Raichle, and B. E. Sobel, "Positron-Emission Tomography," *Scientific American* 243(4) (1980): 171–181.

84. R. A.Brooks, V. J. Sank, W. S. Friauf, S. B. Leighton, H. E. Cascio, and G. Di Chiro, "Design Considerations for Positron Emission Tomography," *IEEE Transactions on Biomedical Engineering* BME-28 (1981): 158–177.

85. A. Abragam, *The Principles of Nuclear Magnetism* (London: Oxford University, 1961).

86. E. R. Andrew, *Nuclear Magnetic Resonance* (Cambridge, England: Cambridge University, 1969).

87. L. A. Shepp, "Computerized Tomography and Nuclear Magnetic Resonance," *Journal of Computer Assisted Tomography* 4 (1980): 94–107.

88. E. R. Andrew, "NMR Imaging of Intact Biological Systems," *Philosophical Transactions of the Royal Society of London,* Series B 289 (1980): 471–481.

89. I. L. Pykett, "NMR Imaging in Medicine," *Scientific American, 246*(5) (May 1982): 78–88.

90. W. S. Hinshaw and A. H. Lent, "An Introduction to NMR Imaging: From the Block Equation to the Imaging Equation," *IEEE Proceedings* (1983): 338–350.

91. J. F. Greenleaf, "Computerized Tomography with Ultrasound," *IEEE Proceedings* 71 (1983): 330–337.

92. K. M. Hanson, "Proton Computed Tomography," in *Computer Aided Tomography and Ultrasonics in Medicine,* ed. J. Raviv, J. F. Greenleaf, and G. T. Herman (Amsterdam, New York, and Oxford: North-Holland, 1979), pp. 97–106.

93. R. A. Koeppe, R. M. Brugger, G. A. Schlapper, G. N. Larsen, and R. J. Jost, "Neutron Computed Tomography," *Journal of Computer Assisted Tomography* (1981): 5 79–88.

94. W. R. Holley, R. P. Henke, G. E. Gauger, B. Jones, E. V. Benton, J. I. Fabrikant, and C. A. Tobias, "Heavy Particle Computed Tomography," in *Proceedings of the Sixth Conference on Computer Applications in Radiology and Computer-Aided Analysis of Radiological Images* (New York: IEEE, 1979), pp. 64–70.

95. L. R. Price, "Electrical Impedance Computed Tomography (ICT): A New CT Imaging Technique," *IEEE Transactions on Nuclear Science* NS-26(2) (1979): 2736–2739.

96. K. A. Dines and R. J. Lytle, "Analysis of Electrical Conductivity Imaging," *Geophysics* 46 (1981): 1025–1036.
97. C. H. Chapman, "A New Method for Computing Synthetic Seismograms," *Geophysical Journal of the Royal Astronomical Society* 54 (1978): 481–518.
98. C. H. Chapman, "On Impulsive Wave Propagation in a Spherically Symmetric Model," *Geophysical Journal of the Royal Astronomical Society* 58 (1979): 229–234.
99. C. H. Chapman, "Generalized Radon Transforms and Slant Stacks," *Geophysical Journal of the Royal Astronomical Society* 66 (1981): 455–460.
100. R. A. Phinney, K. R. Chowdhury, and L. N. Frazer, "Transformation and Analysis of Record Sections," *Journal of Geophysical Research* 86 (1981): 359–377.
101. E. A. Robinson, "Spectral Approach to Geophysical Inversion by Lorentz, Fourier, and Radon Transforms," *IEEE Proceedings* 70 (1982): 1039–1054.
102. P. G. Richards, "Theoretical Seismic Wave Propagation," *Reviews of Geophysics and Space Physics Phys.* 17 (1979): 312–328.
103. G. A. McMechan, "Seismic Tomography in Boreholes," *Geophysical Journal of the Royal Astronomical Society* 74 (1983): 601–612.
104. V. A. Ambartsumian, "On Some Trends in the Development of Astrophysics," *Annual Reviews of Astronomy and Astrophysics* 18 (1980): 1–13.
105. J. H. Thompson and J. E. B. Ponsonby, "Two-Dimensional Aperture Synthesis in Lunar Radar Astronomy," *Proceedings of the Royal Society of London* A303 (1968): 477–491.
106. T. Hagfors, B. Nanni, and K. Stone, "Aperture Synthesis in Radar Astronomy and Some Applications to Lunar and Planetary Studies," *Radio Science* 3 (1968): 491–509.
107. P. A. G. Scheuer, "Lunar Occultation of Radio Sources," *Monthly Notices of the Royal Astronomical Society* 129 (1965): 199–204.
108. R. A. Phinney and D. L. Anderson, "On the Radio Occultation Method for Studying Planetary Atmospheres," *Journal of Geophysical Research and Space Physics* 73 (1968): 1819–1827.
109. W. E. Moore and G. P. Garmire, "The X-Ray Structure of the Vela Supernova Remnant," *Astrophysical Journal* 199 (1975): 680–690.
110. M. D. Altschuler, "Reconstruction of the Global-Scale Three-Dimensional Solar Corona," in *Image Reconstruction from Projections*, ed. G. T. Herman, Topics in Applied Physics, vol. 32 (New York: Springer-Verlag, 1979), pp. 105–145.
111. R. N. Bracewell, "Image Reconstruction in Radio Astronomy," in *Image Reconstruction from Projections*, ed. G. T. Herman, Topics in Applied Physics, vol. 32 (New York: Springer-Verlag, 1979), pp. 81–104.
112. D. J. DeRosier and A. Klug, "Reconstruction of Three-Diomensional Structures from Electron Micrographs," *Nature* 217 (1968): 130–134.
113. B. K. Vainshtein, "Electron Microscopical Analysis of the Three-Dimensional Structure of Biological Macromolecules," in *Advances in Optical and Electron Microscopy*, vol. 7, ed. V. E. Closlett and R. Barer (New York: Academic Press, 1978), pp. 281–377.
114. D. J. DeRosier, "The Reconstruction of Three-Dimensional Images from Electron Micrographs," *Contemporary Physics* 12 (1971): 437–452.

115. R. A. Crowther and A. Klug, "Structural Analysis of Macromolecular Assemblies by Image Reconstruction from Electron Micrographs," *Annual Review of Biochemistry* 44 (1975): 461–482.

116. D. L. D. Caspar and D. J. DeRosier, "The 1982 Nobel Prize in Chemistry," *Science* 218 (1982): 653–655.

117. P. E. Mijnarends, "Reconstruction of Three-Dimensional Distributions," in *Compton Scattering*, ed. B. Williams (New York: McGraw-Hill Book Company, 1977), pp. 323–345.

118. P. E. Mijnarends, "Electron Momentum Densities in Metals and Alloys," in *Positrons in Solids*, ed. P. Hautojärvi (Berlin: Springer-Verlag, 1979), pp. 25–88.

119. B. Williams, ed., *Compton Scattering* (New York: McGraw-Hill Book Company, 1977).

120. P. Hautojärvi, ed., Positrons in Solids, in *Topics in Current Physics*, vol. 12 (New York: Springer-Verlag, 1979).

121. Y. Das and W.-M. Boerner, "On Radar Target Shape Estimation Using Algorithms for Reconstruction from Projections," *IEEE Transactions on Antennas and Propagation* AP-26 (1978): 274–279.

122. W.-M. Boerner, "Development of Physical Optics Inverse Scattering Techniques Using Radon Projection Theory," in *Mathematical Methods and Applications of Scattering Theory*, vol. 130, ed. J. A. DeSanto, A. W. Sàenz, and W. W. Zachary, *Lecture Notes in Physics* (New York: Springer-Verlag, 1980), pp. 301–307.

123. N. N. Bojarski, "N-Dimensional Fast Fourier Transform Tomography for Incomplete Information and Its Application to Inverse Scattering Theory," in *Mathematical Methods and Applications of Scattering Theory*, vol. 130, ed. J. A. DeSanto, A. W. Sàenz, and W. W. Zachary, *Lecture Notes in Physics* (New York: Springer-Verlag, 1980), pp. 277–281.

124. A. J. Devaney, "Inverse Source and Scattering Problems in Optics," in *Optics in 4 Dimensions—1980* ed. M. A. Machado and L. M. Narducci (New York: American Institute of Physics, 1981), pp. 613–626.

125. A. J. Rockmore, R. V. Denton, and B. Friedlander, "Direct Three-Dimensional Image Reconstruction," *IEEE Transactions on Antennas and Propagation* AP-27, (1979): 239–241.

126. C. K. Chan and N. H. Farhat, "Frequency Swept Tomographic Imaging of Three-Dimensional Perfectly Conducting Objects," *IEEE Transactions on Antennas and Propagation* AP-29 (1981): 312–319.

127. C. K. Chan, Analytical and Numerical Studies of Frequency Swept Imaging, Ph.D. thesis, (Philadelphia: University of Pennsylvania, 1978).

128. D. L. Mensa, S. Halevy, and G. Wade, "Coherent Doppler Tomography for Microwave Imaging," *IEEE Proceedings* 71 (1983): 254–261.

129. J. R. Willis, "Interfacial Stresses Induced by Arbitrary Loading of Dissimilar Elastic Half-Spaces Joined over a Circular Region," *Journal of the Institute of Mathematics and Its Applications* 7 (1971): 79–197.

130. J. R. Willis, "The Penny-Shaped Crack on an Interface," *Quarterly Journal of Mechanics and Applied Mathematics* 25 (1972): 367–385.

131. J. R. Willis, "Self-Similar Problems in Elastodynamics," *Philosophical Transactions of the Royal Society of London*, Series A 274 (1973): 435–491.

132. B. P. Hildebrand and D. E. Hufferd, "Computerized Reconstruction of Ultrasonic Velocity Fields for Mapping of Residual Stress," in *Acoustical Holography*, vol. 7, ed. L. W. Kessler (New York: Plenum, 1976), pp. 245–262.

133. S. R. Deans, "Hough Transform from the Radon Transform," *IEEE Transactions of Pattern Analysis and Machine Intelligence* PAMI-3 (1981): 185–188.

134. A. E. Cowart, W. E. Synder, and W. H. Ruedger, "The Detection of Unresolved Targets using the Hough Transform," *Computer Vision, Graphics, and Image Processing* 21 (1983): 222–238.

135. B. W. Stuck, "A New Proposal for Estimating the Spatial Concentration of Certain Types of Air Pollutants," *Journal of the Optical Society of America* 67 (1977): 668–678.

136. R. L. Byer and L. A. Shepp, "Two-Dimensional Remote Air-Pollution Monitoring via Tomography," *Optics Letters* 4 (1979): 75–77.

137. J. P. Barton, *Feasibility of Neutron Radiography for Large Bundles of Fast Reactor Fuel*, Tech. Rept. IRT 6247-004 (San Diego: Instrumentation Research Technology Corporation, 1978).

138. J. G. Sanderson, "Reconstruction of Fuel Pin Bundles by a Maximum Entropy Method," *IEEE Transactions on Nuclear Science* NS-26(2) (1979): 2685–2686.

139. W. A. Ellingson and H. Berger, "Three-Dimensional Radiographic Imaging," in *Research Techniques in Nondestructive Testing*, vol. 4, ed. R.S. Sharpe (New York: Academic Press, 1980), pp. 1–38.

140. B. R. Tittman, "Imaging in NDE," in *Acoustical Imaging*, vol. 9, ed. K. Y. Wang (New York: Plenum, 1980), pp. 315–340.

141. W. G. Wee and T.-T. Hsieh, "An Application of the Projection Transform Technique in Image Transmission," *IEEE Transactions on Systems Management and Cybernetics* SMC-6 (1976): 486–493.

142. H. H. Barrett, "Optical Processing in Radon Space," *Optics Letters* 7 (1982): 248–250.

143. A. F. Gmitro, J. E. Greivenkamp, W. Swindell, H. H. Barrett, M. Y. Chiu, and S. K. Gordon, "Optical Computers for Reconstructing Objects from Their X-Ray Projections," *Optical Engineering* 19 (1980): 260–272.

144. G. R. Gindi, J. Arendt, H. H. Barrett, M. Y. Chiu, A. Ervin, C. L. Giles, M. A. Kujoory, E. L. Miller, and R. G. Simpson, "Imaging with Rotating Slit Apertures and Rotating Collimators," *Medical Physics* 9 (1982): 324–339.

145. W. E. Smith, H. H. Barrett, and R. G. Paxman, "Reconstruction of Objects from Coded Images by Simulated Annealing," *Optics Letters* 8 (1983): 199–201.

146. P. D. Lax and R. S. Phillips, "Scattering Theory," *Rocky Mountain Journal of Mathematics* 1 (1971): 173–223.

147. R. G. Newton, "Three-Dimensional Solitons," *Journal of Mathematical Physics* 19 (1978): 1068–1073.

148. A. M. Cormack and E. T. Quinto, "A Radon Transform on Spheres Through the Origin in R^n and Applications to the Darboux Equation," *Transactions of the American Mathematical Society* 260 (1980): 575–581.

149. V. Guillemin and S. Sternberg, *Geometric Asymptotics*, Mathematical Surveys, No. 14 (Providence, R. I.: American Mathematical Society, 1977).

150. A. S. Cavaretta, Jr., C. A. Micchelli, and A. Sharma, "Multivariate Inter-

polation and the Radon Transform," *Mathematische Zeitschrift* 174 (1980): 263–279.

151. A. S. Cavaretta, Jr., C. A. Micchelli, and A. Sharma, "Multivariate Interpolation and the Radon Transform, Part II, Some Further Examples," in *Quantitative Approximation,* ed. R. A. DeVore and K. Scherer (New York: Academic Press, 1980), pp. 49–62.

152. A. Macovski, "Physical Problems of Computerized Tomography," *IEEE Proceedings* 71 (1983): 373–378.

153. Y. Censor, "Finite Series-Expansion Reconstruction Methods," *IEEE Proceedings* 71 (1983): 409–419.

154. M. E. Davison, "The Ill-Conditioned Nature of the Limited Angle Tomography Problem," *SIAM Journal of Applied Mathematics* 43 (1983): 428–448.

4

The Geometry of Quantum Flow

James V. Lindesay

University of California, Berkeley

Harry L. Morrison

University of California, Berkeley

ABSTRACT: A general set of equations is developed for a stationary physical system with an internal local gauge symmetry. In particular, the equations are related to the phenomenological equations for a multicomponent quantum fluid (e.g., the two-fluid model of liquid ^{4}He). Quantization properties due to the topological structure of the system are examined.

4-1 INTRODUCTION

One of the most successful and intuitively understandable theories for the phenomenology of the superfluid ^{4}He system has been the two-fluid model introduced by Landau.[1] In this model the superfluid system is composed of two weakly interacting fluid components that are not distinguished by atomic content but macroscopically have distinctive thermodynamic properties. The superfluid flow field is obtained from the density of the superfluid component and a superflow velocity field, which of itself is not directly observable.

Previous studies[2,3] have indicated the possibility of describing the superfluid velocity as the gauge vector potential of a Galilean system with an Abelian local symmetry. The topics presently discussed include a general development of these ideas, along with applications to systems with full (in general non-Abelian) local group symmetries. A set of equations is proposed for describing the macroscopic hydrodynamic behavior of the bosonic ^{4}He II system and the fermionic ^{3}He-B system. The exploration of the anisotropic ^{3}He-A system will be presented elsewhere.[4]

4-2   GROUP THEORETIC CONSIDERATIONS

To establish notation, the theory of Lie transformation groups is briefly discussed here. For a more complete discussion, we refer the reader to the texts by Hammermesh[5] and Pontryagin.[6]

Transformation of State Variables

The parameter space for the Lie transformation group is denoted G. For an N-parameter group, the element α represents the parameters

$$\alpha = (\alpha^1, \cdots, \alpha^N).$$

The basis is chosen such that the identity element of the group e can be represented as the origin of G.

$$e = (0, \cdots, 0).$$

For two elements within the group α, $\alpha' \in G$, we define a group multiplication that determines a third element of the group α'',

$$\alpha'' = \phi(\alpha'; \alpha), \qquad \alpha'' \in G,$$

where the function ϕ is an analytic real function of the real parameters α, α'.

The operation $S(\alpha)$ is referred to as forming a representation of the group if for any α, α' the following holds:

$$S(\alpha')S(\alpha) = S(\phi(\alpha'; \alpha)). \tag{4-1}$$

If the state variable ψ transforms under the representation S, then

$$\psi' = S(\alpha)\psi \equiv \psi[\alpha].$$

For this type of representation, the generators of infinitesimal transformation G_r are independent of the state variables

$$G_r \equiv \frac{\partial}{\partial \beta^r} S(\beta)\bigg|_{\beta=e}. \tag{4-2}$$

The group algebra is directly determined from the structure of the group multiplication and representation as

$$[G_r, G_s] = C_{rs}^m G_m,$$

where the structure constants are defined by

$$C_{rs}^m = \frac{\partial}{\partial \alpha^s} \frac{\partial}{\partial \beta^r} [\phi^m(\beta; \alpha) - \phi^m(\alpha; \beta)]|_{\beta = e = \alpha}.$$

Alternatively, if we define the algebraic matrices

$$\Theta_m^s(\alpha) \equiv \frac{\partial}{\partial \beta^m} \phi^s(\beta; \alpha)|_{\beta = e}, \qquad \text{(4-3)}$$

then the structure constants can be represented

$$C_{rs}^m = \Theta_s^n(\alpha) \left(\frac{\partial}{\partial \alpha^n} \Theta_r^k(\alpha) \right) (\Theta^{-1}(\alpha))_k^m$$

$$+ -\Theta_r^n(\alpha) \left(\frac{\partial}{\partial \alpha^n} \Theta_s^k(\alpha) \right) (\Theta^{-1}(\alpha))_k^m.$$

From the group property (4-1) and the definition (4-2) we can determine the useful property

$$\Theta_r^m(\alpha) \frac{\partial}{\partial \alpha^m} S(\alpha) = G_r S(\alpha).$$

Note that by its definition in (4-3), the following equation holds:

$$\Theta_m^s(e) = \delta_m^s.$$

Local Group Transformations

Consider transformations of the state variables characterized as follows:

$$\psi(x : \alpha) = S(\alpha(x))\psi(x), \qquad \text{where} \quad \psi(x) \equiv \psi(x : e).$$

If the representation is a linear M-parameter group, this relation can be expressed

$$\psi^a(x : \alpha) = S_b^a(\alpha(x))\psi^b(x), \qquad a, b = 1, \cdots, M. \qquad \text{(4-4)}$$

Here, and in what follows, repeated indices are summed. If the elements α are independent of $\mathbf{x}$, the transformation is referred to as *rigid*. If the transformation parameters vary with respect to the space-time coordinate $\mathbf{x}$, the transformation is referred to as *local*.

We define linear operators $D_\mu[\alpha]$, referred to as gauge-covariant derivatives, which satisfy

$$D_\mu[\phi(\alpha'; \alpha)]\psi(\mathbf{x} : \phi(\alpha'; \alpha)) = S(\alpha')D_\mu[\alpha]\psi(\mathbf{x} : \alpha). \quad (4\text{-}5)$$

The derivative is covariant in the sense that all derivatives are relative to the same "gauge-fixed frame." The gauge-covariant derivative is expressed in terms of the normal coordinate derivative $\partial_\mu \equiv \partial/\partial x^\mu$, and the gauge fields $W_\mu(\mathbf{x} : \alpha)$ at a point

$$D_\mu[\alpha] = 1 \cdot \partial_\mu - W_\mu(\mathbf{x} : \alpha),$$

where 1 is the identity operation of the representation

$$S(\mathbf{e}) = 1.$$

With these identifications, the "field strengths" can be defined for an integrable set of coordinates

$$F_{\mu\nu}(\mathbf{x} : \alpha) \equiv [D_\mu[\alpha], D_\nu[\alpha]]$$

$$= -\partial_\mu W_\nu(\mathbf{x} : \alpha) + \partial_\nu W_\mu(\mathbf{x} : \alpha)$$

$$+ [W_\mu(\mathbf{x} : \alpha), W_\nu(\mathbf{x} : \alpha)]. \quad (4\text{-}6)$$

These definitions relate the behavior of the operators $F_{\mu\nu}$ under group transformations to the integrability of the parameter space

$$([\partial_\mu, \partial_\nu]S(\alpha')) = -F_{\mu\nu}(\mathbf{x} : \phi(\alpha'; \alpha))S(\alpha') + S(\alpha')F_{\mu\nu}(\mathbf{x} : \alpha).$$

Using (4-5) and (4-6), the following local transformation properties can be determined:

$$F_{\mu\nu}(\mathbf{x} : \alpha) = S(\alpha)F_{\mu\nu}(\mathbf{x} : \mathbf{e})S^{-1}(\alpha), \quad (4\text{-}7)$$

$$[\partial_\mu S(\alpha')] = W_\mu(\mathbf{x} : \phi(\alpha'; \alpha))S(\alpha') - S(\alpha')W_\mu(\mathbf{x} : \alpha). \quad (4\text{-}8)$$

The dependence of the quantities $S(\alpha)$ on $\mathbf{x}$ is only through the parameter α,

$$\partial_\mu S(\alpha) = \partial_\mu \alpha^r \frac{\partial}{\partial \alpha^r} S(\alpha). \quad (4\text{-}9)$$

The functions $A_\mu^r(\alpha)$ are defined through the algebraic relation

$$\partial_\mu \alpha^r = A_\mu^s(\alpha)\Theta_s^r(\alpha). \tag{4-10}$$

Then Equation (4-9) can be expressed

$$\partial_\mu S(\alpha) = A_\mu^s(\alpha)G_s S(\alpha).$$

The expression (4-8) then determines the general transformation properties of the gauge fields:

$$W_\mu(\mathbf{x} : \phi(\alpha'; \alpha)) = A_\mu^s(\alpha)G_s + S(\alpha')W_\mu(\mathbf{x} : \alpha)S^{-1}(\alpha').$$

By the definition of the functions A_μ^r through (4-10), the following properties hold:

$$A_\mu^s(\mathbf{e}) = 0,$$

$$\delta A_\mu^s(\alpha) = \partial_\mu(\delta\alpha^s) + \delta\alpha^m C_{mn}^s A_\mu^n(\alpha). \tag{4-11}$$

To explore additional properties of the fields, the Jacobi identities for the operators D_μ are examined. These identities are represented

$$[D_{\{\alpha}, [D_\mu, D_{\nu\}}1]] = 0$$

where $\{\ \}$ represents the cyclic sum of terms with the indices contained. With the definitions (4-6), these identities give auxiliary equations for the fields

$$\partial_{\{\alpha} F_{\mu\nu\}} = [W_{\{\alpha}, F_{\mu\nu\}}]. \tag{4-12}$$

The conditions (4-8) are quite general, and if satisfied, the gauge-covariant derivative automatically transforms as desired in (4-5). As a specific example, consider the case in which the gauge fields are directly related to the generators through

$$W_\mu(x : \mathbf{e}) \equiv a_\mu^r(\mathbf{x})G_r.$$

Then, generally

$$W_\mu(x : \alpha) = A_\mu^r(\alpha)G_r + a_\mu^r(\mathbf{x})S(\alpha)G_r S^{-1}(\alpha).$$

If we define

$$a_\mu^s(\mathbf{x} : \alpha)G_s \equiv a_\mu^r(\mathbf{x})S(\alpha)G_r S^{-1}(\alpha),$$

then the fields W_μ can be directly related to the generators

$$W_\mu(\mathbf{x} : \boldsymbol{\alpha}) = W_\mu^s(\mathbf{x} : \boldsymbol{\alpha})G_s = [A_\mu^s(\boldsymbol{\alpha}) + a_\mu^s(\mathbf{x} : \boldsymbol{\alpha})]G_s. \quad \textbf{(4-13)}$$

Under a gauge transformation, the function W_μ^s transform inhomogeneously owing to the relation (4-11)

$$\delta W_\mu^s(\mathbf{x} : \boldsymbol{\alpha}) = \partial_\mu(\delta\alpha^s) + \delta\alpha^m C_{mn}^s W_\mu^n(\mathbf{x} : \boldsymbol{\alpha}).$$

In addition, the field strengths $F_{\mu\nu}$ decomposes in a similar way, with

$$F_{\mu\nu}^r(\mathbf{x} : \boldsymbol{\alpha}) = -\partial_\mu W_\nu^r(\mathbf{x} : \boldsymbol{\alpha}) + \partial_\nu W_\mu^r(\mathbf{x} : \boldsymbol{\alpha})$$
$$+ W_\mu^m(\mathbf{x} : \boldsymbol{\alpha})C_{mn}^r W_\mu^n(\mathbf{x} : \boldsymbol{\alpha}).$$

Thus the functions W_μ^s and $F_{\mu\nu}^r$ can be identified to be Yang-Mills[7] potentials and field strengths. The relation (4-7) determines the transformation properties of the field strengths

$$\delta F_{\mu\nu}^r(\mathbf{x} : \boldsymbol{\alpha}) = \delta\alpha^m C_{mn}^r F_{\mu\nu}^n(x : \boldsymbol{\alpha}). \quad \textbf{(4-14)}$$

Integrability and Periodicity

Consider a compact group parameter space with an element of periodicity $\boldsymbol{\Gamma}$. This will mean that for some element (or elements) $\boldsymbol{\Gamma} \neq \mathbf{e}$, the representation satisfies

$$S(\boldsymbol{\phi}(\boldsymbol{\Gamma}; \boldsymbol{\alpha})) = S(\boldsymbol{\alpha}),$$
$$\frac{\partial S(\boldsymbol{\phi})}{\partial\phi^r} = \frac{S(\boldsymbol{\alpha})}{\partial\alpha^r}.$$

Then the group multiplication satisfies

$$\phi^r(\boldsymbol{\Gamma}; \boldsymbol{\alpha}) = \Gamma^r + \alpha^r,$$
$$S(\boldsymbol{\phi}(n\boldsymbol{\Gamma}; \boldsymbol{\alpha})) = S(\boldsymbol{\alpha}) \quad \text{for any integer } n.$$

Suppose that $C[0]$ is a closed curve in which the region enclosed contains no boundaries (see Fig. 4-1). Then $C[0]$ is the boundary $\partial\Sigma$ of a simply connected surface Σ. If we examine the line integral of the gradient of the group parameter $\partial_\mu\alpha^r$ along the curve $C[0]$:

$$\oint_{C[0]} dx^\mu \, \partial_\mu\alpha^r = \oint_{\partial\Sigma[0]} dx^\mu \partial_\mu\alpha^r = \int_{\Sigma[0]} ([\partial_\mu, \partial_\nu]\alpha^r) \, d\Sigma^{\mu\nu}, \quad \textbf{(4-15)}$$

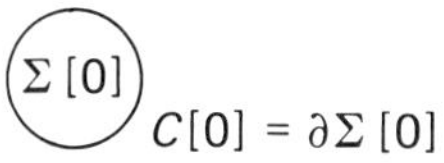

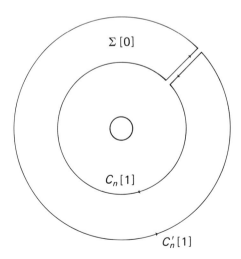

$$C[0] = \partial\Sigma[0]$$

Figure 4-1
Simply connected region.

where the second equality follows by Stokes's theorem. In the case being investigated, the third term in (4-15) vanishes (since curl grad $\alpha^r = 0$), resulting in the following:

$$\oint_{C[0]} dx^\mu \partial_\mu \alpha^r = 0. \tag{4-16}$$

Consider now a multiply connected space that is integrable, except along well-defined boundaries or singular points that break the connectivity. If the curve $C[0]$ can be arbitrarily shrunk to a point without crossing a boundary or singular point, then Equation (4-16) is valid. Figure 4-2 indicates such a curve $C[0]$, which can be constructed of curves $C_n[1]$ and $C'_n[1]$. The region labeled n represents the broken connectivity. Equation (4-16) gives the following relation:

$$\oint_{C[0]} dx^\mu \partial_\mu \alpha^r = 0 = \oint_{C'_n[1]} dx^\mu \partial_\mu \alpha^r - \oint_{C_n[1]} dx^\mu \partial_\mu \alpha^r.$$

Figure 4-2
Multiply connected region.

Therefore, the term $\oint_{C_n[1]} dx^\mu \partial_\mu \alpha^r$ is a topological invariant:

$$\oint_{C=n[1]} dx^\mu \partial_\mu \alpha^r = n\Gamma^r, \qquad (4\text{-}17)$$

where the integer n represents the degree of the map α on $\mathbf{x}$ along $C_n[1]$. If α is a one-parameter characterization of the compact manifold $C_n[1]$ of periodicity of 2π, then n is the winding number of the curve.

In general, if m represents the number of times that a closed curve $C_n[m]$ wraps around the region n, then

$$\oint_{C_n[m]} dx^\mu \partial_\mu \alpha^r = mn\Gamma^r. \qquad (4\text{-}18)$$

For several regions of broken connectivity labeled by k, the relation (4-18) can be generalized:

$$\oint_C dx^\mu \partial_\mu \alpha^r = \oint_C dx^\mu A_\mu^s(\boldsymbol{\alpha}) \Theta^r(\boldsymbol{\alpha})$$

$$= \sum_k m_k n_k \Gamma^r_{(k)}$$

where n_k represents the vortex number of the kth region, and m_k represents the number of times the curve C wraps around the region.

4-3   SYSTEMS WITH LOCAL INTERNAL SYMMETRIES

The considerations of the previous sections are applied to and interpreted within several physical systems in what follows. We assume that there exists some action or thermodynamic potential for which stationarity under variations of field parameters determines the equations of motion or the behavior of the state parameters. In general, dissipative processes are not discussed. Also, the field parameters will be classical, and any quantization of parameters will be owing to the structure of the system of equations.

The action functional takes the form

$$a[\psi, q, \mathbf{a}] = \int d^4x \; \mathcal{F}(\psi, D_\mu \psi, q, \partial_\mu q : F^r_{\mu\nu}),$$

with the covariant derivative D_μ as defined previously. The field parameters

q are taken to transform invariantly under group transformations

$$G_r q = 0, \qquad \text{(4-19)}$$

whereas the field parameters ψ transform according to Equation (4-4).

Stationary of the action under variations of the field parameters results in the Euler-Lagrange equations for the system. Variations on the field ψ give

$$\partial_\mu \Pi^\mu = \Pi^\mu W_\mu(\mathbf{x} : \mathbf{e}) = \frac{\partial \mathfrak{F}}{\partial \psi}, \qquad \text{where } \Pi^\mu = \frac{\partial \mathfrak{F}}{\partial (D_\mu \psi)}. \qquad \text{(4-20)}$$

Stationary of the action with respect to variations on the field q gives

$$\partial_\mu \pi^\mu = \frac{\partial \mathfrak{F}}{\partial q}, \qquad \text{where } \pi^\mu = \frac{\partial \mathfrak{F}}{\partial (\partial_\mu q)}. \qquad \text{(4-21)}$$

Finally, stationarity of the action with respect to the field parameters a_μ^r, which are implicitly contained in the gauge-covariant derivatives and field strengths $F_{\mu\nu}^r$, gives the following set of equations:

$$[\delta_s^r \partial_\mu + W_\mu^m (C_m)_s^r] H_r^{\mu\nu} = J_s^\nu,$$

where

$$H_r^{\mu\nu} \equiv 2 \frac{\partial \mathfrak{F}}{\partial F_{\mu\nu}^r}, \qquad J_s^\nu \equiv \Pi^\nu G_s \psi. \qquad \text{(4-22)}$$

The symbol δ_s^r represents the standard Kronecker delta, and the matrix C_m is in the adjoint representation of the group.

$$(C_m)_s^r = C_{ms}^r.$$

These Euler-Lagrange equations act as source equations for the field parameters.

An equation analogous to the Noether relation for current conservation[8] is obtained by invoking the symmetry condition

$$\frac{\delta E}{\delta \alpha^r} = 0,$$

$$\partial_\mu J_r^\mu = C_{rs}^m (W_\mu^s J_m^\mu - \tfrac{1}{2} F_{\mu\nu}^s H_m^{\mu\nu}). \qquad \text{(4-23)}$$

Thus the non-Abelian character of the equations gives source terms to the currents. A canonical energy-momentum tensor can be reconstructed:

$$T^\mu_\nu \equiv \Pi^\mu D_\nu \psi + \pi^\mu \partial_\nu q + H^{\mu\beta}_r F^r_{\nu\beta} - \delta^\mu_{\nu;} \mathcal{F},$$

$$\partial_\mu T^\mu_\nu = 0.$$

No assumption has been made of the existence of a metric or symmetry under exchange of indices.

For a certain class of densities F a "force" can be defined between components of the system. This class is referred to as a *separable* system, and satisfies

$$\mathcal{F} = \mathcal{F}_{(P)}(\psi, D_\mu \psi, q, \partial_\mu q) + \mathcal{F}_{(G)}(F^r_{\mu\nu}).$$

Then, in addition to the previous equations, we obtain force equations

$$\partial_\mu T^\mu_{(P)\nu} = f_\nu = -\partial_\mu T^\mu_{(G)\nu},$$

where

$$T^\mu_{(P)\nu} \equiv \pi^\mu D_\nu \psi + \pi^\mu \partial_\nu q - \gamma^\mu_\nu \mathcal{F}_{(P)},$$

$$T^\mu_{(G)\nu} \equiv H^{\mu\beta}_r F^r_{\nu\beta} - \delta^\mu_\nu \mathcal{F}_{(G)},$$

$$f_\nu \equiv J^\alpha_r F^r_{\alpha\nu} + \tfrac{1}{2} W^r_\nu C^m_{rs} H^{\alpha\beta}_m F^s_{\alpha\beta}. \tag{4-24}$$

Another class of densities $\mathcal{F}$ has a behavior such that the gauge-covariant derivative appears in a manner that has a geometrical interpretation. These systems are referred to as *substantive* systems, or *material* systems, and satisfy

$$\mathcal{F} = \mathcal{F}_{(M)}(\psi, D\psi, qDq : F^r_{\mu\nu}),$$

where

$$D\psi \equiv U^\mu D_\mu \psi,$$

$$Dq \equiv U^\mu \partial_\mu q \tag{4-25}$$

for some vector function **U**. The substantive derivatives defined above can be interpreted as gauge-covariant directional derivatives along a tangent vector **u**, which characterizes a streamline. For this system, the form of the canonically conjugate fields simplifies

$$\Pi^\mu = U^\mu \frac{\partial F_{(M)}}{\partial (D_\mu)},$$

$$\pi^\mu = U^\mu \frac{\partial F_{(M)}}{\partial (Dq)},$$

$$J_r^\mu = U^\mu \frac{\partial F_{(M)}}{\partial (D\psi)} G_r \psi. \qquad \textbf{(4-26)}$$

The equations developed in this section are next applied within the context of various physical examples. An emphasis is placed upon the possible application of this formalism to the hydrodynamics of the superfluid helium systems. The development remains somewhat general and concise, rather than comprehensive, with references made to similar results from alternative approaches by other authors.

4-4   APPLICATION TO PHYSICAL SYSTEMS

The results of this section are developed by means of various examples of physical systems for which the phenomenology is elegantly modeled using the formalism of the previous sections.

Example 1.   *Identifications for thermal fluid mechanics.* Consider the following action density for a substantive system written in terms of the field parameters ψ and q, which for the present have no internal structure:

$$\mathscr{F} = \mathscr{F}_{(M)}(\psi, D\psi, q, Dq).$$

The substantive derivatives D are as defined in Equation (4-25). For the fluid system, the following identifications can be made:

$$\frac{\partial \mathscr{F}_{(M)}}{\partial (D\psi)} \equiv \rho c, \qquad \frac{\partial \mathscr{F}_{(M)}}{\partial (Dq)} \equiv \zeta c.$$

The form of the conjugate fields follows directly from Equation (4-26):

$$\Pi^\mu = \rho c u^\mu, \qquad \pi^\mu = \zeta c u^\mu,$$

and the equations of motion (4-20) and (4-21) give flux equations for the fluid parameters

$$\partial_\mu(\rho c u^\mu) = \frac{\partial \mathfrak{F}_{(M)}}{\partial \psi},$$

$$\partial_\mu(\zeta c u^\mu) = \frac{\partial \mathfrak{F}_{(M)}}{\partial q}.$$

Characteristic velocity c is included only for dimensional considerations. The canonical energy-momentum tensor has the following components:

$$T^0_0 = u^0(\rho\dot\psi + \zeta\dot q) - \mathfrak{F}_{(M)},$$

$$T^j_0 = u^j(\rho\dot\psi + \zeta\dot q),$$

$$T^0_k = u^0 c(\rho\nabla_k\psi + \zeta\nabla_k q),$$

$$T^j_k = u^j c(\rho\nabla_k\psi + \zeta\nabla_k q) = \delta^j_k \mathfrak{F}_{(M)},$$

where, for a parameter ϕ, the quantity $\dot\phi$ is defined

$$\dot\phi \equiv \frac{\partial\phi}{\partial t}.$$

These identifications are directly applied to a Galilean fluid. For this non-relativistic situation,

$$u^\mu = (1, \bar v_x/c, v_y/c, v_z/c), \quad x^\mu = (ct, x, y, z)$$

$$T^0_0 \equiv u, \qquad \text{energy density}$$

$$T^j_0 \equiv Q_j, \qquad \text{energy flux vector}$$

$$\frac{1}{c} T^0_k \equiv -I_k, \qquad \text{canonical momentum density}$$

$$\frac{1}{c} T^j_k \equiv -\tau_{jk}. \qquad \text{stress-energy tensor} \qquad \textbf{(4-27)}$$

In addition, the substantive chemical potential and substantive temperature are defined by

$$\frac{D\psi}{Dt} = \frac{\partial\psi}{\partial t} + v_j\nabla_j\psi \equiv \mu^*,$$

$$\frac{Dq}{Dt} = \frac{\partial q}{\partial t} + v_j\nabla_j q \equiv T^*.$$

The thermal hydrodynamic equations thus obtained for the Galilean fluid are

$$\frac{\partial \rho}{\partial t} + \sum_{j=1}^{3} \nabla_j(\rho v_j) = \frac{\partial \mathcal{F}_{(M)}}{\partial \psi}, \qquad \text{mass flux}$$

$$\frac{\partial \zeta}{\partial t} + \sum_{j=1}^{3} \nabla_j(\zeta v_j) = \frac{\partial \mathcal{F}_{(M)}}{\partial q}, \qquad \text{entropy flux}$$

$$u = \mu^* \rho + T^* \zeta + v_j I_j - \mathcal{F}_{(M)},$$

$$Q_j = (u + \mathcal{F}_{(M)}) v_j,$$

$$\frac{\partial u}{\partial t} + \sum_{j=1}^{3} \nabla_j Q_j = 0, \qquad \text{energy flux}$$

$$I_k = -\rho \nabla_k \psi - \zeta \nabla_k \mathbf{q},$$

$$T_{jk} = I_k v_j + \delta_{jk} \mathcal{F}_{(M)},$$

$$\frac{\partial}{\partial t} I_k + \sum_{j=1}^{3} \nabla_j T_{jk} = 0. \qquad \text{momentum flux} \qquad \textbf{(4-28)}$$

This set of equations represents results similar to the standard hydrodynamic equations if the action density $\mathcal{F}$ is the density of the thermodynamic potential Ω, which is represented by the pressure P.

$$\mathcal{F} \Leftrightarrow P,$$

action $\Leftrightarrow$ time integral of Ω.

For completeness, the form of the equations for rigid-body rotation is briefly examined. If there are rigid-body rotation and translations, the velocity takes the form

$$\mathbf{v} = \mathbf{v}_T + \boldsymbol{\omega}_R \times \mathbf{r}.$$

Then the energy density u represented in (4-28) can be expressed

$$u = \mu^* \rho + T\zeta + \mathbf{v}_T \cdot \mathbf{g} + \boldsymbol{\omega}_R \cdot \mathbf{l} - P,$$

where T is the canonical temperature conjugate to q defined by

$$T \equiv \frac{\partial q}{\partial t} = T^* - \mathbf{v} \cdot \nabla q.$$

The momentum density **g** is defined by

$$\mathbf{g} = -\rho \nabla \psi,$$

and the angular momentum density **l** is defined

$$\mathbf{l} = \mathbf{r} \times \mathbf{g}.$$

Example 2. *Identifications for systems with general internal symmetries.* These identifications are now generalized to describe a system with internal symmetries. Consider a general action density that is both separable and substantive:

$$\mathfrak{F} = \mathfrak{F}_{(M)}(\psi, D\psi, q, Dq) + \mathfrak{F}_{(G)}(F^r_{\mu\nu}).$$

With the identification previously made for ζ, the condition (4-19),

$$G_r q = 0, \tag{4-19}$$

is interpreted to imply that the gauge fields carry no entropy. Then, if the action density is cyclic in the field parameter q, there is an entropy conservation condition of the form

$$\frac{\partial \zeta}{\partial t} + \nabla \cdot (\zeta \mathbf{v}) = \frac{\partial \mathfrak{F}_{(M)}}{\partial q} = 0. \tag{4-29}$$

Thus the entropy flux will be purely convective through the parameter **v**, and gauge independent.

The identifications for gauge-dependent quantities are explored next. *For dimensional convenience*, we assume that the system contains a scale parameter c_0 that is a velocity. *This parameter has no bearing on the form of the equations*, and can be set to unity if desired. The Galilean parameters are written

$$x^\mu = (c_0 t, \mathbf{x}), \qquad u^\mu = (c_0, \mathbf{v}).$$

With these considerations, the gauge fields and field strengths are identified as follows:

$$\mathbf{W}^r_\mu = (W^r_0, \mathbf{W}^r),$$

$$F^r_{jk} \equiv -\sum_{m=1}^{3} \epsilon_{jkm} \Omega^r_m, \qquad F^r_{0k} \equiv E^r_k,$$

$$\sum_{jk} H_s^{jk} \, \epsilon_{jkl} \equiv 2(\Lambda_s), \qquad H_s^{0k} \equiv (\mathbf{K}_s)_l,$$

$$\mathbf{E}^r = -\frac{1}{c_0} \frac{\partial}{\partial t} \mathbf{W}^r + \nabla \, W_0^r + W_0^s \, C_{sn}^r \mathbf{W}^n,$$

$$\Omega^r = \quad \mathbf{x} \, \mathbf{W}^r - \tfrac{1}{2} C_{sn}^r \mathbf{W}^s \, \mathbf{x} \, \mathbf{W}^n. \tag{4-30}$$

These equations satisfy the auxiliary conditions from Equation (4-12):

$$[\delta_s^r \, \nabla - \mathbf{W}^m (C_m)_s^r] \cdot \Omega^s = 0,$$

$$\left[\delta_s^r \frac{1}{c_0} \frac{\partial}{\partial t} - W_0^m (C_m)_s^r \right] \Omega^s + [\delta_s^r \, \nabla - \mathbf{W}^m (C_m)_s^r] \, \mathbf{x} \, \mathbf{E}^s = 0. \tag{4-31}$$

Next, the dynamic equations are developed for the system. From (4-22), the source equations for the fields are

$$[\delta_s^r \, \nabla + \mathbf{W}^m (C_m)_s^r] \cdot \mathbf{K}_r = -J_s^0,$$

$$- \left[\delta_s^r \frac{1}{c_0} \frac{\partial}{\partial t} \right.$$

$$\left. + W_0^m (C_m)_s^r \right] \mathbf{K}_r$$

$$+ [\delta_s^r \, \nabla + \mathbf{W}^m (C_m)_r^r] \, \mathbf{x} \, \Lambda_r = -\mathbf{J}_s. \tag{4-32}$$

The canonical energy and momentum densities are identified as indicated:

$$T_0^0 \equiv u = u_{(M)} + u_{(G)}, \qquad c_0 T_0^j \equiv Q_j,$$

$$-\frac{1}{c_0} T_k^0 \equiv I_k = I_{(M)k} + I_{(G)k}, \qquad -T_k^j \equiv T_{jk},$$

$$\mathfrak{F} \equiv P - \Lambda_r \cdot \Omega^r.$$

It is convenient to define the quantity γ, which has the units of a free-energy density:

$$\gamma \equiv \pi^0 \frac{1}{c_0} \frac{\partial}{\partial t} \psi.$$

The energy density can now be expressed as

$$u = \gamma + v^r J_r^0 + T\zeta + \mathbf{K}_r \cdot \mathbf{E}^r + \Lambda_r \cdot \Omega^r - P. \tag{4-33}$$

Notice that, as defined, the pressure has a contribution from the gauge field strengths. The energy flux vector satisfies

$$\mathbf{Q} = (u_{(M)} + P)\mathbf{v} + c_0 \mathbf{E}^r \mathbf{x} \, \Lambda_r.$$

The canonical momentum density and flux vector are given as follows:

$$\mathbf{I} = \mathbf{g} - \zeta \nabla q + \mathbf{I}_{(G)} = \mathbf{I}_{(M)} + \mathbf{I}_{(G)},$$

$$T_{jk} = v_j I_{(M)k} - \Omega_j^r(\Lambda_r)_k + \delta_{jk} P, \qquad \textbf{(4-34)}$$

where

$$g \equiv -\frac{1}{c_0} \Pi^0 \mathbf{D} \, \psi,$$

$$\mathbf{I}_{(G)} \equiv \frac{1}{c_0} \mathbf{K}_r \mathbf{x} \, \Omega^r.$$

The current source equation (4-23) can be expressed

$$\left[\delta_s^r \frac{1}{c_0} \frac{\partial}{\partial t} + W_0^m (C_m)_s^r \right] J_r^0 + [\delta_s^r \, \nabla + \mathbf{W}^m (C_m)_s^r] \cdot J_r$$

$$= C_{sm}^r [\Lambda_r \cdot \Omega^m - \mathbf{K}_r \cdot \mathbf{E}^m]. \qquad \textbf{(4-35)}$$

To complete this set of equations, we identify the forces acting between components. Using relation (4-24), the following identifications can be made:

$$f_0 = -\mathbf{J}_r \cdot \mathbf{E}^r + W_0^m C_{ms}^r (\mathbf{K}_r \cdot \mathbf{E}^s - \Lambda_r \cdot \Omega^s),$$

$$\mathbf{f} = J_r^0 \mathbf{E}^r + \mathbf{J}_r \mathbf{x} \, \Omega^r + \mathbf{W}^m C_{ms}^r (\mathbf{K}_r \cdot \mathbf{E}^s - \Lambda_r \cdot \Omega^s). \qquad \textbf{(4-36)}$$

The first of Equations (4-36) represents a material energy dissipation into gauge-field energy density. The second equation represents a force density on the material component owing to the presence of the gauge fields. The set of equations presented in this example is examined in more detail in the following examples.

Before proceeding, it is interesting to note the similarity of these equations with those obtained by Khalatnikov and Lebedev[9] using Poisson bracket relationships for the various state parameters. The correspondence between the approach employed here and that of Khalatnikov and Lebedev has been explored by the authors and found to be consistent.

The goal of the remainder of this section is the application of the formalism developed above to quantum fluids. In this spirit, we assume that the system has an effective mass scale m, along with the parameter $\hbar$. Using these parameters, dimensional fields can be defined for the system:

$$\mu^r \equiv -\frac{\hbar c_0}{m} W_0^r, \qquad \text{gauge potential}$$

$$w^r \equiv \frac{\hbar}{m} \mathbf{W}^r, \qquad \text{gauge velocity}$$

$$\omega^r \equiv \frac{\hbar}{m} \Omega^r, \qquad \text{vorticity}$$

$$\epsilon^r \equiv \frac{\hbar}{m} \mathbf{E}^r, \qquad \text{field strength}$$

$$\rho_r \equiv \frac{m}{\hbar c_0} \frac{\partial \mathfrak{F}_{(M)}}{\partial(D\psi)} C_r \psi, \qquad \text{component densities}$$

$$\kappa_r \equiv \frac{m}{\hbar} \mathbf{K}_r, \qquad \text{conjugate fields}$$

$$\lambda_r \equiv \frac{m}{\hbar} \Lambda_r.$$

Example 3. *Abelian symmetries and the two-fluid model.* If the gauge group algebra is Abelian, then the structure constants c_{ms}^r vanish. In particular, a one-parameter group of transformations will be explored in relation to a two-fluid model for superfluid ^{4}He.

Recall the relations (4-34). The vector $\mathbf{g}$ represents the mass momentum density for the fluid:

$$\mathbf{g} = -\frac{1}{c_0} \Pi^0 \mathbf{D}\psi = \mathbf{p}_{(n)} + \rho_{(s)}\mathbf{w},$$

where

$$\mathbf{p}_{(n)} \equiv -\frac{1}{c_0} \Pi^0 \boldsymbol{\nabla}\psi,$$

$$\rho_{(s)} \equiv \frac{m}{\hbar c_0} \Pi^0 G\psi.$$

G is the generator of the one-parameter group of transformations. Also define

$$\mu_{(s)} \equiv -\frac{\hbar c_0}{m} W_0.$$

Here the parenthetically enclosed subscripts refer to normal or superfluid components of the system.

The set of hydrodynamic equations for the system can now be directly generated. The Euler-Lagrange equations are

$$\frac{\partial}{\partial t} \Pi^0 + \boldsymbol{\nabla} \cdot (\Pi^0 \mathbf{v}) - (\mu_{(s)} - \mathbf{v} \cdot \boldsymbol{\omega})G$$

$$= \frac{\partial \mathcal{F}_{(M)}}{\partial \psi}, \qquad\qquad \text{field equation}$$

$$\frac{\partial}{\partial t} \zeta + \boldsymbol{\nabla} \cdot (\zeta \mathbf{v}) = \frac{\partial \mathcal{F}_{(M)}}{\partial q}, \qquad\qquad \text{entropy density flux}$$

$$\boldsymbol{\nabla} \cdot \boldsymbol{\kappa}(\boldsymbol{\epsilon}, \boldsymbol{\omega}) = -\rho_{(s)} c_0,$$

gauge field strength source
equations

$$-\frac{1}{c_0} \frac{\partial}{\partial t} \boldsymbol{\kappa}(\boldsymbol{\epsilon}, \boldsymbol{\omega}) + \boldsymbol{\nabla} \times \boldsymbol{\lambda}(\boldsymbol{\epsilon}, \boldsymbol{\omega}) \qquad\qquad \textbf{(4-37)}$$

$$= -\rho_{(s)} \mathbf{v}.$$

The symmetry condition (4-35) results in a statement of substantive superfluid current conservation:

$$\frac{\partial}{\partial t} \rho_{(s)} + \boldsymbol{\nabla} \cdot (\rho_{(s)} \mathbf{v}) = 0.$$

The subsidiary conditions (4-30) or (4-31) can be expressed in the equivalent ways

$$\boldsymbol{\epsilon} = -\frac{1}{c_0} \left[\boldsymbol{\nabla} \mu_{(s)} + \frac{\partial}{\partial t} \boldsymbol{\omega} \right],$$

$$\boldsymbol{\omega} = \boldsymbol{\nabla} \times \mathbf{w},$$

or, alternatively,

$$\frac{1}{c_0} \frac{\partial}{\partial t} \boldsymbol{\omega} + \boldsymbol{\nabla} \times \boldsymbol{\epsilon} = 0,$$

$$\boldsymbol{\nabla} \cdot \boldsymbol{\omega} = 0. \qquad\qquad \textbf{(4-38)}$$

Note that from condition (4-14), the quantities ϵ and ω are gauge invariant, and therefore so are the conjugate quantities κ and λ.

The form of the energy density follows from Equation (4-33),

$$ u = \gamma + \mu_{(s)}\rho_{(s)} + T\zeta + \kappa \cdot \epsilon + \lambda \cdot \omega - \rho. $$

Thus, if any two parameters ζ, ϕ are canonically conjugate, then

$$ \frac{\partial u}{\partial \zeta} = \phi, \qquad \zeta \text{ extensive; } \phi \text{ intensive.} $$

The remaining flux parameters are given by

$$ \mathbf{Q} = (u_{(M)} + \rho)\mathbf{v} + c_0 \, \epsilon \times \lambda, \qquad \text{energy flux vector} $$

$$ \mathbf{I} = \mathbf{g} - \zeta \nabla q + \frac{1}{c_0} \, \kappa \times \omega, \qquad \text{canonical momentum density} $$

$$ T_{jk} = v_j(g_k - \zeta \nabla_k q) - \omega_j \lambda_k + \delta_{jk}\rho. \qquad \text{momentum flux tensor} $$

The conservation laws are

$$ \frac{\partial u}{\partial t} + \nabla \cdot \mathbf{Q} = 0, $$

$$ \frac{\partial}{\partial t} I_k + \nabla_j T_{jk} = 0. \qquad\qquad \textbf{(4-40)} $$

Notice, however, that the set of equations (4-39) and (4-40) does allow for exchange of energy-momentum between the two components of the super-fluid system. The energy-density dissipation and force density for such exchanges are given by

$$ f_0 = -\rho_{(s)} \, \mathbf{v} \cdot \epsilon, $$

$$ \mathbf{f} = \rho_{(s)} \, [c_0 \, \epsilon + \mathbf{v} \times \omega]. $$

Note that if there are no material forces acting on the superfluid compo-nent, an effect occurs within the bulk fluid analogous to the Hall effect for currents in conductors in the presence of magnetic fields. The superfluid component then distributes itself in that region such that

$$ \epsilon = -\frac{1}{c} \, \mathbf{v} \times \omega. $$

The physical effects of the equations developed above will be interpreted next. First, note that any quantity representing a physical observable (e.g., u, $\mathbf{g}$, ζ) is gauge invariant. In particular, examine the momentum density $\mathbf{g}$:

$$\mathbf{g} = p_{(n)} + \rho_{(s)}\boldsymbol{\omega} = -\frac{1}{c_0}\Pi^0\boldsymbol{\nabla}\psi + \rho_{(s)}\boldsymbol{\omega}.$$

The terms are not separately invariant under local gauge transformations, whereas the physical momentum density is gauge invariant. Thus the current $\mathbf{g}$ is constructed from two-component gauge-dependent currents $\mathbf{p}_{(m)}$ and $\rho_{(s)}\boldsymbol{\omega}$.

Consider next the canonical momentum flux equation in (4-40). This equation can be written as follows:

$$\left[\frac{\partial}{\partial t}g_k + \nabla_j(v_j g_k)\right] + \left[\frac{1}{c_0}\frac{\partial}{\partial t}(\kappa \times \omega)_k - \nabla_j(\omega_j\lambda_k)\right]$$

$$= \zeta\frac{D}{Dt}(\nabla_k q) - \nabla_k P. \tag{4-41}$$

In what follows this equation is examined in two frames of reference.

The local frame for which $v = 0$ is referred to as the substantive frame and, for a particular choice of gauge, corresponds to the Galilean frame that is comoving with the normal component. In this frame, all substantive parameters are identical to the corresponding canonical parameters; for example, $T^* = T$. Equation (4-41) is represented in this frame by

$$\frac{\partial}{\partial t}\left(\mathbf{g} + \frac{1}{c_0}\boldsymbol{\kappa} \times \boldsymbol{\omega}\right) = \zeta\boldsymbol{\nabla}T - \boldsymbol{\nabla}P + (\boldsymbol{\omega}\boldsymbol{\cdot}\boldsymbol{\nabla})\boldsymbol{\lambda}.$$

If in this frame of reference, the field equations (4-37) imply the irrotationality condition $\boldsymbol{\omega} = 0$, then the equation further simplifies:

$$\frac{\partial}{\partial t}\mathbf{g} = \zeta\boldsymbol{\nabla}T - \boldsymbol{\nabla}P.$$

This equation apparently predicts a temperature gradient associated with pressure gradients if there is no fluid flow ($\mathbf{g} = 0$) and a fountain effect if there is no pressure gradient.

Another relevant frame is the rest frame of the fluid, for which $\mathbf{g} = 0$; that is, there is no net momentum flux. Since the density $\mathbf{g}$ is constructed from two terms, and the entropy flux equations (4-29) depend only on the Galilean parameter $\mathbf{v}$, there can be a net entropy flux within the system. The momentum flux equation (4-41) can be expressed as indicated if the equations are cyclic in the heat parameter q:

$$\frac{1}{c_0} \frac{\partial}{\partial t} (\, \kappa \times \omega \,) = \zeta \frac{D}{Dt} \nabla q - \nabla P + (\, \omega \cdot \nabla \,) \lambda \qquad (4\text{-}42)$$

if q is a cyclic parameter. The term involving the heat parameter can be expressed in alternative forms,

$$\frac{D}{Dt} \nabla q = \nabla T + (\mathbf{v} \cdot \nabla) \nabla q = \nabla T^* - \sum_{j=1}^{3} (\nabla v_j)(\nabla_j q). \qquad (4\text{-}43)$$

For instance, for a stationary system with uniform flow, the system satisfies

$$0 = \nabla T^* - \nabla P + (\, \omega \cdot \nabla \,) \lambda \,.$$

If there do exist gradients in the parameter $\mathbf{v}$, Equations (4-42) and (4-43) will imply the formation of gradients in other physical parameters. This result is a consequence of the substantive form of the equations, and is analogous to a stationary gauge wheel effect[10] (without dissipative terms).

In general, the material force equation for the system can be determined directly. With the following identifications:

$$\mathbf{I}_{(M)} \equiv \mathbf{g} - \zeta \nabla q \,,$$

$$T_{(M)jk} \equiv v_j I_{(M)k} + \delta_{jk} \mathcal{F}_{(M)} \,,$$

the force density equation for the system is

$$\frac{\partial}{\partial t} I_{(M)k} + \nabla_j T_{(M)jk} = f_k = \rho_{(s)} c_0 \, [(\, \epsilon + \mathbf{v}/c_0 \times \omega \,)]_k \,.$$

This momentum flux is "radiated" via the gauge fields, since total momentum density is conserved. This equation is valid for compressible flow and for general equations of state.

Experiments have indicated that in the frame of reference for which the "normal field" is locked, the flow of the superfluid system is irrotational:

$$\omega = \nabla \times \mathbf{w} = 0 \,.$$

Therefore, in order to satisfy the experimental criteria, Equations (4-37), which govern the creation of gauge field strengths, should have irrotationality as a solution in this frame. Since other experiments indicate the lack of entropy flow in this frame of reference, it is apparently parameterized by $\mathbf{v} = \mathbf{0}$.

For the case of irrotational superflow, if the group is periodic in the sense discussed previously, quantization conditions emerge. Examine Equations (4-13) as applied to this situation:

$$\mathbf{w}(\mathbf{x}{:}\alpha) = \frac{\hbar}{m} [\nabla\alpha + \mathbf{a}(\mathbf{x}{:}\alpha)].$$

Notice that, for an Abelian system, the parameter $\mathbf{a}$ is independent of the gauge parameter α.

$$\mathbf{w}(\mathbf{x}{:}e) = \mathbf{w}(\mathbf{x}) = \frac{\hbar}{m} \mathbf{a}(\mathbf{x}),$$

$$\mathbf{w}(\mathbf{x}{:}\alpha) = \frac{\hbar}{m} \nabla\alpha + \mathbf{w}(\mathbf{x}).$$

If $\boldsymbol{\omega} = \nabla \times \mathbf{w}(\mathbf{x}) = 0$, then there exists a gauge transformation parameter $-\alpha_0(\mathbf{x})$ that will eliminate the vector components of the gauge fields:

$$\mathbf{w}(\mathbf{x}{:}{-}\alpha_0) = 0 = -\frac{\hbar}{m} \nabla\alpha_0 + \mathbf{w}(\mathbf{x}).$$

Therefore, the circulation can be represented

$$\oint \mathbf{w} \cdot d\mathbf{l} = \frac{\hbar}{m} \oint \nabla\alpha_0 \cdot d\mathbf{l}.$$

The result (4-17) then implies the following condition:

$$\oint_{C_n[1]} \mathbf{w} \cdot d\mathbf{l} = 2\pi n \frac{\hbar}{m}. \tag{4-44}$$

The integer n represents the degree of the map $C_n[1]$ on the local group parameter α_0 with periodicity 2π.

To complete the discussion of the two-fluid model, the phenomenon of critical flow is briefly discussed. The mass flux through a surface Σ can be expressed

$$\frac{\text{mass}}{\text{time}} = \iint_\Sigma \mathbf{g} \cdot d\mathbf{A}.$$

Experimental results are typically expressed in terms of the volumetric flow rate σ_V given by

$$\sigma_V = \int\int_\Sigma \mathbf{g} \cdot d\mathbf{A}.$$

For critical flow, the "normal fluid" is typically locked, and the superfluid flow is given by

$$\mathbf{g}_c = \rho_{(s)} \mathbf{W}_{\text{crit}}.$$

As a mechanism for the onset of dissipative superflow, consider Equation (4-37) for conjugate field creation in a stationary system:

$$\nabla \times \boldsymbol{\lambda} = -\rho_{(s)} \mathbf{v}.$$

In the superfluid rest frame, dissipation sets in when the parameter $\mathbf{v} = -\mathbf{w}_{\text{crit}}$. Then the volumetric flow rate satisfies

$$\sigma_{V\,\text{crit}} = \frac{1}{\rho} \int\int_\Sigma \nabla \times \boldsymbol{\lambda}_c \cdot d\mathbf{A} = \frac{1}{\rho} \oint_{\partial\Sigma} \boldsymbol{\lambda}_c \cdot d\mathbf{l}.$$

If the system has cylindrical symmetry about the flow direction, then

$$\sigma_{V\,\text{crit}} = \frac{\lambda_c(R)}{\rho}\, 2\pi R,$$

$$\lambda_c(r) = \frac{w_{\text{crit}}}{r} \int_0^r r' \rho_{(s)}(r')\, dr', \qquad \text{azimuthally directed,} \qquad \textbf{(4-45)}$$

where R is the maximum radius of the channel. One may expect the formation of a vortex ring when the circulation ω around a loop of the size of a vortex core takes on a quantized value as determined by (4-44)

$$\int \mathbf{w} \cdot d\mathbf{l} = \int\int_\Sigma \boldsymbol{\omega} \cdot d\mathbf{A} \simeq \omega(R)\pi a^2 = \frac{h}{m}$$

$$\text{vortex core radius} = a \ll R.$$

For example, suppose that the action is like the Maxwell action $F_{(G)} = \boldsymbol{\omega} \cdot \boldsymbol{\omega}/2\xi$, where ξ is a coupling constant of dimension length/mass. Then, from

Equation (4-45).

$$\omega_c(R) = \xi \, \frac{w_{crit}}{R} \int_0^R r' \rho_{(s)}(r') \, dr'.$$

As a first approximation, if the superfluid is incompressible, the form for $\sigma_{V \, crit}$ is determined to be

$$\rho_{(s)} \omega_{crit} \simeq \frac{4\hbar}{\xi m a^2} \frac{1}{R},$$

$$\sigma_{V \, crit} = \frac{\rho_{(s)}}{\rho} \, \omega_{crit} R^2 \simeq \left(\frac{4\pi\hbar}{m a^2 \rho} \right) R.$$

Improvements can be made on these predictions using compressional equations of state. Experiments[11] indicate that the actual behavior of the critical flow rate with channel size is

$$\sigma_{V \, crit} \sim R^{3/4}$$

so that the behavior indicated would need be only slightly refined by the equation of state.

Example 4. *Extension to systems with "spin densities."* For the case that follows, it is assumed that the integral symmetry group directly factorizes into a group of spin rotations and an Abelian one-parameter group. The order parameter is represented

$$\psi = \phi_{spin} \times \phi_{(0)}$$

and satisfies Equation (4-20). The generators for the groups will be, respectively, E_m and $G_{(0)}$, for $m = 1$, 2, or 3. The algebra for the operators will be as follows:

$$[\Sigma_m, \Sigma_n] = \sum_{r=1}^{3} \epsilon_{mnr} \Sigma_r,$$

$$[\Sigma_m, G_{(0)}] = 0, \quad \text{all } m,$$

with ϵ_{mnr} representing the standard fully antisymmetric tensor that takes on values $+1$, -1, or 0 according to the cyclicity of the values of the indices.

The order parameter transforms according to

$$\delta\psi = \delta\alpha G_{(0)}\psi + \delta\alpha'\Sigma_r\psi.$$

Again, the gauge fields are parameterized for dimensional convenience:

$$W_\mu \equiv \left(-\frac{m}{\hbar c_0}\mu_{(0)}, \frac{m}{\hbar}\boldsymbol{\omega}\right),$$

$$W^r_\mu \equiv \left(-\frac{v^r}{c_0}, \mathbf{W}^r\right).$$

The densities are represented

$$\rho_{(0)} \equiv \frac{m}{\hbar c_0}\Pi^0 G_{(0)}\psi, \qquad \text{mass density}$$

$$s_r \equiv \frac{1}{c_0}\Pi^0 E_r\psi, \qquad \text{spin density}$$

$$\mathbf{j}_{(0)} \equiv \frac{m}{\hbar}\Pi G_{(0)}\psi, \qquad \text{mass current}$$

$$\mathbf{j}_r \equiv \Pi \Sigma_r\psi. \qquad \text{spin current}$$

With these identifications, the hydrodynamic equations follow directly from the results in Example 2. From Equation (4-33), the energy density is expressed

$$u = \gamma + \mu_{(0)}\rho_{(0)} + v^r s_r + T\zeta + \boldsymbol{\kappa} \cdot \boldsymbol{\epsilon}$$
$$+ \boldsymbol{\lambda} \cdot \boldsymbol{\omega} + \boldsymbol{\kappa}_r \cdot \boldsymbol{\epsilon}^r + \boldsymbol{\lambda}_r \cdot \boldsymbol{\omega}^r - p,$$

and the momentum density is expressed

$$g = \mathbf{P}_{(n)} + \rho_{(s)}\mathbf{w} + s_r\mathbf{W}^r.$$

The current symmetry condition (4-35) is expressed

$$\frac{\partial}{\partial t}\rho_{(0)} + \boldsymbol{\nabla} \cdot \mathbf{j}_{(0)} = 0,$$

$$\frac{\partial}{\partial t}s_n + \boldsymbol{\nabla} \cdot \mathbf{j}_n = \epsilon_{nmr} [\boldsymbol{\lambda}_r \cdot \boldsymbol{\omega}^m - \boldsymbol{\kappa}_r \cdot \boldsymbol{\epsilon}^m - s_r v^m + \mathbf{j}_r \cdot \mathbf{W}^m].$$

These relations are similar to current equations presented in Khalatnikov and Lebedev.[9] Results for the field strengths and force densities can be directly obtained by replacing the structure constants c_{mn}^s with ϵ_{mns} in Equations (4-30), (4-32), and (4-36).

4-5   A SIMPLE EXAMPLE

To illustrate the methods presented in the previous section, a relatively simple, yet somewhat general, system is examined. Consider first the system determined by a free-energy density given here:

$$\mathfrak{F} = i\frac{\hbar}{2m}[\psi^\dagger\partial_0\psi - (\partial_0\psi)^\dagger\psi] - \frac{\hbar}{2m^2}(\nabla\psi)^\dagger\cdot(\nabla\psi)$$

$$- V(\psi^\dagger\psi, q, \dot{q}, \nabla q), \quad \textbf{(4-46)}$$

where q is a real scalar field and

$$\partial_0 q = \dot{q} = \frac{\partial q}{\partial t}.$$

Stationarity of the free energy under variations in ψ and $\psi^\dagger$ result in two equations of state for these fields. One of the equations can be expressed as

$$\frac{\partial}{\partial t}(\psi^\dagger\psi) + \nabla\cdot\left[\frac{\hbar}{2im}(\psi^\dagger\nabla\psi - (\nabla\psi)^\dagger\psi)\right] = 0. \quad \textbf{(4-47)}$$

In addition, there is another second-order differential equation relating the fields ψ to V. From the form of Equation (4-47) it is tempting to make the following identifications:

$$\psi^\dagger\psi \equiv \rho, \quad \text{density}$$

$$\frac{\hbar}{2im}[\psi^\dagger\nabla\psi - (\nabla\psi)^\dagger\psi] \equiv \mathbf{j}. \quad \text{current density} \quad \textbf{(4-48)}$$

Then Equation (4-47) represents a continuity equation for the density ρ.

Stationarity of the free energy under variations in q results in an additional equation. Following the identifications in Example 1 of Section 4-4,

$$\dot{q} \equiv T,$$

$$-\frac{\partial V}{\partial \dot{q}} \equiv \zeta, \quad -\frac{\partial V}{\partial \nabla q} \equiv \pi,$$

the stationarity condition is

$$\frac{\partial \zeta}{\partial t} + \nabla \cdot \pi = -\frac{\partial V}{\partial q}. \tag{4-49}$$

The parameter q is interpreted as a canonical heat parameter, and if $\mathfrak{F}$ is cyclic in q, Equation (4-49) represents conservation of entropy density with entropy flux vector given by π.

The canonical energy and momentum densities can be directly obtained using (4-24). Using the identifications (4-27), the energy density is given by

$$u = \frac{\hbar^2}{2m^2} (\nabla \psi^\dagger) \cdot (\nabla \Psi) + V(\rho, q, T, \nabla q) + T\zeta.$$

The energy flux vector Q is given by

$$Q = -\frac{\hbar^2}{2m^2} [\dot{\psi}^\dagger \nabla \psi + (\nabla \psi)^\dagger \dot{\psi}] + T\pi. \tag{4-50}$$

For this system of equations, the canonical momentum density is expressed

$$\mathbf{I} = \mathbf{j} - \zeta \nabla q \tag{4-51}$$

and the momentum flux tensor is

$$\tau_{ik} = \frac{\hbar^2}{2m^2} [(\nabla_j \psi)^\dagger (\nabla_k \psi) + (\nabla_k \psi)^\dagger (\nabla_j \psi)] - \pi_i \nabla_k q + \delta_{ik} \mathfrak{F}. \tag{4-52}$$

For further illustration, consider the case in which V is independent of q and ∇q. The flux vector π vanishes identically, and Equation (4-49) reads

$$\frac{\partial \zeta}{\partial t} = 0.$$

Furthermore, the complex fields $\Psi(x, t)$ can be expressed as follows:

$$\psi(\mathbf{x}, t) = |\psi(\mathbf{x}, t)| e^{i\theta(\mathbf{x}, t)} = \sqrt{\rho(\mathbf{x}, t)}\, e^{i\theta(\mathbf{x}, t)}.$$

For this form, the current is expressible from (4-48) as

$$\mathbf{j} = \rho \cdot \left(\frac{\hbar}{m} \nabla \theta\right). \tag{4-54}$$

Thus, the current density is a measure of the gradient of the phase of the field ψ. The canonical energy density u is expressed as

$$u = \frac{1}{2}\rho \left| \frac{\hbar}{2m} \nabla \log \rho \right|^2 + \frac{1}{2}\rho \left| \frac{\hbar}{m} \nabla \theta \right|^2 + V(\rho, T) + T\zeta.$$

The energy flux vector in (4-50) becomes

$$\mathbf{Q} = -\left(\frac{\hbar}{2m} \frac{\partial}{\partial t} \log \rho \right) \rho \left(\frac{\hbar}{2m} \nabla \log \rho \right) + \left(\frac{\hbar}{m} \dot{\theta} \right) \mathbf{j}.$$

The momentum flux equation, using the forms in (4-51) and (4-52), is expressible as

$$\frac{\partial}{\partial t}\mathbf{j} + \nabla \cdot \tau^{(1)} = \zeta \nabla T + \nabla \mathcal{F},$$

$$\tau_{ik}^{(1)} = \rho \left(\frac{\hbar}{2m} \nabla_i \log \rho \right) \left(\frac{\hbar}{2m} \nabla_k \log \rho \right) + \left(\frac{\hbar}{m} \nabla_i \theta \right) j_k.$$

Thus these relations form a plausible system of equations for a simple hydrodynamics.

Equation (4-46) for the density $\mathcal{F}$ has a global gauge invariance

$$\psi(x) \rightarrow \psi'(\mathbf{x}) = e^{i\alpha}\psi(\mathbf{x}),$$
$$\mathcal{F}(x) \rightarrow \mathcal{F}'(x) = \mathcal{F}(x).$$

The equation will have local gauge invariance if the derivatives $\partial_\mu \psi$ are replaced by gauge covariant derivatives $D_\mu \psi$ given by

$$D_\mu \psi \equiv (\partial_\mu - iW_\mu \mathcal{G})\psi.$$

Here $W_\mu(x, t)$ is a real field, and $\mathcal{G}$ is the (Hermitian) generator of the phase symmetry transformation. We assume that q is invariant under the transformation, and therefore the gauge fields will not carry quantities with q content. This assumption means that the gauge-derived fields will carry no entropy or heat flux. The form of the equation for $\mathcal{F}$ that has local gauge invariance is

$$\mathcal{F} = \frac{i\hbar}{2m} [\psi^\dagger (D_0 \psi) - (D_0\psi)^\dagger \psi] - \frac{\hbar^2}{2m^2} (\mathbf{D}\psi)^\dagger - (\mathbf{D}\psi)$$

$$- V(\psi^\dagger \psi, q, \dot{q}, \nabla q) + \mathcal{F}_G(\epsilon, \omega). \quad \textbf{(4-55)}$$

The gauge-independent quantities ϵ and $\boldsymbol{\omega}$ are as defined in Equation (4-38). One can directly verify the invariance

$$\psi(\mathbf{x},\, t) \rightarrow \psi'(\mathbf{x},\, t) = e^{i\alpha(\mathbf{x},\, t)}\psi(\mathbf{x},\, t),$$

$$W_\mu(\mathbf{x},\, t) \rightarrow W'_\mu(\mathbf{x},\, t) = W_\mu(\mathbf{x},\, t) + \partial_\mu\alpha(\mathbf{x},\, t),$$

$$\mathfrak{F}(\mathbf{x},\, t) \rightarrow \mathfrak{F}'(\mathbf{x},\, t) = \mathfrak{F}(\mathbf{x},\, t). \qquad (4\text{-}56)$$

The stationarity of Equation (4-55) under variations of the field ψ results in a modified equation of the form of (4-47):

$$\frac{\partial}{\partial t}\, (\psi^\dagger\psi) + \boldsymbol{\nabla} \cdot \left[\frac{\hbar}{2im}\, (\psi^\dagger\boldsymbol{\nabla}\psi - (\boldsymbol{\nabla}\psi)^\dagger\psi) - \frac{\hbar}{m}\, \psi^\dagger\psi W \right] = 0. \qquad (4\text{-}57)$$

The following identifications are made:

$$\rho \equiv \psi^\dagger\psi, \qquad \text{gauge-invariant density}$$

$$\rho(s) \equiv \psi^\dagger\psi,$$

$$\mathbf{j} \equiv \frac{\hbar}{2im}\, [\psi^+\boldsymbol{\nabla}\psi - (\psi)^+\psi],$$

$$\mathbf{w}_{(s)} \equiv -\frac{\hbar}{m}\, \mathbf{W}.$$

Then, the continuity equation (4-57) has the form

$$\frac{\partial\rho}{\partial t} + \boldsymbol{\nabla} \cdot [\mathbf{j} + \rho_{(s)}\mathbf{w}_{(s)}] = 0. \qquad (4\text{-}58)$$

Since the parameter q is invariant under $\mathcal{G}$, the stationarity condition (4-49) does not change:

$$\frac{\partial\zeta}{\partial t} + \boldsymbol{\nabla} \cdot \boldsymbol{\pi} = -\frac{\partial V}{\partial q}. \qquad (4\text{-}49)$$

The equation (4-58) is interpreted as a mass flux equation for a two-component system, and Equation (4-49) indicates that one of the components carries no entropy. Since the quantities $\mathbf{j}$ and $\rho_{(s)}\mathbf{w}_{(s)}$ are not gauge invariant (although the sum *is* gauge invariant), the decomposition is unique only up to a gauge transformation.

The phase symmetry conditions (4-56) allow the construction of a conserved current using (4-23). For this system the current takes the form

$$\mathcal{J} = \frac{\hbar}{2im} [\psi^\dagger \mathcal{G} \nabla \psi - (\nabla \psi)^\dagger \mathcal{G} \psi] + \psi^\dagger \mathcal{G}^2 \psi \mathbf{w}_{(s)}.$$

The current is conserved in the sense

$$\frac{\partial \rho_{(s)}}{\partial t} + \nabla \cdot \mathcal{J} = 0. \tag{4-59}$$

Since the form of the density $\mathcal{F}$ in (4-55) is separable, the canonical energy-momentum flux parameters can be expressed in terms of the interaction between "material" component and gauge fields as indicated in Equation (4-24). For the material component, the energy density is represented

$$U_{(p)} = \frac{\hbar}{2m^2} (\mathbf{D}\psi)^\dagger \cdot (\mathbf{D}\psi) + V(\rho, q, T, \nabla q) + T\varsigma.$$

By directly substituting the form (4-54) for the gauge-covariant derivatives, this can be reexpressed:

$$U_{(p)} = \frac{\hbar^2}{2m^2} (\nabla \psi)^\dagger \cdot (\nabla \psi) + \mathbf{w}_{(s)} \cdot \mathcal{J} + V(\rho, q, T, \nabla q) + T\varsigma. \tag{4-60}$$

Thus there is an additional term present due to the velocity $\mathbf{w}_{(s)}$ and current $\mathcal{J}$, which is of the form one would expect if $\mathbf{w}_{(s)}$ were a true velocity. The energy flux vector is given by

$$\mathbf{Q}_{(p)} = -\frac{\hbar^2}{2m^2} [(\mathbf{D}\psi)^\dagger (\mathbf{D}_0\psi) + (\mathbf{D}_0\psi)^\dagger (\mathbf{D}\psi)] + T\boldsymbol{\pi}.$$

The canonical momentum is expressed

$$\mathbf{I}^{(p)} = \mathbf{j} + \rho_{(s)}\mathbf{w}_{(s)} - \varsigma \nabla q. \tag{4-61}$$

Again, this expression is indicative of a two-component system, although it is obtained in an entirely different way from Equation (4-58). The canonical momentum flux tensor has the form

$$\tau_{ik}^{(p)} = \frac{\hbar^2}{2m^2} [(D_i\psi)^\dagger (D_k\psi) + (D_k\psi)^\dagger (D_i\psi)] - \pi_i \nabla_k q + \delta_{ik}\mathcal{J}^{(P)}.$$

The coupling between material and gauge fields is expressed in (4-24).

For this system of equations, the expressions obtained are

$$\frac{\partial U_{(P)}}{\partial t} + \nabla \cdot \mathbf{Q}_{(P)} = - \ \mathfrak{J} \cdot \boldsymbol{\epsilon} \ ,$$

$$\frac{\partial I_k^{(P)}}{\partial t} + \nabla_j \ \tau_{jk}^{(P)} = \rho_{(s)}(\boldsymbol{\epsilon} \ )_k + (\mathfrak{J} \times \boldsymbol{\omega})_k \ . \qquad \textbf{(4-62)}$$

The first of equations (4-62) represents an energy dissipation if $\mathfrak{J}$ and $\boldsymbol{\epsilon}$ have nonvanishing parallel components, while the second represents a force on the material system owing to the presence of the gauge fields.

To complete the discussion, the form of the currents is explored when the complex fields ψ are expressed as in (4-53),

$$\psi(\mathbf{x}, t) = |\psi(\mathbf{x}, t)| e^{i\theta(\mathbf{x}, t)} = \sqrt{\rho(\mathbf{x}, t)} \ e^{i\theta(\mathbf{x}, t)} \ . \qquad \textbf{(4-53)}$$

The form of the continuity equation (4-58) is

$$\frac{\partial \rho}{\partial t} + \nabla \cdot \left[\left(\frac{\hbar}{m} \ \nabla \theta \right) + \rho_{(s)} \mathbf{W}_{(s)} \right] = 0 \ . \qquad \textbf{(4-63)}$$

In addition, suppose that the operator $\mathfrak{G}$ serves only as a projection into the space of $\rho_{(s)}$.

$$\mathfrak{G}^2 = \mathfrak{G} \ .$$

Then the current $\mathfrak{J}$ can be expressed

$$\mathfrak{J} = \rho_{(s)} \left(\mathbf{W}_{(s)} + \frac{\hbar}{m} \ \nabla \theta \right) \Big]$$

and the gauge symmetry condition (4-59) is

$$\frac{\partial \rho_{(s)}}{\partial t} + \nabla \cdot \left[\rho_{(s)} \left(\mathbf{W}_{(s)} + \frac{\hbar}{m} \ \nabla \theta \right) \right] = 0 \ . \qquad \textbf{(4-64)}$$

It is very tempting to make the following identifications:

$$\rho_{(n)} \equiv \rho - \rho_{(s)} \ , \qquad \mathbf{v}_{(n)} \equiv \frac{\hbar}{m} \ \nabla \theta \ ,$$

$$\mathbf{v}_{(s)} \equiv \mathbf{W}_{(s)} + \frac{\hbar}{m} \ \nabla \theta \ .$$

For these identifications, the expressions (4-63) and (4-64) give apparent two-fluid relationships:

$$\frac{\partial \rho}{\partial t} + \boldsymbol{\nabla} \cdot [\rho_{(n)}\mathbf{v}_{(n)} + \rho_{(s)}\mathbf{v}_{(s)}] = 0 \,,$$

$$\frac{\partial \rho_{(s)}}{\partial t} + \boldsymbol{\nabla} \cdot (\rho_{(s)}\mathbf{v}_{(s)}) = 0 \,,$$

$$\frac{\partial \rho_{(n)}}{\partial t} + \boldsymbol{\nabla} \cdot (\rho_{(n)}\mathbf{v}_{(n)}) = 0 \,.$$

Note that as defined, the densities ρ, $\rho_{(s)}$, and $\rho_{(n)}$ are gauge invariant. The canonical energy and momentum densities (4-60) and (4-61) can be expressed

$$U_{(P)} = \frac{\hbar^2}{2m^2} (\boldsymbol{\nabla}\psi)^\dagger \cdot (\boldsymbol{\nabla}\psi) + (\mathbf{v}_{(s)} - \mathbf{v}_{(n)}) \cdot \mathbf{\mathfrak{J}} + V(\rho, q, T, \boldsymbol{\nabla}q) + T\varsigma$$

$$\mathbf{I}_{(P)} = \rho_{(s)}\mathbf{v}_{(s)} + \rho_{(n)}\mathbf{v}_{(n)} - \varsigma\boldsymbol{\nabla}q \,.$$

4-6   CONCLUSIONS

The formalism presented describes many of the macroscopic and phenomenological properties of certain superfluid systems. The formalism can be directly generalized to describe systems with partial symmetries. The form of the equations is similar to that of electromagnetism, thus allowing direct use of developed analytic techniques for the solutions of the hydrodynamic equations. In addition, these similarities provide an intuitive link to the understanding of the phenomenology and critical behavior of quantum fluids.

ACKNOWLEDGMENTS

J. V. L. wishes to acknowledge the support of the Chancellor's Distinguished Fellowship of the University of California, Berkeley. H. L. M. would like to acknowledge the encouragement of Arnold P. Jones.

REFERENCES

1. L. D. Landav, "The Theory of Superfluidity of Helium,"*Journal of Physics* 5 (1971): 71–90.

2. H. L. Morrison and J. V. Lindesay, "Galilean Presymmetry and the Quantization of Circulation," *Journal of Low Temperature Physics* 26 (1977): 899–907.

3. J. Chela-Flores, "Gauge Theory of Superfluidity," *Journal of Low Temperature Physics* 21 (1975): 307–319.

4. J. V. Lindesay and H. L. Morrison, "Symmetry Properties of Anisotropic Superfluids," (1984): preprint.

5. M. Hamermesh, *Group Theory* (Reading, Mass.: Addison-Wesley, 1962).

6. L. S. Pontryagin, *Topological Groups* (New York: Gordon and Breach, 1966).

7. C. N. Yang and R. L. Mills, "Conservation of Isotopic Spin and Isotopic Gauge Invariance," *Physical Review* 96 (1954): 191–195.

8. E. Noether, *Göttinger Nachrichten* 2 (1918): 235.

9. I. M. Khalatnikov and V. V. Lebedev, "Equation of Hydrodynamics of Quantum Liquids in the Presence of Continuously Distributed Singular Solitions," *Supplement of the Progress of Theoretical Physics*, no. 69 (1980): 269–280.

10. T. L. Ho and N. D. Mermin, "Gauge Wheel of Superfluid ^{4}He," *Physical Review Letters* 44 (1980): 330–333.

11. W. M. Van Alphen, G. J. Van Haasteren, R. De Bruyn Ouboter and K. W. Taconis, "The Dependence of the Critical Velocity of the Superfluid on Channel Diameter and Film Thickness," *Physics Letters* 20 (1966): 474–475.

5

Classical Chaos, the Geometry of Phase Space, and Semiclassical Quantization

William P. Reinhardt

*University of Colorado and National
Bureau of Standards*

ABSTRACT: Basic concepts relating to the onset of chaos in deterministic dynamics are illustrated using conservative Hamiltonian dynamics in two degrees of freedom—that is, dynamics in a four-dimensional phase space. The correspondence between regular and chaotic dynamics and concepts of integrability and nonintegrability are established, and semiclassical quantization (i.e., determination of estimates of energy eigenvalues of the corresponding Schrödinger operator using only classical dynamical information as input) is carried out utilizing the invariant tori of integrable (regular) classical dynamics. The origins of classical chaos are briefly discussed in terms of orbit bifurcation and in terms of the Painlevé analysis of the complex time singularities of the orbits themselves. Expectations of quantum correspondence principle ramifications of classical chaos are discussed in terms of Percival's "irregular spectrum" and in terms of the statistics of nearest-neighbor level spacings. It is pointed out that although onset of classical chaos and "irregular spectra" often occurs in parallel in numerical experiments, mere presence of classical chaos does not imply quantum chaos. A geometric motivation for this observation is given in terms of a qualitative analysis of the fragmentation of manifold structure as a system becomes nonintegrable. This analysis focuses on the remaining time-independent structure of phase space rather than dynamics of individual trajectories, and indicates that observations of chaotic trajectories need not imply gross destruction of the manifold structure; rather, small patches of strong nonanalyticity may develop, leaving large fragments of the tori intact. This phenomenon is investigated using Padé tables to accelerate convergence of the Birkhoff-

Gustavson normalization. Quantization on fragmented tori is discussed for the Hénon-Heiles problem and for two-dimensional rational billiards. In the latter case the mechanism of fragmentation is completely understood. In conclusion, it is suggested that, owing to remnants of tori, classical chaos and irregular spectra are not in one-to-one correspondence, and that integrable approximations to the classical dynamics can yield irregular quantum spectra, provided that appropriate uniform approximations are made.

5-1   INTRODUCTION AND PHENOMENOLOGY

Introduction and Scope

Deterministic classical chaos is a striking phenomenon in conservative and dissipative dynamics in that systems of a few nonlinear equations, or the simple nonlinear mappings abstracted from them, can often show sudden and unmistakable transitions between regular motion, implying existence of constants of motion, and chaotic behavior, implying absence of such constants. Such transitions can occur as a function of initial conditions or on variation of external parameters.[1] The aim of this chapter is to explore the phenomenology of classical chaos and the possibility of related quantum phenomena, induced by the presence of such underlying chaos, as they occur in conservative Hamiltonian systems. Examples are restricted to systems of two degrees of freedom; that is, to discussion of the geometry of four-dimensional phase space. This limitation may seem severe, but such systems show a major fraction of the complexity of the more general case and are certainly not fully understood. Restriction to conservative systems makes possible a parallel discussion of quantum systems and classical-quantum correspondence, within the framework of "ordinary" nonrelativistic wave mechanics.

The chapter begins in a very elementary manner with illustrations of the phenomenon of classical chaos using a simple and traditional example.[2] Such illustrations are based on numerically determined classical trajectories, obtained by direct integration of the Hamiltonian equations of motion. The techniques of Poincaré section and Fourier analysis are seen to be sensitive indicators of the onset of "macroscopic" classical chaos. In particular, the possible fractal[3] nature of the power spectrum of a chaotic classical observable is pointed out. Section 5-1 continues with discussion of angle action variables,[4,5] the invariant tori of integrable classical dynamics,[5-8] and the Einstein-Brillouin-Keller (EBK) semiclassical quantization[9-11] on these tori. Various approaches to the origin of chaos, from a trajectory-based point of view, are then reviewed: local exponentiation[12-15] and negative curvature[16]; bifurcations, and period multiplying[17-21]; and, finally, the rapidly developing

Painlevé[22-28] approaches that suggest fractal boundaries of analyticity for trajectories, as a function of complex time.

In discussing quantized versions of the same dynamical systems, care must be taken to distinguish the problems of: (1) defining quantum analog(s) of classical chaos (i.e., is there any such phenomenon as "quantum chaos"?); and (2) providing necessary and sufficient conditions (if, indeed, there are such) for predictions of relationships between classical chaos and such suitably defined "quantum chaos." Many conjectures, with respect to both (1) and (2) have been made,[29,30] a general sentiment being[30]:

In particular, any transition of classical dynamics from regular to random ought to be reflected in the form of the wave function for quantum states, and in the distribution of quantum energy levels, especially in the semiclassical limit (i.e., $\hbar \rightarrow 0$, *Ed.*) where these are densely populated. (P. 18)

The various conjectures evolving from and around such an expectation are outlined, illustrated, and discussed in Section 5-2.

With the exception of the discussion of the Painlevé analysis later in this section, much of the material in Sections 5-1 and 5-2 has been discussed and reviewed many times within the last few years. The reader is referred to references 29–40 for such reviews and to references 41–49 for collections of articles. Monographs,[7,8,50-52] and more recently even undergraduate texts,[53-55] are available for more detailed expositions of much of the purely classical dynamics. Recent primary and secondary articles and collections discussing closely related topics, not discussed here in any detail, are concerned with "quantum integrability,"[56-58] the bifurcation, or period doubling, path(s) to chaos in dissipative and conservative systems,[45,59,60] and in particular the use of renormalization of techniques in these areas.[61-64]

The distinction between classically regular and classically chaotic deterministic motion made in Sections 5-1 and 5-2, and in references 2, 7, 8, and 12–49, most often is based on the quite distinctive behavior of individual trajectories in the two regimes. In Section 5-3 a rather different view of the mechanism of the onset of classical chaos is taken. This view is largely intuitive and is based on work carried out in the author's research group. It is proposed that, in many classes of systems, good local approximate constants of motion exist, and often destruction of classical tori may be a highly nonuniform phenomenon in that fragments of the invariant manifold structure remain.[65-70] This proposal in turn suggests that for the purposes of semiclassical quantization rather than for following the dynamics of individual chaotic trajectories, partially intact tori might be used in an Einstein-Brillouin-Keller-type quantization in the presence of classical chaos.[67-70] This suggestion is explored and illustrated in Section 5-4. Section 5-5 contains a summary.

Regular and Chaotic Classical Dynamics

The most studied of the differential conservative dynamical systems in two degrees of freedom is undoubtedly that of Hénon and Heiles,[2, 12–49, 68, 69, 71] which grew[72, 73] out of the study of families of Hamiltonian systems of interest in the astrophysics of galactic and planetary dynamics.[74–79] We use the Hénon-Heiles system to illustrate regular and chaotic classical dynamics. Calling the coordinates $\bar{q} = (q_1, q_2)$ and the momenta $\bar{p} = (p_1, p_2)$, the Hénon-Heiles Hamiltonian is[2]

$$H(\bar{p}, \bar{q}) = \tfrac{1}{2}(p_1^2 + q_1^2) + \tfrac{1}{2}(p_2^2 + q_2^2) + q_1^2 q_2 - \tfrac{1}{3}q_2^3, \qquad (5\text{-}1)$$

which is a system of two 1 : 1 degenerate harmonic oscillators (i.e., the oscillator frequencies ω_1, ω_2 are equal) with a cubic polynomial coupling. The Hénon-Heiles potential energy surface is illustrated in Figure 5-1. The time evolution of the four phase-space variables (q_i, p_i; $i = 1, 2$) is given by Hamilton's canonical equations:[4, 5]

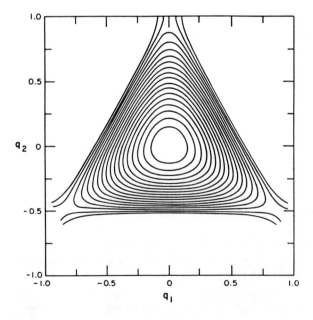

Figure 5-1
Equipotentials for the Hénon-Heiles problem. The surface has a local minimum at $q_1 = q_2 = 0$ where $E = 0$. Classical dissociation can occur via three equivalent cols at $E = 1/6$. All of the classical motion considered here is that trapped in the well near $q_1 = q_2 = 0$, with energies $0 \leq E < 1/6$.

$$\dot{p}_i = -\frac{\partial H}{\partial q_i}, \qquad i = 1, 2, \tag{5-2a}$$

$$\dot{q}_i = \frac{\partial H}{\partial p_i}, \qquad i = 1, 2. \tag{5-2b}$$

Typical coordinate space motion of a regular (or quasi-periodic) trajectory is shown in Figure 5-2, where $x \equiv q_1$, $y \equiv q_2$. Similar trajectory plots could be just as easily displayed as trajectories in other planes, for example, (q_1, p_2) or (p_1, p_2). A question, closely related to problems in the foundations of classical statistical mechanics, is whether the trajectory fills the volume defined by energy conservation in the phase space—and if it does, whether the trajectory is ergodic; that is, whether time averages are equal to phase volume averages. To examine this question, a technique of Poincaré is particularly appropriate for systems of two degrees of freedom, with an energy-conserving four-dimensional phase space. If E is fixed, only three independent phase variables remain for a conservative system. Motion in this three-dimensional subspace is easily projected into two dimensions via

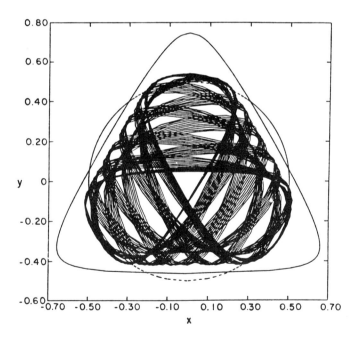

Figure 5-2
A quasi-periodic trajectory for the Hénon-Heiles problem. The triangular boundary is the equipotential at the energy of the trajectory. $x \equiv q_1$, $y \equiv q_2$.

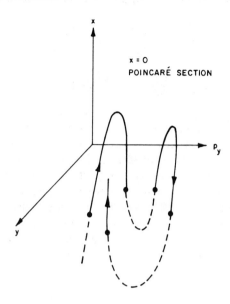

Figure 5-3
Definition of the $x = 0$ Poincaré section.

the Poincaré section technique. In Figure 5-3 the definition of the Poincaré section is illustrated; it arises by taking points along the trajectory where $x(t) = q_1(t) = 0$ and displaying the resulting intersection in the $(y, p_y) = (q_2, p_2)$ plane as shown. The resulting Poincaré section corresponding to the coordinate space trajectory of Figure 5-2 is shown in Figure 5-4. Although such a finite precision/finite time computer study does not immediately lead to rigorous mathematical deduction, it certainly appears that:

> The motion on the energy shell (i.e., in the energy-conserving subspace of the four-dimensional phase space) is not ergodic. Rather, the motion seems to be confined to a two-dimensional surface, which intersects the $x = q_1 = 0$ plane in two closed curves. Ergodic or quasi-ergodic motion would, of necessity, fill the energy shell or a portion of it, which would correspond to an area on the Poincaré surface.
>
> Correspondingly, a constant of motion, in addition to $E = H(\bar{p}, \bar{q})$, must exist. It is this constant that reduces the dimensionality of the subspace of the trajectory from the three-dimensional energy shell to a two-dimensional surface embedded in the four-dimensional phase space.

How are we to understand these statements for this nonseparable system?

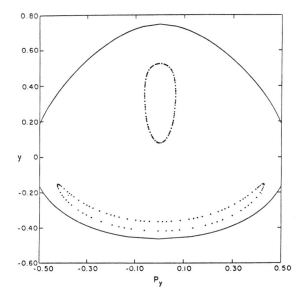

Figure 5-4
Poincaré section for the quasiperiodic trajectory of Figure 5-2. Were the trajectory ergodic—or quasi-ergodic—it would fill an area on the sectioning surface. This does not appear to be the case, suggesting existence of a dynamical constant of motion in addition to $H(\bar{p}, \bar{q})$ itself.

What is the topology of the apparent confining surface? How is this surface related to the new constant of motion?

To visualize the topology of the surface confining the trajectory of Figure 5-4 it is useful to consider the general uncoupled problem obtained by neglecting the cubic terms in the Hamiltonian of Equation (5-1).

$$H_{\text{osc}}\,(\bar{p}, \bar{q}) = \tfrac{1}{2}\,(p_1 + \omega_1^2\,q_1) + \tfrac{1}{2}\,(p_2^2 + \omega_2^2\,q_2). \qquad \textbf{(5-3)}$$

In this case, total energy is conserved, and additionally, energy in each of the two oscillators is separately conserved. The topology of the phase space in this simple case is illustrated in Figure 5-5, where the topological Cartesian product of two circles is a torus.[80] In Figure 5-5(b) a trajectory confined to a torus is shown, and it is not difficult to imagine that an arbitrary slice through the torus would produce a Poincaré section similar to that of Figure 5-4. What is suggested by the results of Figure 5-4, then, is that the particular trajectory examined is on a two-dimensional torus. Since such a torus is a two-dimensional surface embedded in a four-dimensional phase space,

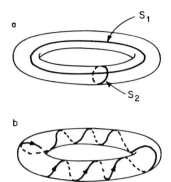

Figure 5-5
(*a*): A torus defined as the Cartesian product $S_1 \times S_2$ of the respective phase spaces of two oscillators. (*b*): A trajectory on a torus (embedded in the four-dimensional phase space) will be periodic if the frequencies ω_1 and ω_2 are commensurate. Otherwise the trajectory will be ergodic on the torus.

its various projections and sections into the plane are not always immediately recognizable, a point discussed in great detail by de Almeida and Hannay.[81]

Because motion of the uncoupled system [H of Eq. (5-3)] is also on a torus, the effect of the coupling terms in Equation (5-1) distorts the confining surface in phase space, but without changing its topology. A quantum-mechanical analogy would be to consider a perturbation distorting an unperturbed, nodeless ground-state wavefunction into a perturbed, nodeless ground-state function: Much distortion may take place, but the mathematical character of the wavefunction has not really changed (e.g., it is still single-valued, strictly positive in the finite plane, twice differentiable, etc.). Figure 5-6 indicates a *composite* Poincaré surface of section for the Hénon-Heiles problem, obtained by running several trajectories (all at the same total energy, $E = 0.10$) to give an overall view of the geometric organization of phase space. The phase space appears to consist of nested tori, indicating that the trajectory of Figure 5-4 is not an isolated accident.

However, the suggestion of a phase space with organizational regularity of Figure 5-6 is immediately shattered by examination of the trajectory of Figure 5-7(*a*) and (*b*), shown both in coordinate space and in Poincaré section. What has happened? The trajectory of Figure 5-7 does not seem to be confined to a "smooth" (i.e., many times differentiable) surface. Does this imply that the torus structure has disappeared or perhaps that the tori are now continuous but nondifferentiable, as illustrated in Figure 5-8? Or is it the case that the trajectory of Figure 5-7 just has not been integrated long enough to allow it to display its regularity? Perhaps there no longer is any underlying structure in phase space. In the absence of any simple answers

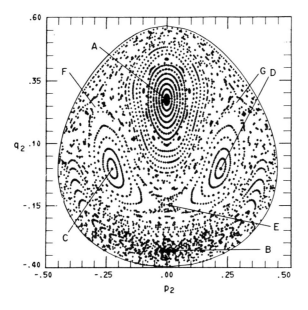

Figure 5-6
A composite of Poincaré sections of approximately 20 trajectories at $E = 1/10$ for the Hénon-Heiles problem. The dynamics is dominated by families of tori surrounding fundamental stable periodic orbits labeled A, B, C, and D, respectively. Points E, F, and G are unstable fixed points. It is evident that at least most of phase space is filled with tori.

to the above questions, and to note that the trajectory of Figure 5-7 does seem to differ qualitatively from the regular trajectory of Figure 5-4 the former is referred to as *chaotic*. Such trajectories are also often referred to as "stochastic," but this term tends to suggest that the motion is nondeterministic, which is not the case at all. Figure 5-9 shows a composite Poincaré section, illustrating the existence of both regular and chaotic motion at the same energy ($E = 0.13$) for the Hénon-Heiles problem. It was this mix of regular and chaotic behavior that so startled early workers.[2] The possibility, indeed prevalence, of such mixed behavior is the content of the celebrated Kolmogorov-Arnold-Moser (KAM) theory,[7, 8, 82–84] whose central tenet is that weak perturbations do not destroy all tori, but rather tori are destroyed in increasing measure as a function of perturbation strength or distance from stable periodic orbits. This theory is illustrated in Figure 5-10. The KAM theory does not, however, discuss possible structure in the chaotic volumes of phase space.

The Poincaré surface provides an excellent visualization of the transition to chaos in systems of two degrees of freedom. A second diagnostic, which

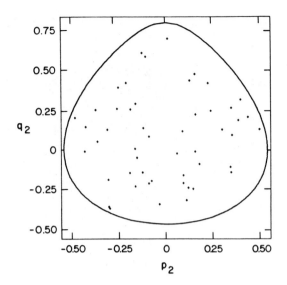

Figure 5-7 (a)
A chaotic trajectory for the Hénon-Heiles problem

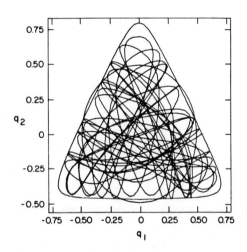

Figure 5-7 (b)
Poincaré surface of section for the trajectory of Figure 5-7(a). It is evident, on comparison of this section with that of Figure 5-4, that a dramatic change has occurred. Hints of such chaotic motion can be observed near the unstable fixed points in Figure 5-6.

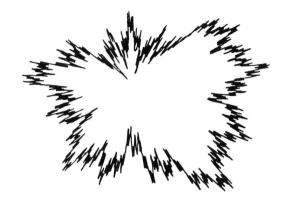

Figure 5-8
Cross section of an everywhere continuous but nondifferentiable torus. (*After J. D. Farmer, E. Ott, and J. A. Torke, "The Dimension of Chaotic Attractors," Physica 7D [1983] 176.*)

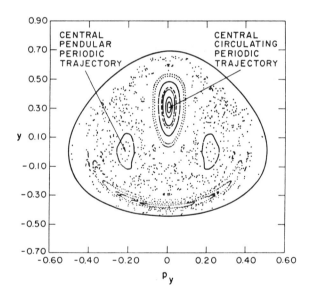

Figure 5-9
A composite Poincaré section for the Hénon-Heiles problem at $E \sim 0.13$. Volumes of chaos and regular motion coexist in large measure. The stable periodic orbits (*A, B, C, D* of Figure 5-6) have apparently stabilized the neighboring dynamics. (*From R. B. Shirts and W. P. Reinhardt, "Approximate Constants of Motion for Classically Chaotic Vibrational Dynamics: Vague Tori, Semiclassical Quantization, and Classical Intramolecular Energy Flow," Journal of Chemical Physics 77 [1982]: 5205.*)

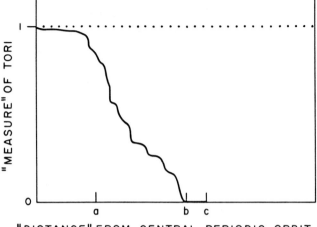

Figure 5-10
Measure of phase space occupied by tori as a function of distance from a stable periodic orbit; see Figures 5-6 and 5-9. For systems of two degrees of freedom chaos will be difficult to detect numerically until point c, beyond which there are no tori, as even presence of tori in measure 0 (as between b and c) will trap chaotic orbits between concentric tori, making their numerical identification difficult. (*After R. B. Shirts and W. P. Reinhardt, "Formal Integrals for a Nonintegrable Dynamical System: Preliminary Report," in Quantum Mechanics in Mathematics, Chemistry, and Physics, ed. K. Gustafson and W. P. Reinhardt [New York: Plenum, 1981], p. 586.*)

also shows a striking qualitative change and, further, is applicable to systems of higher dimensionality, involves examination of the frequency content of individual trajectories or their ensemble averages. Following the work of Noid, Koszykowski, and Marcus,[85] Figure 5-11 shows the power spectrum

$$I(\omega) = \frac{1}{2\pi} \lim_{\tau \to \infty} \frac{1}{\tau} \left| \int_{-\tau}^{\tau} f(t)e^{-i\omega t}\, dt \right|^2 \tag{5-4}$$

for a chaotic trajectory on the Hénon-Heiles surface for the choice $f(t) = x(t) + y(t)$. The spectrum is obviously broad banded and complex, as opposed to the equivalent power spectrum of a quasi-periodic trajectory, these latter showing a small number of distinct frequencies, with bandwidths apparently only limited by the finite size of the range $(-\tau \to +\tau)$ in Equation (5-4).

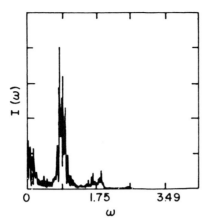

Figure 5-11
Power spectrum of a chaotic orbit. (*From D.W. Noid, M. L. Koszykowski, and R. A. Marcus, "A Spectral Analysis Method of Obtaining Molecular Spectra from Classical Trajectories," Journal of Chemical Physics 67 [1977]: 408.*)

However, the possibility exists that spectra of the type shown in Figure 5-11 contain a quantitative measure of the extent to which the dynamics generating them is chaotic. Knauf[86] has recently analyzed power spectra of the type of Figure 5-11, using a Hamiltonian dynamics abstracted from the N-vortex model[87,88] of flow in compressible fluids. He finds that the power spectra have a fractal[3] dimension an observation that may be the frequency space signature of the type of phase space behavior so imaginatively indicated in Figure 5-8. Figure 5-12 illustrates a sequence of curves (the Koch curves[89]) whose limiting length increases without bound. To define a measure of such curves (and many others) Mandelbrot[3] has suggested that a function $L(\epsilon)$, the length as a function of an elementary increment a "ϵ" considered as $\epsilon \to 0$, determining the fractal dimension D, by the relationship

$$L(\epsilon) = \epsilon^{1-D}. \tag{5-5}$$

To apply this to the Koch curve, illustrating its utility, we paraphrase Mandelbrot:[3] Taking ϵ equal to the elementary length defined by the longest straight line at a given stage of the sequence, it is evident that in going to the next stage, $\epsilon \to \epsilon/3$, and that $L \to 4/3\ L$. Thus

$$L(\epsilon/3) = \tfrac{4}{3} L(\epsilon).$$

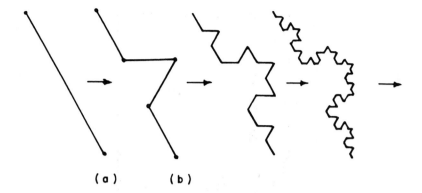

Figure 5-12
Sequence of Koch curves of increasing length. Continuing the sequence leads everywhere to an nondifferentiable curve of infinite length. As discussed in the text, such a curve may be usefully thought of as having a fractal dimension $D > 1$. In this case $D = \log 4/\log 3$.

If this equation is to retain the form of Equation (5-5), it must be the case that

$$(\epsilon/3)^{1 - D} = \tfrac{4}{3} (\epsilon)^{1 - D}$$

or,

$$3^{D - 1} = \tfrac{4}{3},$$

giving $D = \log 4/\log 3 \approx 1.26$ as the dimension of the limiting Koch curve.

Alternatively, D can be determined from $L(\epsilon)$ directly by using a finer and fine-measuring scale "ϵ" and plotting $\log L(\epsilon)$ versus $\log \epsilon$, and noting that the slope is $(1 - D)$. Such a log-log plot, obtained by Knauf[86] for the "length" of the power spectrum of a vortex model Hamiltonian system, is shown in Figure 5-13 as a function of ϵ. The fractal dimension, from Equation (5-5), is ~ 1.24. As the power spectrum analyzed in Figure 5-13 (not shown, see reference 85) looks, on inspection, like that of Figure 5-11, it would not be surprising to find that the curve of Figure 5-11 had a fractal dimension.

The implications of the observation of a fractal power spectrum are not entirely clear. A reasonable suggestion would be that rational frequencies in the spectrum are suppressed and that the large amplitudes correspond to those irrationals that are least well rationally approximated. This hypothesis

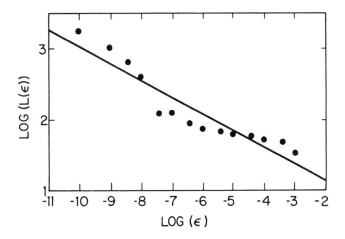

Figure 5-13
Richardson log-log plot (see Mandelbrot[3]) of the length $L(\epsilon)$ of a power spectrum of classically chaotic motion versus the ruler length ϵ. The nonunit slope of this curve indicates that the power spectrum $D > 1$. (*After A. Knauf, "Das No-Vortex-System der Klassischen Mechanik," thesis, Free University of Berlin [Berlin, 1983].*)

is closely related to the more usual KAM assumptions about the destruction of tori (see the section on orbit-based pictures of the origins of chaos). Related Fourier spectra associated with chaotic point maps are discussed in references 21, 90, and 91.

Integrability and Quantization on Tori

One-dimensional bound classical motion may be approximately quantized by use of the Bohr condition:

$$\frac{I}{2\pi} = \frac{1}{2\pi} \oint p \, dq = \left(m + \frac{\alpha}{4} \right) \hbar. \tag{5-6}$$

Here p, q are the conjugate classical momentum and position, m the integer quantum number, and α the Keller-Maslov index.[29] Equation (5-6) is referred to as a semiclassical result in that it attempts to take purely classical dynamical information as the input to a quantization rule. The cyclic integral is about one classical period and determines the value of the action I. Einstein[9] first suggested, in 1917, that the appropriate generalization of Equation (5-6), the nonseparable n-dimensional case for regular classical motion, was what is now referred to as the Einstein-Brillouin-Keller (EBK)[9-11] quantum condition. An approximation to a quantized energy is obtained at an energy E if the n independent quantum conditions

$$\frac{I_i}{2\pi} = \frac{1}{2\pi} \oint_{\mathcal{C}_i} \bar{p} \cdot d\bar{q} = \left(m_i + \frac{\alpha_i}{4} \right) \hbar, \qquad i = 1, 2, \ldots, n, \qquad (5\text{-}7)$$

are simultaneously satisfied. In Equation (5-2),

$$\bar{p} \cdot d\bar{q} \equiv \sum_{i=1}^{n} p_i \, dq_i$$

is the invariant one-form of Poincaré,[6] and the quantization condition has an elegant geometric interpretation in that the n paths $\mathcal{C}_i$ are the n independent cycles on the surface of an invariant (i.e., time independent) torus of dimension n, embedded in the $2n$-dimensional phase space.[9,29,30] These tori are, of course, the tori corresponding to the regular dynamics of the previous section. This geometric quantization rule is illustrated in Figure 5-14 for the case $n = 2$, where it is noted that as the torus is a Lagrangian manifold,[6]

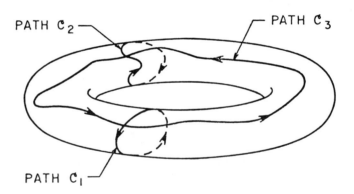

PATH $\mathcal{C}_2$ PATH $\mathcal{C}_3$

PATH $\mathcal{C}_1$

INVARIANT TORUS

Figure 5-14
EBK quantization on an invariant torus. The paths $\mathcal{C}_1$ and $\mathcal{C}_3$ are the representation of the two fundamental topologically independent one-cycles on the torus. They cannot be continuously deformed into one another without leaving the surface. As the torus is a Lagrangian manifold, path integrals of the invariant one-form $\bar{p} \cdot d\bar{q}$ are invariant to continuous path distortion on the non-simply connected surface. Thus paths $\mathcal{C}_1$ and $\mathcal{C}_2$ give an equivalent quantization condition, and either may be used. Note that none of $\mathcal{C}_1$, $\mathcal{C}_2$, $\mathcal{C}_3$ are classical trajectories.

continuous distortion of the paths $\mathcal{C}_i$, leaves the EBK quantization conditions invariant.

An alternative, and completely algebraic, quantization method follows from the fact that if the phase space consists *only* of invariant tori, a Hamiltonian of form

$$H(\bar{p}, \bar{q}) = H(p_1, \ldots, p_n; q_1, \ldots, q_n) \qquad (5\text{-}8)$$

may be written as a function of the n-actions, I_i, of Equation (5-7).[6] That is, a canonical transformation may be found such that the I_i of Equation (5-7) become the new variables:

$$H_{\text{new}} = \tilde{H}(\bar{I}) = \tilde{H}(I_1, \ldots, I_n) , \qquad (5\text{-}9)$$

the I_i now being generalized momenta, the actions, conjugate to the angles θ_i. As the θ_i do not appear in H_{new}, they are referred to as *ignorable*.[4] Such a system is said to be *integrable* if the n independent I_i have vanishing Poisson brackets with one another and are analytic and single-valued functions of $(\bar{p}, \bar{q})$.[6, 30, 82] The I_i $(\bar{p}, \bar{q})$ themselves are then said to be in *involution*. Note that although all separable systems are integrable, the converse is false. The EBK quantization condition, Equation (5-7), in terms of the new Hamiltonian, $\tilde{H}$, in the new variables, I_i, is

$$E_{m_1, m_2, \ldots, m_n} = \tilde{H}\left[\left(m_1 + \frac{\alpha_1}{4}\right)\hbar, \left(m_2 + \frac{\alpha_2}{4}\right)\hbar, \ldots, \left(m_n + \frac{\alpha_n}{4}\right)\hbar\right],$$

implying that if the transformation to the form $\tilde{H}(\bar{I})$ can be analytically carried out, quantization follows at once.

The fact that the θ_i are ignorable in $\tilde{H}(\bar{I})$ implies that as

$$\dot{I}_i = -\frac{\partial \tilde{H}(\bar{I})}{\partial \theta_i} = 0,$$

the actions are constants of motion as befits their association with the time-independent torus. The angles θ_i evolve at constant frequency

$$\omega_i(\bar{I}) = \frac{\partial \tilde{H}(\bar{I})}{\partial I_i} \qquad (5\text{-}10)$$

and provide a coordinatization of a trajectory as it evolves, confined to the torus. In the language of angle action variables, an integrable system corresponds to existence of only regular motion, and the orbits on the tori are periodic (quasi-periodic) corresponding to ratios of the $\omega_i(\bar{I})$ being rational (irrational). If, for a system of n degrees of freedom, there are fewer than n such constants (including $\tilde{H}$!) the system is nonintegrable—or chaotic.

This short discussion indicates that the semiclassical determination of energy levels is often, at least conceptually, straightforward in regions of phase space where the dynamics are integrable (regular). The reader is referred to the literature for numerical implementation of these ideas for quantization of nonseparable systems.[68, 92-99] In his 1917 paper, Einstein clearly noted that Equation (5-7) is not immediately applicable to quantization of chaotic, or nonintegrable, classical dynamics, since the invariant tori no longer exist. Einstein did not suggest an alternative method suitable to this latter case. One alternative is to search for a semiclassical scheme that is *not* based on the existence of tori. Quantization schemes based on the orbits themselves, rather than on the assumption of an underlying phase space structure, would seem in order. Gutzwiller[100, 101] has developed such methods based on the stationary phase approximation to the Feynmann path integral. This approach yields a quantization condition requiring enumeration of all periodic orbits. The relation of Gutzwiller's approach to EBK quantization, in the case in which all orbits are regular, has been discussed by Berry and Tabor[102-104]; other discussions of orbit based rather than torus based quantization are indicated in references 105–111. A quite different approach is taken in Sections 5-3 and 5-4.

Orbit-Based Pictures of the Origins of Chaos

The generic nonseparable nonlinear Hamiltonian for $n \geq 2$ is noninte-grable, although the KAM analysis (see Figure 5-9) does not preclude exis-tence of tori in nonintegrable systems. It might seem that there is a paradox: If *any* tori exist, the n constants $I_i(\bar{p}, \bar{q})$ must exist on the tori. Why doesn't this imply integrability? For nonintegrable systems, the $I_i(\bar{p}, \bar{q})$ are no longer analytic (in the sense of Cauchy) functions of the $(\bar{p}, \bar{q})$. In fact, when an integrable system is perturbed by a generic perturbation, periodic orbits are destabilized,[112] resulting in destruction of those tori where the $\omega_i(\bar{I})$ are rationally related; that is, those tori that support periodic orbits. The tori that survive most robustly are those where the frequency ratios are least well approximated by rationals. This case implies a highly complex interleaving of tori and destroyed tori. This qualitative discussion suggests that even if the $I_i(\bar{p}, \bar{q})$ exist on a torus, differentiability as a function of the $(\bar{p}, \bar{q})$ is unlikely in a neighborhood (as is essential for Cauchy analyticity) because an arbitrary neighborhood will contain destroyed tori. However, $I_i(\bar{p}, \bar{q})$ could well be analytic when such neighborhoods are appropriately restricted in the sense of Borel (as opposed to Cauchy) analyticity; the Borel[113] concept of analyticity is more general in that it does not require existence of a deriv-ative in an arbitrary neighborhood but allows use of analyticity in domains defined as complements of what may be very complex sets. Analyticity on the resulting swiss cheese-like domains may be a useful, formal way of describing the complexity of the phase space of nonlinear systems, but it has not yet been applied.[114]

How are we to understand the mechanism for such "destabilization" of periodic orbits? In this section this problem of the origin of chaos is discussed qualitatively in terms of (1) exponentiation of trajectories and suggested local origins of chaos, and (2) accumulation of bifurcations.

Local exponentiation of trajectories

How can trajectories be ergodic? This very old question, related to the foundations of classical statistical mechanics, was answered by Hopf[16] in 1934 by showing that if *all* neighboring trajectories diverged exponentially in time, ergodicity was implied; and, more importantly, that at least one such system existed, namely, free motion on a Riemannian surface of constant negative curvature.[115] That this result is nontrivial can be understood by noting that for $n = 2$, existence of *a single torus* immediately divides the three-dimensional energy shell into disconnected volumes "inside" and outside the two-torus, and thus precludes ergodicity. For $n \geq 3$, the situation is more complex because an n-dimensional surface (the n-torus) does not divide the $2n\text{-}1$ dimensional energy shell into disjoint parts, which leads to the possibility of Arnold diffusion.[116, 117]

Although Hopf's particular example seems of limited interest in the analysis of generic dynamical systems, which are neither ergodic nor integrable, the concept of exponentiation of trajectories provides both a possible measure of the presence and extent of chaos and a possible predictive tool. Figure 5-15 shows the time dependence of the separation of initially neighboring trajectories for regular and chaotic dynamics on the Hénon-Heiles surface. It is apparently the case that exponential divergence characterizes chaos, and linear divergence characterizes regular motion. Several authors[12-15, 35, 36] have attempted to use these ideas as a simple predictive tool for identification of integrable systems and for prediction of the values of the energy (or coupling strength) at which large-scale chaos sets in. More specifically, consider the dynamics of difference trajectories, which is also a Hamiltonian system. If

$$\eta_i(t) = p_i(t) - p_i^0(t), \qquad i = 1, \ldots, n,$$
$$\xi_i(t) = q_i(t) - q_i^0(t).$$

$\bar{p}^0(t), \bar{q}^0(t)$ being a "reference" trajectory, then, by writing $p_i(t) = p_i^0(t) + \eta_i(t)$, and $q_i(t) = q_i^0(t) + \xi_i(t)$ and keeping *only first-order terms* in the

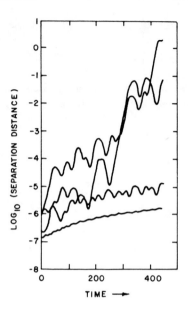

Figure 5-15
Exponential separation of chaotic trajectories (upper curves) compared with linear separation of neighboring quasi-periodic trajectories (lower curves). (*From J. Ford, "The Transition from Analytic Dynamics to Statistical Mechanics," Advances in Chemical Physics 24 [1973]: 176.*)

difference trajectory, $(\eta_i(t),\ \xi_i(t))$, we have (in matrix notation) the coupled first-order differential equations

$$\begin{pmatrix} \dot{\eta} \\ \dot{\xi} \end{pmatrix} = \begin{pmatrix} 0 & A \\ M & 0 \end{pmatrix} \begin{pmatrix} \eta \\ \dot{\xi} \end{pmatrix}, \qquad (5\text{-}11)$$

where

$$A_{i,j+n} = -\left. \frac{\partial^2 V}{\partial q_i^0\, \partial q_j^0} \right|_{\overline{q}^0(t)}$$

$$M_{i+n,\, j} = \frac{\delta_{ij}}{m_i},$$

m_i being the mass of the ith degree of freedom. Toda,[12] Brumer and Duff,[13] and Cerjan and Reinhardt[14] have analyzed the stability of Equation (5-11) adiabatically in the sense that $q^0(t)$ was held fixed in time. Since the matrix A is essentially a local curvature tensor, stability changes from oscillatory (stability) to growing exponential (instability) as the curvature, which is proportional to and has the sign of det(A), becomes negative as a function of $(\bar{q}^0)$. Kosloff and Rice[15] have appended the idea that only exponentiation *transverse* to the phase flow is likely to lead to chaotic behavior. Much of this qualitative analysis is illustrated[68, 118] for the Hénon-Heiles problem in Figure 5-16. Examination of Figure 5-16 indicates that this type of local analysis, closely related to Hopf's idea of chaos being generated by negative curvature, gives a good qualitative, and even semiquantitative, prediction of the minimal energy necessary to ensure substantial chaos, and once this energy has been achieved, prediction of which trajectories will exponentiate, and which parts of phase space will retain tori in large measure. Although this rather simple and nonrigorous analysis works well for model systems such as Hénon-Heiles[2] and Barbanis[77] and for coupled Morse[119, 120] oscillators, it can fail to detect integrable systems, and certainly is incorrect in that it gives the impression that no chaos exists below a critical energy. The latter is not the case: Below the critical energy, chaos is simply of small measure and very difficult to detect numerically. However, the suggestion of

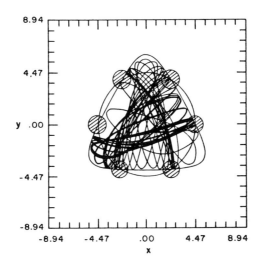

Figure 5-16

Marked local instability of trajectories occurs when trajectories enter one or more of the small shaded circles superimposed on the triangular equipotential. This was observed empirically by Jaffé,[118] and is consistent with the analysis of references 12–15, which focus on analytic location of regions of exponential separation of trajectories. (*From C. Jaffé, "Semiclassical Quatization of Nonseparable Hamiltonian Systems," Ph.D. thesis, University of Colorado [1979], p. 101.*)

a local origin of chaos will carry over to the discussion of Sections 5-3 and 5-4.

Chaos through bifurcations and period multiplication

The Poincaré surfaces of Figures 5-6 and 5-9 were generated through laborious and time-consuming stepwise integration of Hamilton's equations [Eqs. (5-2a) and (5-2b)]. Both from the point of view of numerical reliability (which can be a very serious problem for strongly exponentiating trajectories[121, 122]) and speed, many advantages accrue from the use of recursive point mappings[2, 19, 123] of the form

$$\begin{pmatrix} P_n \\ q_n \end{pmatrix} = T \begin{pmatrix} P_{n-1} \\ q_{n-1} \end{pmatrix}. \tag{5-12}$$

Such mappings can effectively directly generate Poincaré surfaces, showing all of the phenomenology of, say, Figure 7(b), or Figure 7-9 for a composite of many mappings. If such maps are properly chosen they can model both dissipative and conservative systems; and, if integer arithmetic can be used, roundoff and truncation induced instabilities are avoided. Such maps are mentioned in the present context because an extensive literature has grown up in the process of understanding: (1) bifurcation of unstable periodic orbits, and the subsequent period doubling[19, 62, 63]; (2) universal scaling behavior[19] (and accompanying use of renormalization techniques)[61, 64] associated with such period doubling bifurcation, leading to a detailed fleshing out of the introductory remarks of this section, based on a detailed analysis of behavior of periodic orbits [i.e., fixed points of Eq. (5-12)] as a function of an external parameter defining T.

The only detailed example of such analysis, which is far less complete than the analysis of point maps, applied to a differentiable dynamical system with a four-dimensional phase space, is that of Bountis[124] and Helleman and Bountis,[17] which extensively elaborates the work of Rod[125] on the periodic orbits of the Hénon-Heiles system. Such orbits are rather more difficult to find than the fixed points of a point map. Bountis and Helleman have proceeded variationally, and their results are of interest in that they appear to be tied very closely to the results of the local stability picture of Figure 5-16.

Using a variational technique, Bountis and Helleman[17] found periodic orbits of varying period, τ_r, and winding number $\sigma = m_2/m_1$,

where

$$2m_1 = \text{the number of zeroes of } q_1(t) = x(t) \text{ in the period } \tau_r,$$

$$2m_2 = \text{the number of zeroes of } q_2(t) = y(t) \text{ in the period } \tau_r,$$

Results are shown in Figure 5-17 *for those trajectories that start on an equipotential.* This constraint does not allow examination of all periodic trajectories, but allows easy visualization, since the curves, labeled by the various relatively prime winding numbers, $\sigma = 1, 2/5, 3/4, 6/7, \ldots$, may be indicated by their $\dot{x}(0) = \dot{y}(0) = 0$ origin on the potential surface. The results of Figure 5-17, although restricted to a subclass of periodic orbits, are entirely consistent with the local stability analysis summarized in Figure 5-16, in that periodic orbits of different winding numbers, and thus very different periods,

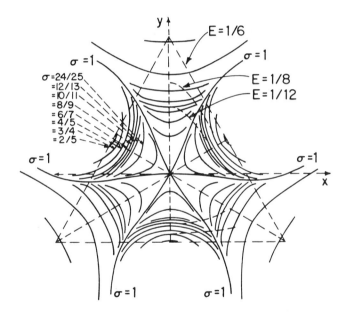

Figure 5-17
Loci of points on the Hénon-Heiles equipotentials corresponding to periodic orbits of fixed winding number σ (see text). Note that the loci accumulate on the $\sigma = 1$ line, which pass through the shaded circles of Figure 5-16. Both the winding number analysis and local instability analysis have located the same regions of strong sensitivity to initial conditions. (*From R. H. G. Helleman and T. Bountis, "Periodic Solutions of Arbitrary Period, Variational Methods," Lecture Notes in Physics 93 [1979]: 357.*)

accumulate along the $\sigma = 1$ lines, which run precisely through the small circles of Figure 5-16. The local stability analysis of references 12 through 15 thus seems to have located the regions of strongest sensitivity to initial conditions, at least as measured by the winding number criterion.

Analytic Properties of Trajectories: Painlevé Analysis

Within the past few years very rapid progress has been made in understanding the complex time singularities associated with integrable and nonintegrable Hamiltonian dynamics for $n = 2$. By assuming that the only "movable" singularities are poles, new classes of integrable systems have been discovered, including and extending the classes of known systems with constants of motion that are low-order polynomials in the momenta.[126, 127] Further, in the case of nonintegrable systems the movable singularities suggest a natural boundary of analyticity of a fractal nature, whose signature in the frequency domain might well be the fractal power spectrum discussed earlier. The section ends with a brief mention of the problem of intermittency and the corresponding complex time analysis.

Movable singularities in the solutions of nonlinear differential equations are those that depend on the arbitrary constants (e.g., initial conditions) rather than the location of a singularity in the equation itself. An excellent textbook discussion is that of Hille,[128] who gives the simple examples

$$\frac{dx(t)}{dt} + e^{x(t)} = 0$$

with solution

$$x(t) = \log\left[(t - c)^{-1}\right] \tag{5-13}$$

and

$$\frac{dx(t)}{dt} + x(t)\log^2 x(t) = 0 \tag{5-14}$$

with solution

$$x(t) = e^{(t - c)^{-1}}. \tag{5-15}$$

In each case, c is an arbitrary constant in the solution, but the locations of the logarithmic branch point of Equation 5-13 and the essential singularity of Equation (5-15) are determined by the value of c, which does not appear in the equations themselves. Thus the singularities of Equations (5-13) and (5-15) are "movable." S. Kovalevskaya[129] and P. Painlevé (see references 22–28 and 128) investigated the forms of nonlinearity possible if the prior assumption is made that the only movable singularities are poles. This is referred to as the Painlevé property. Following the work of Kovalevskaya, and much more recently, Ablowitz, Ramani and Segur,[22] the following KP (Kovalevskaya-Painlevé) conjecture has emerged: If a system of coupled nonlinear ordinary differential equations has the Painlevé property they are equivalent to a partial differential equation soluble by an inverse scattering technique, or in the case of a system of Hamilton's equations the system is integrable. Working with a generalized Hénon-Heiles Hamiltonian,

$$H_{A,\,B,\,C,\,D}\,(\bar{p},\,\bar{q})\;=\;\frac{1}{2}\,p_1^2\;+\;\frac{A}{2}\,q_1^2\;+\;\frac{1}{2}\,p_2^2$$

$$+\;\frac{B}{2}\,q_2^2\;+\;Cq_1^2\,q_2\;-\;\frac{D}{3}\,q_2^3\,. \qquad \textbf{(5-16)}$$

Chang, Tabor, Weiss and Corliss;[23] Bountis, Segur, and Vivaldi;[27] Ramani, Povizzi and Grammaticos;[26] and Tabor[24] have investigated the Hamiltonian of Equation (5-16). These authors have found that enforcing the Painlevé property (or, as in references 24 and 26, an "extended" or "weak" Painlevé property that allows *rational* branch singularities in addition to poles, but excludes "higher" types of singularities, including irrational branch points [e.g., $(t - t_*)^\alpha$ with α not rational]) gives rise to specific values of the constants A, B, C, D such that the Hamiltonian is integrable.

Perhaps even more important than the fact that the Painlevé property seems to be the beginning of a unification of integrability of systems of nonlinear ordinary and partial differential equations[22,28] is the insight it suggests in the analysis of the singularity structure of the dynamics of regular and chaotic trajectories for nonintegrable systems. Such an analysis suggests a possible relationship to the fractal power spectra of the discussion on regular and chaotic classical dynamics, and the possibility of fractal tori alluded to in Figure 5-8 and to be discussed in Section 5-3.

Following the analysis of Chang et al.[23] for the usual Hénon-Heiles case, where $A = B = C = D = 1$ in Equation (5-16), the Painlevé analysis suggests examination of the leading order singularity for a time near a (complex) singularity at t_*), by writing

$$x(t) \;=\; q_1(t) \;=\; a(t - t_*)^\alpha\,, \qquad\qquad \textbf{(5-17)}$$

$$y(t) \;=\; q_2(t) \;=\; b(t - t_*)^\beta\,, \qquad\qquad \textbf{(5-18)}$$

Substitution of (5-17) and (5-18) into Hamilton's equations for the Hénon-Heiles problem, and equating the most singular terms, gives two cases for the exponents, α, β, which determine the type of singularity, at t_*:

Type 1 singularity: $\alpha = \beta = -2$ (double poles).

Type 2 singularity: $\alpha = \frac{1}{2} \pm i\,(\sqrt{47}/2)$, $\beta = -2$.

Note that $\alpha = \frac{1}{2} \pm i\,(\sqrt{47}/2)$ gives a singularity of a type consistent with lack of integrability of the Hénon-Heiles system. Detailed numerical studies[23] allowed location of the values of t_* corresponding to the singularities of types 1 and 2. The nearest such singularities to the real axis are illustrated in Figure 5-18 for regular and chaotic trajectories. These "nearest" singularities appear to be less regularly arranged for the chaotic trajectory. However, in both the regular and chaotic cases Chang et al.[23, 25] located an

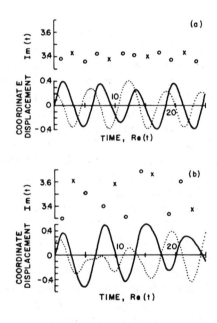

Figure 5-18
Nearest Painlevé complex time singularities along (a) a quasiperiodic Hénon-Heiles trajectory and (b) along a chaotic trajectory. The singularities in (b) are more irregular, but no detailed analysis of the relationship of singularity structure to chaos has yet been given; the singularities of (a) are certainly not *completely* regular. (*After Y. F. Chang, M. Tabor, J. Weiss, and G. Corliss, "On the Analytic Structure of the Henon-Heiles System," Physics Letters 85A [1981]: 212.*)

infinite number of other singularities lying on sequence of self-similar polygons closely related to the Koch curves of Figure 5-12. These cases are illustrated in Figure 5-19.

Chang et al., have thus identified a natural boundary of analyticity, and found that that boundary was a fractal curve of dimension $D \approx 1.14$ (see Fig. 5-19). This result is consistent with fractal natural boundaries found by Greene and Percival[130] and Percival[131] in studies of point maps, in an analysis independent of a Painlevé-type-ansatz such as that of Equations (5-17) and (5-18). The existence of such a fractal boundary of nonanalyticity ought to correlate directly with the analysis of power spectra discussed earlier. What property of the fractal boundary gives a simple ($D = 1$) spectrum in the regular case a $D > 1$ power spectrum in the chaotic case?

As a parting example[132] of the relationship between complex time singularity structure and complexity of dynamics,[133] consider the correlation indicated in Figure 5-20. In this case, rather than a Hamiltonian system, the high-frequency component of $v(t)$, an intermittent solution of the nonlinear Langevin equation

$$m\dot{v}(t) = -\gamma v(t) - v(t)^3 + f(t),$$

where $m > 0$, $\gamma > 0$, and $f(t)$ is the random force giving rise to the friction constant γ, through the fluctuation dissipation theorem. We return to a discussion of this figure in Section 5-3.

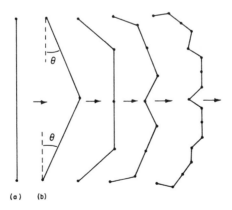

(a)   (b)

Figure 5-19
Generation of self-similar natural boundary of analyticity corresponding to the Painlevé singularities of the Hénon-Heiles problem. The sequence of curves, with line segments connecting Painlevé singularities, has a fractal limit with $D \sim 1.14$. See references 23 and 25 for further discussion.

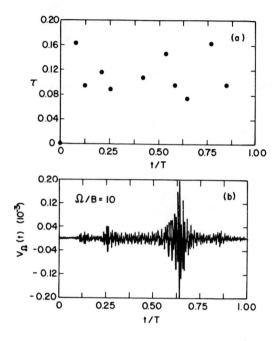

Figure 5-20
(a): Correlation of complex time singularities [$\tau = \text{Im}(t/T)$] with (b): bursts in the frequency projected Langevin velocity $V_a(t)$. (*From U. Frisch and R. Morf, "Intermittency in Nonlinear Dynamics and Singularities at Complex Times," Physical Review A 23 [1981]: 2684.*)

5-2 EXPECTATIONS: CLASSICAL CHAOS AND THE CORRESPONDENCE PRINCIPLE

What is the quantum analog or analogs of classical chaos? Or, asking a different question, For a fixed small value of $\hbar$, when and precisely how do we expect to see quantum systems reveal to us that their underlying classical dynamics is chaotic? Are there necessary and sufficient conditions relating classical and quantum dynamics that allow a unique statement of a correspondence principle? A reasonable observer today must conclude that not enough is known to give a complete answer to any of these questions. On the other hand, the field is very rapidly developing, and within a few years many of the questions may have partial answers, or perhaps it will be realized that there are different questions that can be answered definitively. In the sections that follow early expectations are reviewed and results of numerical experiments that seem to support them are presented, followed by a critique. We will see that we have restricted attention to discussion of level spacing distributions and the form of individual quantum wave functions. This is owing to limitations of space, and excludes the whole area of classical and quantum wave packet dynamics,[134-142] probed experimentally by low-resolution spectroscopy and mathematically in terms of time evolution of overlaps,[143-153] $|<\psi(0), \psi(t)>|^2$ and other correlation functions.[154-161]

The Irregular Spectrum

General expectations

The concepts of regular and irregular spectra to be associated (for high enough quantum level) with regular and chaotic dynamics were introduced by Percival.[162, 163] Put briefly, the states of the regular spectrum are labeled by association of the quantum numbers m_i with classical actions I_i, and thus with tori via Equations (5-8) and (5-10). Thus each state is labeled by n quantum numbers. Classical frequencies on quantized tori are related to quantum energy differences via the correspondence principle.[164] This relation is in strong contrast to the irregular spectrum. Paraphrasing Percival,[162, 163] in the irregular spectrum:

P1. As there are not n actions, it may not be possible to label a given quantum state in terms of n quantum numbers. States will be labeled only by remaining global constants such as energy or, say, angular momentum.

P2. As many frequencies (see Figs. 11 and 13) are present in a chaotic trajectory. . . . The quantum distribution of levels could appear random.

P3. The energies of the irregular spectrum will be highly sensitive to perturbations.

These statements are referred to as P1–P3 in what follows. Only those aspects of Percival's characterization relevant to the present discussion are summarized in P1–P3. More extended discussion appears in references 29, 30, and 38.

Wave functions

What would the wave function for an individual stationary state in the irregular spectrum look like? Berry[165] has made specific predictions for both of the wave functions corresponding to the regular and irregular spectrum. (Additionally, he has predicted properties of its Wigner transform,[166, 167] not discussed here.) Many of these predictions are best illustrated by example. Figure 5-21(a) shows a chaotic trajectory for free motion in a

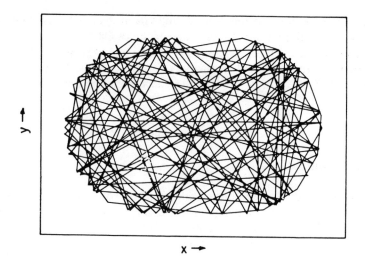

y →

x →

Figure 5-21 (a)
Chaotic trajectory within the two-dimensional stadium boundary. Motion is free in the interior, with specular reflection taking place at the hard boundary. (*From M. V. Berry, "Semiclassical Mechanics of Regular and Irregular Motion," in Proceedings of the 1981 "Les Houches" Summer School, vol. 36, Chaotic Behavior in Deterministic Systems [Amsterdam: North-Holland, 1983], p. 183.*)

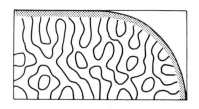

Figure 5-21 (b)
Nodal patterns [i.e., lines where $\psi(x, y) = 0$] corresponding to a high-lying quantum eigenstate of the stadium billiard. Corresponding to the "random" orientation of momenta in (a), the wave fronts of (b) appear random (only $\frac{1}{4}$ of the pattern is shown—the rest is symmetrically related). (*From S. W. McDonald and A. N. Kaufman, "Spectrum and Eigenfunctions for a Hamiltonian with Stochastic Trajectories," Physical Review Letters 42 [1979]: 1190.*)

two-dimensional hard-wall enclosure, consisting of parallel sides of length L, capped with semicircular ends. This "stadium billiard" is a strongly chaotic classical system,[168] which leads to the fact that a typical (i.e., non-periodic) trajectory not only fills configuration space [see Figure 5-21(a)], but, when the trajectory returns to the neighborhood of a point where it previously passed, its momentum will not be correlated with the previous value.

If we have a localized wave packet centered on such a trajectory, the packet will constantly interfere (owing to both random direction and phase) with "itself" on these return visits, as long as the return visits are within a typical de Broglie wavelength. The result of such interference is intuitively a wavefunction spread more or less uniformly over the whole configuration space, with more-or-less randomly directed wave fronts (and thus more-or-less random nodal patterns). This result is in contrast to the regular case, where the wavefunction is localized on the coordinate space projection of the corresponding torus, with amplitude maxima on the classical caustics.[165, 169] A high-lying quantum state of the stadium billiard, as determined numerically by McDonald and Kaufman,[170] is as shown in Figure 5-21(*b*) and seems to have many of these expected properties. This brief discussion is greatly amplified in reference 165; numerical investigations of properties of regular and irregular wavefunctions are now numerous.[38, 169, 171–176]

Statistics of Nearest-Neighbor Level Spacings

Berry and Tabor,[177] and Berry[178] have made substantial progress in quantifying Percival's suggestion that the distribution of levels be different in the regular and irregular spectrum. If it is assumed that all quantum energy levels are accurately given by EBK quantization on tori, Berry and Tabor have shown that level clustering dominates the level spacing distribution. More specifically, the level spacing probability distribution, $P(S)$, for nearest-neighbor spacing, normalized to a unit average spacing, is Poisson. That is,

$$P(S) = e^{-S}. \tag{5-19}$$

(Note that exponentially small tunneling corrections of order $\exp(-1/\hbar)$ have been neglected compared to an average level spacing of $(2\pi\hbar)^n$.)

In a generic system of quantum levels of the same symmetry, rather than the degeneracy suggested by Equation (5-19) being probable, "avoided crossings" are the rule—that is, the *Wigner–von Neuman noncrossing rule*.[179] By an avoided crossing is meant that when a single parameter in a quantum Hamiltonian is varied (for example, an external field, the value of $\hbar$, or the ratio of length to width of a confining potential) degeneracies are produced with zero probability measure, simply because degeneracy of eigenvalues requires that *two* conditions be simultaneously met. The simplest example shows this to be the case: For the eigenvalues of the 2×2 (real symmetric) matrix

$$\begin{pmatrix} a_{11} & a_{12} \\ a_{21} & a_{22} \end{pmatrix} \tag{5-20}$$

to be degenerate it must be true that

$$(a_{11} - a_{22})^2 + 4(a_{12})^2 = 0$$

and thus that

$$a_{11} = a_{22}$$

and

$$a_{12} = a_{21} = 0.$$

Note that if a *symmetry* requires that $a_{12} = a_{21} = 0$ then only one additional condition need be met and variation of a single parameter, say the value of a_{11}, can produce degeneracy. Berry[178] has carried out an analysis of the expected level statistics in the case in which j parameters need to be varied to ensure degeneracy (j is the codimension) and finds that, *as $S \to 0$,*

$$P(S) \xrightarrow[\lim\ S \to 0]{} \sigma S^{j-1}, \tag{5-21}$$

σ being a constant. The case $j = 1$ corresponds to EBK quantization on tori, and we regain the small S behavior of Equation (5-19). The usual Wigner–von Neuman case [as would apply to the matrix of Equation (5-20)] corresponds to $j = 2$, and suggests a linear behavior for small S. However, it is important to note that j is not a measure of the "strength" of classical chaos, and that Berry's arguments leading to Equation (5-21) are really an analysis of the geometry of the sheet structure of det $[E - A(j)]$, $A(j)$ being an Hermitian matrix, as a function of j-parameters. *Thus neither σ, nor the range of validity of the limiting law of Equation (5-21) are predicted, nor is their relation to any "measure" of chaos indicated.* (An earlier, now discredited, analysis of Zaslavskii[180] had suggested a limiting law of form σS^ν, where ν was related to the rate of exponentiation of chaotic trajectories. This analysis at least gave a quantum distribution law that depended on a measure of the strength of the classical chaos.)

As an example of the lack of a specific connection of Equation (5-21) with classical chaos [see parenthetical statement following Equation (5-19)], note that if the exponentially small tunneling corrections are included in the analysis of the Berry-Tabor[177] derivation of the Poisson law [Eq. (5-19)] for EBK quantization of integrable systems, the distribution will "turn over," as $S \to 0$, in accord with the limiting law of Equation (5-21), but on a scale too small to be seen on a linear plot of $P(S)$ versus S. The crucial question is then, What is the relationship (if any!) between the range of validity of Equation (5-21) and some measure of the strength and/or prevalence of classical chaos? Numerical experiments have been carried out in attempts to assess the validity of Equation (5-21), and we now proceed to discuss some of them.

Numerical Experiments

Extensive numerical work has ensued following Percival's 1973 suggestion[162] of existence of an irregular spectrum. Pomphrey[181] investigated the quantum levels of Hénon-Heiles-like Hamiltonian

$$H_\alpha(\bar{p}, \bar{q}) = \tfrac{1}{2}(p_1^2 + q_1^2) + \tfrac{1}{2}(p_2^2 + q_2^2) + \alpha(p_1^2 q_2 - \tfrac{1}{3}q_2^3) \quad (5\text{-}22)$$

by examining the sensitivity of numerically computed eigenvalues to the parameter α. [See Percival's P3 (p. 23).] Pomphrey investigated the second differences, Δ_i^2,

$$\Delta_i^2 \equiv [(E_i(\alpha + \Delta\alpha) - E_i(\alpha)) - (E_i(\alpha) - E_i(\alpha - \Delta\alpha))]$$

near $\alpha = 0.088$ with $\Delta\alpha = \pm 0.001, \pm 0.002$; he found that unusually large second differences correlated well with the onset of classical chaos, as measured by the fraction of the surface of section occupied by chaotic trajectories (see Fig. 5-9, where it is clear that this area may be reasonably, if not exactly, estimated, the difficulty being apparent on examination of Fig. 5-10). In another investigation, Noid et al.,[182] did not observe any such "anomalous" second differences. However, their quantum results assumed $\alpha \cong 0.1118$ (rather than 0.088), which had the effect of giving many fewer, bound states in the energy region in which the classical phase space corresponds to bound motion, that is, below the dissociation threshold. Thus it has been argued[183] that the "correspondence" limit had not yet set in. Stated another way, $\hbar$ was large enough in the calculations of reference 105 to mask the onset of an "irregular spectrum." (The fact that for "large" $\hbar$ essentially the same quantum levels can be obtained from potential energy surfaces showing strikingly different classical behavior (with respect to chaos) is the substance of an interchange between Lehmann, Scherer, and Klemperer,[184, 185] and Farrelly and Reinhardt.[186])

Pomphrey's results have been confirmed by Pullen and Edmonds.[187] A correlation between classical chaos and the onset of large second differences has been observed, also by Pullen and Edmonds,[188] for the potential $\tfrac{1}{2}(q_1^2 + q_2^2) + \alpha q_1^2 q_2^2$, which has numerical advantages, since tunneling cannot occur.

What gives rise to the anomalously large second differences discussed above? A more global view of large numbers of levels over a wide range of

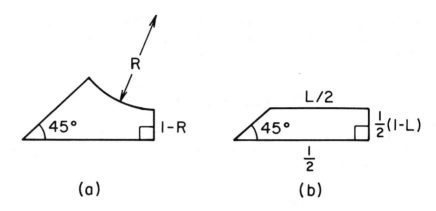

Figure 5-22
(a): Truncated Sinai billiard[189] (b): Truncated square.[190] The dynamics of interest is free motion in two dimensions within these hard-wall enclosures.

α is illuminating. This has unfortunately not been carried out for the Hénon-Heiles because of expense and because the level of effort and care needed to ensure sufficient numerical quality of results in calculations of this type. For nonintegrable billiard problems in two dimensions, Berry[189] and Richens and Berry[190] have successfully reduced the dimensionality of the eigenvalue determination for the billiards shown in Figure 5-22. These are, respectively, the truncated Sinai torus billiard[189] and the truncated square torus billiard,[190] chosen to give eigenstates with no geometrically imposed degeneracy. The energy levels for the truncated Sinai problem are shown in Figure 5-23, as a function of R, and for the truncated 45° right triangle in Figure 5-24 as a function of L. It is evident in both of these classically nonintegrable systems that abundant (sometimes multiple) avoided crossings dominate much of the spectra. A histogram of the level spacing distribution derived from Figure 5-23 is shown in Figure 5-25(a). A similar histogram for the level distribution of Figure 5-24 is shown in Figure 5-25(b). For comparison, the level distribution for the distribution of eigenvalue spacings corresponding to an EBK quantization of an integrable system is shown in Figure 5-25(c). Those distributions (and that of the stadium; see reference 170) not only indicate (as they must) the $P(S) \rightarrow \sigma S$ expected behavior [Eq. (5-21)], but the observed distributions peak *at a value very near to the average level spacing*. We refer to this type of distribution (i.e., peak ≈ average) as one displaying *robust* avoided crossing, as opposed to, say, the *exponentially small* avoided crossings owing to penetration through "large" barriers. The onset of such robust avoided crossings is of course completely consistent with Percival's P1–P3, as characterizing the irregular spectrum: P1 follows as if levels are always in a state of entering or leaving a crossing region.

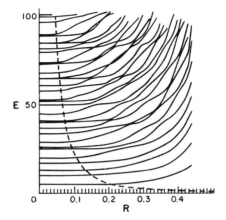

Figure 5-23
Quantum levels for the truncated Sinai billiard of Figure 5-22(*a*) as a function
of the truncation radius *R*. (*From M. V. Berry, "Quantizing a Clasically Ergodic
System, Sinai's Billiard and the KKR Method," Annals of Physics 131 [1981]: 172.*)

They are a superposition of several "unperturbed" states and thus are not
labeled by a single set of quantum numbers; P2 is satisfied because the his-
tograms of Figures 5-25(*a*), (*b*) closely resemble the Dyson-Wigner distribution[191–195]

$$P(S) = \frac{\pi}{2} S \exp\left(-\pi S^2/4\right)$$

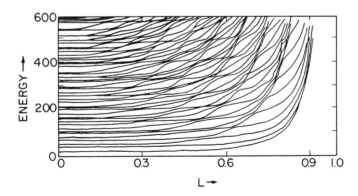

Figure 5-24
Quantum levels for the truncated square torus billiard of Figure 5-22(*b*) as a
function of the truncation length *L*. (*From P. J. Richens and M. V. Berry, "Pseudointegrable Systems in Classical and Quantum Mechanics," Physica 2D [1981]:
502.*)

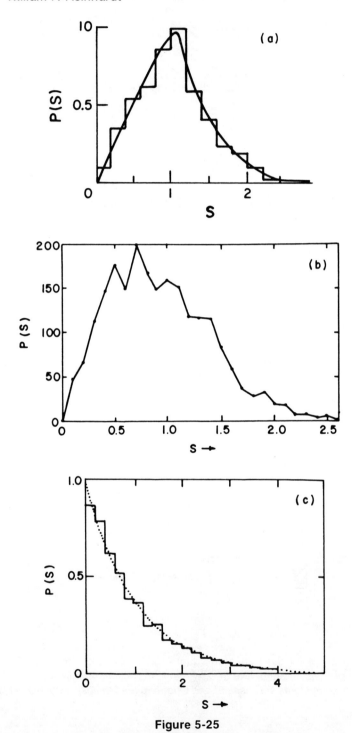

Figure 5-25

expected for an ensemble of random matrices, and P3 is evident from examination of the spectra themselves.

Discussion and Critique

Comparison of the empirical results of the previous section with the discussion of the first two parts of Section 5-2 indicates that Percival's irregular spectrum has been identified. However, and this is the theme of the rest of the present chapter, there is actually little evidence that, although Percival's P1–P3 were suggested by correspondence principle arguments, any necessary and/or sufficient relationships are known relating classical chaos and "quantum" chaos as defined by P1–P3. Rather, we suggest that the irregular spectrum is simply the generic quantum behavior of a *nonseparable* system, in the same sense that a generic classical nonseparable (and nonlinear) system is nonintegrable. (The theme of most of the papers of references 171–176, especially that of Hose and Taylor,[175] is that quantum chaos is related to one's inability to find a separable basis giving an adequate description of each quantum level. This assumption is perfectly consistent with the present remarks: It is the implicit presumption of these authors that this is, of necessity, related to a classical chaos that is questioned here.)

Genericity may well imply that the quantum and classical phenomena of "irregular spectrum" and "classical chaos" are often seen to occur in parallel, but as will be seen, there is no necessity of a one-to-one relationship.

That classical chaos does not imply an irregular spectrum is immediately clear on examination of the quadratic Zeeman problem. A numerical spectrum, calculated by Zimmerman, Cash, and Kleppner[196] is shown in Figure 5-26 as a function of external field *B*. A typical composite surface of section [197, 198] is shown in Figure 5-27, indicating the generically expected mixture of classical chaos and classical tori. The dashed line[199] in Figure 5-26 indicates the boundary between "purely integrable" dynamics, determined

---

Figure 5-25 (at left)
Level spacing distributions *P(S)*, (*a*): for the spectrum of Figure 5-23. (*From M. V. Berry, "Quantizing a Classically Ergodic System, Sinai's Billiard and the KKR Method," Annals of Physics 131 [1981]: 179.*) (*b*): for the spectrum of Figure 5-24. (*From P. J. Richens and M. V. Berry, "Pseudointegrable Systems in Classical and Quantum Mechanics," Physica 2D [1981]: 503.*) (*c*): a model integrable problem. (*From M. V. Berry and M. Tabor, "Closed Orbits and the Regular Bound Spectrum," Proceedings of the Royal Society of London 349 [1976].*) It is evident that the truncated Sinai billiard and truncated torus billiard have irregular spectra in the sense that the spectra are dominated by robust avoided crossings. These results contrast strongly with the EBK *P(S)* of (*c*).

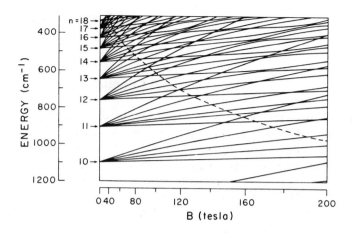

Figure 5-26
Quantum levels for the H atom in an external *B* field as calculated in reference 196. The dashed line indicates the boundary between classically regular and generic KAM dynamics, as illustrated in Figure 5-27. Regular motion is to the *left,* that is, at low fields, or small *n.* (*From D. Farrelly and W. P. Reinhardt, "Highly Excited States of HCN: The Probable Applicability of Classical Dynamics," Journal of Chemical Physics 78 [1983]: 607.*)

visually, and the onset of macroscopic chaos, as illustrated in Figure 5-27. Histograms of $P(S)$ following from Figure 5-26 in the regular and chaotic/regular regimes are shown in Figures 5-28(a), (b). It is evident that in both cases the distributions are Poisson-like.[186, 198] [The recent suggestion of an observation of a quantum irregular spectrum for the Zeeman problem made by Harada and Hasegawa[200] is at $B = 4.7$ tesla, and involves levels with quantum numbers substantially higher (up to $\sim n = 34$) than those shown in Figure 5-26. These authors are using the $|m| = 0$ levels of Clark and Taylor,[201] who also suggested a possible onset of an irregular spectrum. In both papers only visual analysis of a single oscillator strength distribution (at a field of 4.7 tesla) were used to impute the possible onset of irregularity. Results of the type shown in Figure 5-26 should be extended to much higher n values to investigate the spectrum systematically at the higher quantum numbers involved here. Also, note that the $|m| = 1$ spectra of Clark and Taylor[202] appear to show only a regular spectrum up to $n \approx 34$, in conflict with the classical work of Robnik[203] if one believes that classical chaos always implies an irregular spectrum.] As opposed to the study of Noid et al.[182] of the Hénon-Heiles problem with $\alpha = 0.1118$, which resulted in levels being too sparse to show avoided crossings, in Figure 5-26 avoided crossings are everywhere, including many "triple" crossings. These crossings are decidedly not "robust" in either the regular or chaotic regimes.

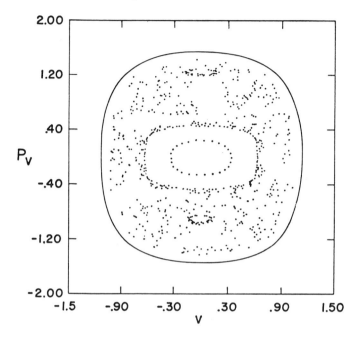

Figure 5-27
Composite Poincaré surface for the Zeeman problem of Figure 5-26, showing
the generically expected mixture of tori and chaos. It is this type of phase space
that is characteristic of the classical dynamics to the right of the dashed line in
Figure 5-26. (*From W. P. Reinhardt and D. Farrelly, "The Quadratic Zeeman Effect
in Hydrogen: An Example of Semi-Classical Quantization of a Strongly Non-Separable but Almost Integrable System," Journal de Physique 43 colloque supplement C2 [1982]: 38.*)

This fact has led Zimmerman et al.[196] to suggest that the system is almost
integrable *quantum mechanically*.[56–58] We will argue in Section 5-3 that in
spite of the appearance of classical chaos in the surface of section of Figure
5-27, the system is almost integrable[197] classically. This discussion suggests
that mere observation of classical chaos *need not* imply an irregular spectrum.[186] Further, in Section 5-4 it will be argued that the robust avoided
crossings of Figure 5-25 do not necessarily imply failure of an appropriately
modified EBK quantization on the tori of a "nearby" integrable classical
system. That is, it will be argued that quantum chaos and classical chaos do
not imply each other, but since both are generic they often accidently occur
together once the system is no longer approximately separable.

A related view is held by Marcus, who[204, 205] has noted that a possible *analog* of the classical chaos produced by overlapping (or multiple) classical

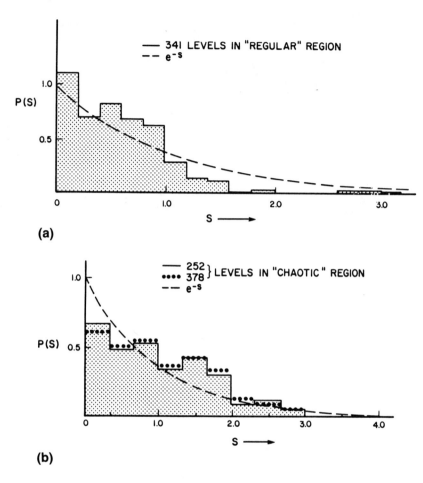

Figure 5-28
$P(S)$ distributions using level spacings to the *left* and *right,* respectively, of the dashed line in Figure 5-26. (*a*) indicates the expected Poisson distribution characteristic of a regular spectrum. (*b*) indicates that classical chaos has *not* created an irregular spectrum. The avoided crossings are still exponentially small in the "chaotic" region, in strong contrast to expectations of a necessary relationship between classical chaos and an irregular spectrum.

nonlinear resonances[39] is a definition of quantum chaos involving multiple avoided crossing, as illustrated in Figure 5-29. Occurrence of many such robust overlapping crossings is consistent with Percival's P1–P3, and, intuitively at least, with Berry's criteria for an irregular wavefunction. Indeed, Marcus[206] has noted that the "irregular" nodal patterns of Stratt, Handy, and Miller[173] and those of Feit, Fleck, and Steiger[176] can be understood on

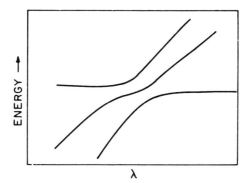

Figure 5-29
Overlapping avoided crossings used by Marcus in his characterization of quantum chaos.

such a basis, and are thus "complex" but not mysterious. More to the point, Marcus[204, 205] has noted that classical chaos is certainly not sufficient to generate such multiple avoided crossings—a conclusion upheld by the several triple crossings of Figure 5-26.

We now begin to discuss a possible physical basis for this lack of necessity in the relation between classical chaos and the onset of a Wignerian level spacing distribution.

5-3 CHAOS AND THE FRAGMENTATION OF TORI

Qualitative Arguments

Consider the trajectory of Figure 5-30(a), which shows the Poincaré section of a chaotic trajectory for the Hénon-Heiles Hamiltonian, run for a long enough time to see that a substantial volume of the phase space is filled in a more or less uniform manner. The only clearly visible "structures" of this chaotic phase space are a set of disconnected, but smooth, inner boundaries (filled with tori, impenetrable for this $n = 2$ problem; see Fig. 5-9) and the outer boundary of the surface of section controlled by the total energy. However, if the time development of this surface of section is examined in a sequential sectioning, as carried out by Shirts and Reinhardt,[65] as is done in Figures 5-30(b), (c), and (d), respectively, a suggestion of structure immediately appears. Figures 5-30(b) and 5-30(d), consisting of the first and last thirds of the points in the section of Figure 5-30(a), both suggest an underlying approximate constant of motion, referred to as a vague torus in reference 65. The trajectory of Figure 5-30(c) may then be thought of as being

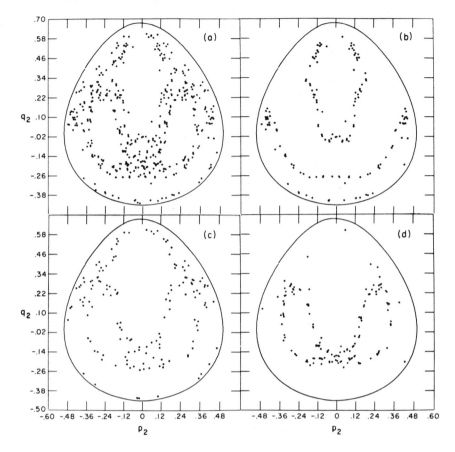

Figure 5-30

(a) displays a long-time Poincaré section (~225 points) of a chaotic Hénon-Heiles trajectory. The trajectory appears to fill an area, with approximate uniformity. Perhaps time averages equal phase averages in this subvolume of phase space. However, as clearly indicated in the *sequential* Poincaré section, which shows the first 75 intersections, second 75, and final 75 as (b), (c), and (d), respectively, there is substantial intermediate time regularity. In fact, the motion can be viewed as a transition between two approximate (or, vague) tori. See reference 65 for further discussion. (*From Quantum Mechanics in Mathematics, Chemistry, and Physics, ed. K. Gustafson and W. P. Reinhardt [New York: Plenum, 1981], p. 5210.*)

in a transition region. Referring back to Figure 5-16, this transition region corresponds to the trajectory with a turning point in or near the six shaded circles associated with local instability, which then allow a "transition" between vague tori to take place.[65] It would be most interesting to investigate the sequence of Figures 5-30(b), (c), (d) in terms of the possible changes in Painlevé singularity structure of Figure 5-18. There are many other examples of short-term surfaces of section showing behavior suggesting that a remnant of a torus might exist: See Figure 12 of reference 65, Figure 6 of the original Hénon-Heiles paper, reference 2, and a related but not quite so chaotic sequential section calculated by Channon and reported by Lebowitz.[207] Is there any way to quantify this intuitive idea?

The Birkhoff-Gustavson Normal Form: Semiclassical Quantization

To obtain analytic insight into possible phase space structure in a chaotic volume of phase space, we turn from trajectory-based methods, such as the Poincaré section and studies of bifurcations of periodic orbits, to time-independent algebraic techniques for direct determination of the tori themselves. Many such techniques using Fourier expansions and iterative solution of nonlinear matrix equations have been introduced to identify "good" action variables, and implicitly the invariant tori.[208-210] We focus here on the Birkhoff-Gustavson[50, 211] technique, introduced in the context of semiclassical quantization by Swimm and Delos.[98] This is a purely algebraic method, which is suitable for computational implementation in symbolic computer languages such as MACSYMA. [Deprit[212] has suggested that it may be advantageous to use a LISP-type language, combined with a Lie transformation-based algorithm (for example, Deprit[213]) rather than MACSYMA, as used by Swimm and Delos,[98] or in reference 68, or an arithmetic implementation (see, for example, Giorjelli[214]) of the normalization.]

The Birkhoff-Gustavson technique assumes a Hamiltonian of the form

$$H = \sum_{j=0}^{\infty} \epsilon^j H_{j+2}, \qquad H_2 = \sum_{i=1}^{N} \frac{\omega_i}{2} (p_i^2 + q_i^2),$$

where H_k is a homogeneous polynomial in p_i and q_i of order k and ϵ is a formal perturbation parameter. The lowest-order term in the Hamiltonian H_2 is a sum of harmonic oscillators each with its own frequency, ω_i. The perturbation terms $Hlyk$ ($k > 2$) contain all anharmonic and coupling terms. The object of normal form theory is to find a canonical transformation S such canonical transformation S such that the transformed Hamiltonian is a function of a new set of only the N variables $\tilde{\pi}_i$. Specifically for the non-resonant case, denoting a "new" variable, we want S such that

$$H(p_i, q_i) \xrightarrow{S} \tilde{H}(\tilde{\pi}_i),$$

where

$$\tilde{\pi}_i = \tfrac{1}{2} (\tilde{p}_i^2 + \tilde{q}_i^2).$$

Use of Hamiltonian equations in the $\tilde{p}_i$, $\tilde{q}_i$ variables immediately establishes that $\dot{\tilde{\pi}}_i(\tilde{p}_i, \tilde{q}_i) = 0$, and, because the $\tilde{\pi}_i$ may now be identified with the actions, L_i, of Equations (5-7) and (5-8), quantization proceeds at once. The algebra of the determination of the transformations, S, is quite involved,[211, 213] and rests on determination of the null spaces of the "normal" operator

$$D = \{ \; , H_2\} = \sum_{i=1}^{n} \omega_i \left(p_i \frac{\partial}{\partial q_i} - q_i \frac{\partial}{\partial p_i} \right)$$

and its adjoint. In the case of a nonresonant problem, functions of the basis $\pi_i = \frac{1}{2}(p_i^2 + q_i^2)$ span this space. Gustavson[211] showed that for resonant systems, additional basis functions besides the π_i are needed to span the null space of D. Although the members of this extended set of null space basis functions are not necessarily independent or in involution, they may be used to find the independent constants of motion that are in involution. For example, for the Hénon-Heiles system $n = 2$, $\omega_1 = \omega_2 = 1$ (a $1:1$ resonance), a possible null space basis set is [68]

$$\pi_0 = \frac{1}{2}(p_1^2 + x^2) + \frac{1}{2}(p_2^2 + y^2),$$

$$\pi_1 = \frac{1}{2}(p_1^2 + x^2) - \frac{1}{2}(p_2^2 + y^2),$$

$$\pi_2 = p_1 y - x p_2,$$

$$\pi_3 = p_1 p_2 + xy.$$

For the $1:1$ resonant Hénon-Heiles problem the Birkhoff-Gustavson procedure produces, through the order in ϵ transformed, a new Hamiltonian of the form

$$H^{\text{new}} = H_\epsilon(I_1, I_2, \theta_2),$$

with $H_\epsilon(I_1, I_2, \theta_2)$ and I_1 being constants of motion. I_1 is obtained as a power series

$$I_1(\bar{p}, \bar{q}) = \sum_{j=2}^{\infty} \epsilon^j I_1^{(j+2)}(\bar{p}, \bar{q}),$$

where the $I_1^{(j+2)}(\bar{p}, \bar{q})$ are homogeneous polynomials in (p_1, p_2, q_1, q_2). Note that if the process is truncated at finite order, H^{new} is always integrable.

Swimm and Delos[98] normalized the $1:1$ resonant Hénon-Heiles Hamiltonian through sixth order and obtained excellent semiclassical energies for those parts of the Hénon-Heiles phase space where classical motion is regular in large measure. However, they also obtained *some* semiclassical quantum results that were in good agreement with exact quantum eigenvalues, even though they corresponded to parts of phase space where the classical dynamics was chaotic. Because the sixth-order truncated normalization of the Hénon-Heiles is an integrable approximation to the actual chaotic dynamics, how can it be that an integrable approximation to the dynamics gives a good semiclassical (EBK) result? The suggestion is, of course, that, as already indicated earlier in the first part of Section 5-3, actual dynamics might well be much less chaotic than indicated in the long-time Poincaré section.

Jaffé and Reinhardt[68] carried out normalization, also through sixth order in ϵ, finding explicitly that

$$H = H_1(J_1) + H_2(J_1, J_2, \theta_2),$$

where

$$H_1(J_1) = J_1 - \frac{5\epsilon^2}{12} J_1^2 - \frac{67\epsilon^4}{432} J_1^3 - \frac{42{,}229\ \epsilon^6}{155{,}520} J_1^4,$$

and

$$
\begin{aligned}
H_2(J_1, J_2, \theta) = {} & \frac{7\epsilon^2}{12} J_2^2 - \frac{7\epsilon^4}{144} J_1 J_2^2 + \frac{7\epsilon^4}{18} (J_1^2 - J_2^2)^{3/2} \\
& \times \cos 6\theta_2 - \frac{\epsilon^6 J_2^2}{155{,}520} [458{,}682\ J_1^2 - 575{,}855\ J_2^2] \\
& - \frac{35\epsilon^6}{648} J_1 J_2^2 (J_1^2 - J_2^2)^{1/2} \cos 2\theta_2 + \frac{35\epsilon^6}{1458} J_2^2 (J_1^2 - J_2^2) \\
& \times \cos 4\theta_2 + \frac{2093\epsilon^6}{2160} J_1 (J_1^2 - J_2^2)^{3/2} \cos 6\theta_2.
\end{aligned}
$$

In this new set of variables Jaffé and Reinhardt obtained the complete set of quantum levels for the Hénon-Heiles problem [for $\alpha = (80)^{-1/2} \approx 0.1118$; see Eq. (5-22)] including *all* levels in the chaotic parts of phase space. (See Table 5-1.) [The differences between the Swimm-Delos work and that of Jaffé and Reinhardt are in the choice of the transformation of Eq. (5-23). See reference 68 for a full discussion.] The evident success of this procedure then suggests that since quantum mechanics is a smoothing operation, perhaps the Birkhoff-Gustavson procedure gives an integrable approximation to the actual Hamiltonian, which is then an adequate approximation for semiclassical quantization via the EBK method. This seemingly plausible idea, however, strongly conflicts with the association of chaos with the irregular

Table 5-1

High-lying Quantum Levels for the Hénon-Heiles Problem as Obtained by Numerical Solution of the Schrödinger Equation (QM),[b] Uniform Semiclassical Quantization (JR)[c] and Primitive Semiclassical Quantization (SD).[d] All of These Levels Correspond to Classically Chaotic Phase Space.

Quantum Numbers[a]	Symmetry of Quantum States[a]	QM[b]	JR[c]	SD[d]
12 ± 0	A	11.966	12.011	11.864
12 ± 2	E	11.968	12.017	—
12 ± 4	E	12.206	12.217	—
12 ± 6	A*	12.777	12.274	12.310
		12.334	12.332	
12 ± 8	E	12.480	12.490	12.491
12 ± 10	E	12.712	12.749	12.750
12 ± 12	A*	13.077	13.0975	13.097
		13.087	13.0976	
13 ± 1	E	12.762	12.824	—
13 ± 3	A*	12.748	12.827	—
		13.032	13.060	
13 ± 5	E	13.081	13.104	—
13 ± 7	E	13.233	13.238	—
13 ± 9	A*	—	13.4397	—
		—	13.4474	
13 ± 11	E	—	13.7351	—

[a] D. W. Noid and R. A. Marcus[94]
[b] D. Noid, as reported in Swimm and Delos[98]
[c] C. Jaffé and W. P. Reinhardt[68]
[d] R. Swimm and J. Delos[98]

spectrum! Can it be that classical chaos actually has little to do with quantum mechanics? Before reaching a conclusion, however, note that for $\alpha \cong 0.1118$ the exact quantum spectrum[182] itself did *not* appear to be irregular (see again the discussions of reference 107), a point to which we will return in Section 5-4.

Convergence of the Birkhoff-Gustavson Expansion

In addition to allowing quantization of $\tilde{H}(I_1, I_2, \theta_2)$, the Birkhoff-Gustavson procedure gives the power series expansion for $I_1(\bar{p}, \bar{q})$. Shirts and Reinhardt[65] examined the truncated expansion for the original Hénon-Heiles problem ($\alpha = 1$), an 11th-order approximate constant in the form

$$I_1^{approx}(\bar{p}, \bar{q}) = I_1(p, q) \equiv \epsilon^2 I_1^{(4)} + \epsilon^3 I_1^{(5)} + \cdots + \epsilon^{11} I_1^{(13)}, \quad (5\text{-}24)$$

where the superscripts denote the order of the homogeneous polynomial in $(\bar{p}, \bar{q})$. $I_1^{approx}(\bar{p}, \bar{q})$ is an approximate constant of motion in the sense that

$dI/dt = \{I, H\}$ is zero through order ϵ^{11}. Because the $I_4, \ldots, I_{13}$ were available as closed algebraic expressions, many questions could be asked: How good is the resulting constant I_1^{approx}? At very low energies, $0 \leq E \leq 1/10$, the level surfaces of $I_1^{\text{approx}}(\bar{p}, \bar{q})$ appear to suggest the same constants of motion as numerically determined by surface of section techniques. (Compare Figs. 5-6 and 5-31). At this energy ($E = 0.10$) the results are a

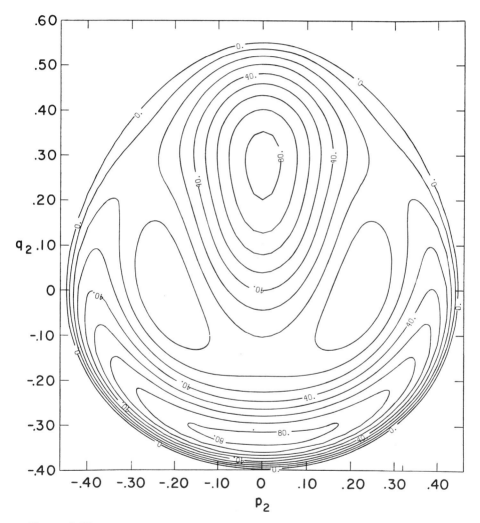

Figure 5-31
Level surfaces of the constant $I_1^{(13)}$, equation, at $E = 1/10$. These surfaces may be compared with those of Figure 5-6. It is evident that the truncated power series has captured the essence of the time-independent phase space structure.

substantial qualitative and quantitative improvement over the original work of Gustavson,[211] who only had calculated $I_1(\bar{p}, \bar{q})$ through ϵ^6 (see reference 211, especially Figs. 10 and 12). Another measure of the "goodness" of $I_1^{13}(\bar{p}, \bar{q})$ of Equation (5-24) is to evaluate it "along" trajectories governed by the exact dynamics. At $E = 1/8$, two such evaluations, both corresponding to quasi-periodic motion, are shown as the dashed curves in Figure 5-32. In one case a respectable constant has been obtained; in another it is evident that although some progress has been made, a good constant has not been obtained. In this latter case, there are two possibilities: Either a constant does not exist, *or* the truncated power series expansion of $I^{(13)}(\bar{p}, \bar{q})$ does not adequately represent the constant, *even though it exists.*

For the Hénon-Heiles problem we do not expect $I_1(\bar{p}, \bar{q})$, analytic in $(\bar{p},$

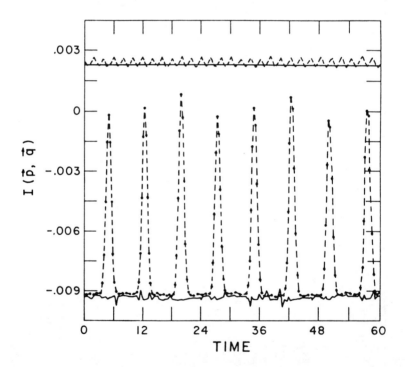

Figure 5-32
Approximate constant $I_i^{(13)}(\bar{p}, \bar{q})$ of Equation (5-24) evaluated along actual trajectories of the Hénon-Heiles problem (see reference 65 for an extended discussion). The dashed lines resulted from use of the truncated series for two different periodic trajectories. The solid line is the [5/4] Padé of the truncated series, evaluated as a function of $(\bar{p}, \bar{q})$ along the same trajectories, and has resulted in greatly improved constants. (*From Quantum Mechanics in Mathematics, Chemistry, and Physics, ed. K. Gustafson, and W. P. Reinhardt [New York: Plenum, 1981], p. 5213.*)

$\bar{q}$), to exist in the sense of Cauchy (see Fig. 5-10 and the discussion on integrability and quantization on tori in Section 5-1). But, suppose, for a moment, that we did expect an analytic constant but that Equation (5-24) didn't give a good representation—perhaps we are not taking enough terms, or the radius of convergence of the Taylor expansion has been exceeded. What could then be done to obtain more information from the Taylor coefficients I_4 through I_{13}, all of which are analytic functions of $(\bar{p}, \bar{q})$? Given information in the form of Taylor coefficients, a usual procedure would be to use the Taylor coefficients to form the Padé[215,216] table and to attempt to extract a better converged representation of the function.

The Padé idea is simple: Given $f(z)$, analytic at z_0, with a unique truncated Taylor representation

$$f(z) = \sum_{n=0}^{N+M+1} a_n(z_0)(z - z_0)^n,$$

the members of the Padé table, labeled $[N/M]$, are defined as

$$[N/M](z) \equiv \frac{P_N(z)}{Q_M(z)}, \tag{5-25}$$

where $P_N(z)$, $Q_M(z)$ are the polynomials

$$P_N(z) = p_0 + p_1(z - z_0)^1 \cdots p_N(z - z_0)^N,$$

$$Q_M(z) = 1 + q_1(z - z_0)^1 \cdots q_M(z - z_0)^M,$$

and the $N + M + 1$ coefficients $p_0, p_1, \ldots, P_N, q_1, \ldots, q_M$ are determined by requiring that $[N/M](z)$ have an identical Taylor expansion (about z_0) as does $f(z)$, through order $N + M + 1$. The rational expression of Equation (5-25) allows a representation that is often valid in a far greater domain about z_0 than the Taylor expansion, whose domain of convergence is a disk centered at z_0 and of radius determined by the nearest singularity to z_0. The Padé rational expression is not deterred by polar singularities, and in fact often has no real difficulty in coping with even branch point singularities as long as the Padé representation of associated cuts is well controlled.[215-219]

Figure 5-32 (solid lines) indicates the result of evaluation of a new approximate constant $I^{\text{Padé}}(\bar{p}, \bar{q})$ along trajectories generated by the actual dynamics. It is seen that the Padé technique has generated a substantially improved constant of motion. Encouraged by this ability to accelerate the convergence of the power expansion of Equation (5-24), the same technique was applied as a function of $(\bar{p}, \bar{q})$, at $E = 0.10$, where the classical phase space shows hints of the onset of chaos, as well as regular motion, resulting in the level surfaces of Figure 5-33. To a large extent, Figure 5-33 resembles Figure 5-31,

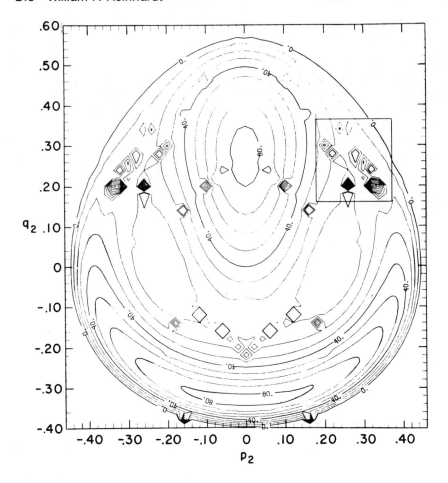

Figure 5-33
Level surfaces of the [5/4] Padé arising from the truncated power series of
Equation (5-24) at $E = 1/10$. comparison with Figure 5-31 indicates convergence
over much of the phase space. However, in the areas represented by the dense
diamonds and squares, the Padé has poles, indicating lack of convergence. The
fact that many members of the Padé table show clusters of zero and poles in
the same location is interpreted as a sign of strong, but highly localized nonanalyticity, and correlate this with local destruction of the manifold (see Fig. 5-
35). (*From Quantum Mechanics in Mathematics, Chemistry, and Physics, ed. K.
Gustafson and W. P. Reinhardt [New York: Plenum, 1981], p. 5211.*)

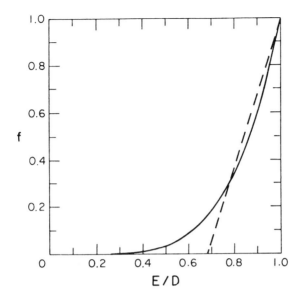

Figure 5-34
Fraction of area of surface of section occupied by chaos (expressed as E/D, D being the classical dissociation energy $\frac{1}{6}$) as determined empirically by Hénon and Heiles,[2] and as a function of the energy-dependent residues of the poles of the [5/4] Padé, as shown in Figure 5-33. It is clear that the Padé measure of strength of nonanalyticity correlates well with the onset and growth of chaos. See reference 65 for a full discussion. (*From Quantum Mechanics in Mathematics, Chemistry, and Physics, ed. K. Gustafson and W. P. Reinhardt [New York: Plenum, 1981], p. 5213.*)

indicating that in some parts of phase space the truncated Taylor series [Eq. (5-24)] and the Padé (the [5/4] is shown) summation give similar results, indicating convergence of both. In other parts of phase space, the Padé has poles and thus diverges strongly from the approximate constant suggested by the power series expansion. Are these poles an artifact? We do not believe so. Many members of the Padé tables showed poles in similar regions of the Poincaré section, and these regions are in reasonable proximity to those corresponding top the regions of high instability in the local exponentiation analysis of Section 5-1. In Figure 5-34 the *residues* of the poles of the [5/4] Padé are seen to correlate well with the fractional volume of chaotic phase space as a function of energy.

Thus although there is no possibility of existence of a global analytic second constant $I(\bar{p}, \bar{q})$ for the Hénon-Heiles problem (if there were, Hénon-Heiles would be integrable), even in the presence of chaos, the strongest nonanalyticity of $I(\bar{p}, \bar{q})$ appears to be highly localized, and we conjecture

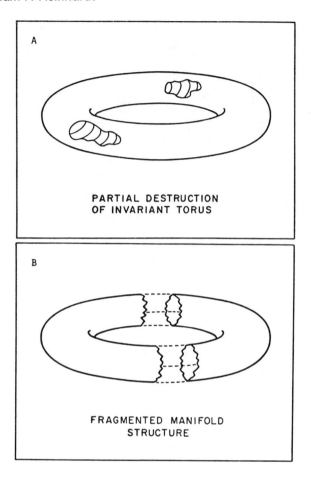

Figure 5-35
Heuristic drawings indicating the idea of local destruction of manifold surfaces. In (*A*) two patches of the torus are shown as missing. If a trajectory enters this part of the phase space it can hop to another (fragmented) torus (see Fig. 5-30). In (*B*) the fragmentation is such that individual trajectories would not show the intermediate term stability of say Figure 5-30(*b*), even though most of the manifold surface remains. (*From C. Jaffé and W. P. Reinhardt, "Uniform Semi-classical Quantization of Regular and Chaotic Classical Dynamics on the Hénon-Heiles Surface," Journal of Chemical Physics 77 [1982]: 5195.*)

that it is this strongest nonanalyticity that is sensed by the Padé approximants. In other parts of the chaotic phase space, reasonable approximate local constants of motion exist, even though the truncated power series, Equation (5-24), does not always represent them. These ideas are summarized in Figures 5-35(*a*) and (*b*), where they re seen to be in complete accord with the discussions of Sections 5-1 and 5-3, in that a local origin of chaos implies a highly nonuniform destruction of the underlying manifold structure. At least to the present author this accord is highly reassuring in that it suggests that the growth of chaos does not require global and uniform nonanalytic behavior; the invariant manifolds can be eaten away little by little, as nonanalyticity grows. The sequence of Figures 5-30(*b*), (*c*), and (*d*) is then interpreted within the framework of Figures 5-35(*a*), (*b*) as a hopping[220-223] between approximate constants, or from one partially intact manifold to another. The same situation appears to hold in the Zeeman[197] case, indicating that the surface of section of Figure 5-27 again far overemphasizes the extent of the destruction of the manifold structure of phase space.

A possible picture of the onset of such destruction is an outgrowth of the studies of Schenker and Kadanoff[224, 225] (following the work of Greene[226, 227]), on the destruction of the one-dimensional (KAM) manifolds in two-dimensional point mappings—A sequence of KAM curves is shown in Figure 5-36, where the presumed limit would be a continuous but nondifferentiable curve in the plane. It would likely have a fractal dimension greater than 1. Although no direct evidence is currently at hand, the arguments made here suggest that the analogous "last" two-dimensional torus (i.e., the one on the edge of chaos) for the Hénon-Heiles problem would be a surface, some part or parts of which would have nonuniform fractal dimension $D > 2$. The nonanalyticities suggested in Figure 5-35(*a*), (*b*) would then grow out of these localized fractal patches as a function of appropriate parameter or

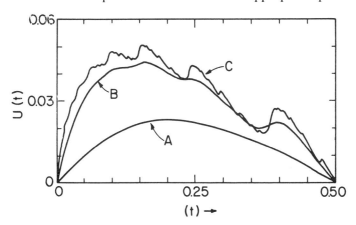

Figure 5-36

The sequence of curves $A \rightarrow B \rightarrow C$ correlates with the onset of destruction of KAM surfaces in the point mapping work of Schenker and Kadanoff.[224] The KAM surface becomes (at stage *C*) a continuous nondifferentiable function before it is destroyed. It is conjectured that the analog of such behavior is the precursor of the destroyed patches of Figure 5-35. (*After S. J. Shenker and L. P. Kadanoff, "Critical Behavior of a KAM Surface: I. Empirical Results" Journal of Statistical Physics 27 [1982]: 643.*)

parameters. Numerical techniques to check this conjecture are currently under development.

5-4   SEMICLASSICAL QUANTIZATION ON FRAGMENTED TORI

The suggestion, summarized in Figure 5-35(*a*), (*b*), that the onset of classical chaos, which appears so dramatically in individual trajectories, appears to be a much more gradual onset of nonanalyticity when viewed in terms of local constants of motion, allows a simple interpretation of the results of Swimm and Delos[98] and Jaffé and Reinhardt:[68] If "enough" of the manifold structure is intact, one can still attempt to carry out an EBK quantization on the remnants of the torus.

The question then arises, Are there cases in which the mechanism of fragmentation of the torus can be understood in detail and at least an attempt made to quantify such arguments? To this end, we consider the family of problems referred to as rational billiards in two dimensions. This discussion is followed by that of crossings and avoided crossings in integrable systems.

Rational Billiards and Semiclassical Quantization

Billiards[168, 170, 189, 190, 223, 228–232] in two dimensions refers to the class of dynamical systems consisting of free motion in a finite region, and hard-wall interactions at the boundary producing specular reflection. Two such problems have already been considered: the *stadium billiard*[168, 170] [see Fig. 5-22(*a*)] and the *truncated Sinai billiard*[189] [see Fig. 5-22(*b*)]. In each of these cases, the hard-wall enclosure is defined by line segments and segments of circular arcs. The rational billiards[190, 231, 232] are the subclass of billiard problems in which all boundaries are line segments, and all angles are *rational* multiples of π, that is, of the form $(n\pi)/m$, with n, m being integers. Simple examples are the rectangle, equilateral triangle, 45°-45°-90° right triangle, 30°-60°-90° right triangle, and parallelogram with one rational angle. The motion of a point mass in a rational enclosure has the special property that, given an initial momentum vector $\mathbf{p}$ in the plane, not only is $|\mathbf{p}|$ conserved by the specular collisions with the walls, but rationality of the angles implies that, independent of the number of collisions with the walls, only a *finite number* of new vector directions $\mathbf{p}_{\text{new}}$ are produced. Does this imply integrability? In the sense that the rational billiards are analytically *soluble* the answer is yes, and Eckhardt and Ford[231, 232] have concluded that solutions are of the same algebraic complexity[232, 233] as usual integrable systems. However, in the sense of integrability being associated with motion on an invariant torus, most rational billiards are nonintegrable!

To see how this comes about we follow the arguments of Keller and Rubinow;[228] Richens and Berry;[190] and Eckhart, Vivaldi, and Ford:[231,232] To visualize a trajectory in a rational billiard, rather than following the reflected trajectory in the fixed interior, follow reflections of the enclosure: The trajectory then is a straight line. This situation is illustrated in Figure 5-37(a). Without discussing the rigorous demonstration, it is now quite simple to state the requirement for integrability of a rational billiard: Consider the set

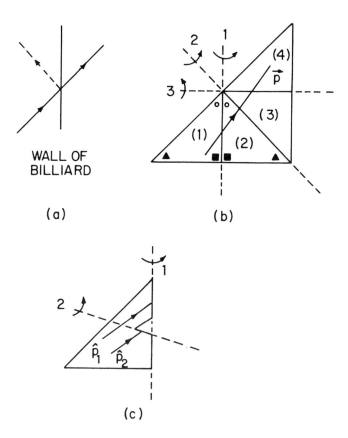

Figure 5-37
(*a*): Specular reflection at a hard wall. The dashed line shows the actual reflected trajectory. The solid line shows the same trajectory, reflected about the in-plane axis of the wall. Thus, if the enclosure is reflected, all trajectories will be straight lines. (*b*): Example of the integrable 45° right triangle billiard. Reflections generated by the set of all initial **p** vectors will exactly tesselate the plane. The ∘, △, serve to label the orientation of the triangle in that if different sequence reflections lead to two triangles occupying the same location, the orientation must be preserved. (*c*): A nonintegrable rational billiard. Various reflections will partially overlap; holes will occur: The plane will not be tesselated.

of all reflections of the billiard generated by choosing the "generating" vector, **p,** in all directions, and with all displacements of initial origin. If the original billiard plus the set of all reflections uniquely fills the plane in a one-to-one manner, the system is integrable. This situation is illustrated in Figures 5-37(*b*) and (*c*), where the 45°-45°-90° right triangle is seen to be integrable, and the indented triangle is seen to be nonintegrable. In the non-integrable case, the set of all reflections in general overlap one another in arbitrary ways, but since each momentum generates only a finite number of new momenta, the motion can be described as taking place on a surface of genus n, $n \geqslant 2$. Genus 1 is a torus (a sphere with one handle); genus n is a sphere with n handles. (See Figure 5-38.) Richens and Berry[190] refer to these cases as pseudointegrable systems.

A simple nonintegrable rational billiard is the truncated 45°-45°-90° right triangle considered by Richens and Berry[190] [Fig. 22(*b*)]. This pseudo-integrable system has a phase space of genus 2 and is illustrated in Figure 5-39. Richens and Berry point out that as this genus 2 surface admits *four* independent cycles that it is unclear how to proceed in a semiclassical quantization based on the four resulting actions. They numerically quantized the system, obtaining the levels (as a function of L) shown in Figure 5-24. A corresponding $P(S)$ distribution is shown in Figure 5-25(*b*). Again, lack of integrability seems to lead to strong level repulsion—but in this case the classical "chaos" is mild, as most neighboring trajectories do not exponentiate.

However, in this case one can make some intuitive guesses as to how things should proceed semiclassically.[198] For small L (i.e., for very little truncation) the motion is on an almost completely intact torus (see Fig. 5-

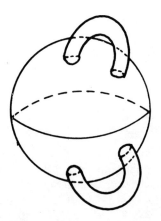

Figure 5-38
Sphere with two handles (genus 2).

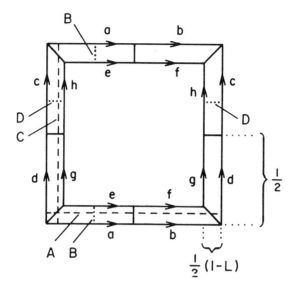

Figure 5-39
Phase space of the truncated square torus billiard of Figure 5-22(b), as discussed by Richens and Berry. The identically labeled directed lines are to be identified, giving a genus 2 surface. Genus 2 implies four independent one-cycles, which are labeled A, B, C, D. (*From J. Richens and M. V. Berry, "Pseudointegrable Systems in Classical and Quantum Mechanics," Physica 2D [1981]: 500.*)

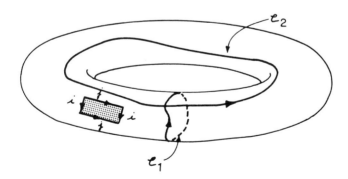

Figure 5-40
For small L, the segments e, f, g, h, in Figure 5-39 are short compared to $\frac{1}{2}$. Identification of the segments a, b, c, d thus gives the torus shown, with a "small" hole. The hole is shaded, and ignored in the quantization on paths $\mathcal{C}_1$ and $\mathcal{C}_2$. The hole is actually a handle on the torus, as may be visualized by realizing that the directed lines (i), (j) are to be identified.

40), and we can make the ansatz that the "hole" in the torus should not have much effect on the quantum levels. Inspection of the exact energy levels (Fig. 5-24) indeed indicates surprisingly little sensitivity to the value of L, in the small L limit. In the alternative limit $L \to (1 - \epsilon)$, the motion is effectively trapped in a rectangle of dimension $(\epsilon/2)$ by $\frac{1}{2}(1 - \epsilon)$, and one can imagine that this trapped motion can occasionally switch sheets[234] [in the spirit of the transition from one vague torus to another in Figs. 5-30(*a*), (*b*), (*c*)]. If this situation is represented heuristically by the genus 2 surface of Figure 5-41, the indicated quantization is suggested, and should be satisfactory unless the equivalent quantization of the "other" fragmented torus also predicts a quantum state at the *same* energy, in which case uniform approximation would be required.

Implementation of this idea suggests a semiclassical quantization of the truncated 45°-45°-90° triangle via the paths A and B of Figure 5-39, with the resultant levels being given by the simple expression

$$E_{n_x n_y} = \frac{4n^2 y}{(1 - L)^2} + 4n_x^2 \quad \text{with} \quad n_x > n_y \geq 1 , \qquad \textbf{(5-26)}$$

where the requirement $n_x > n_y$ gives the exact result in the $L \to 0$ limit. The levels given by the ansatz of Equation (5-26) are shown in Figure 5-42 for $n_y = 1, 2, 3, 4$, and $n_x = n_y, \cdots, 10$. It is evident that the general structure of Figure 5-24 has been captured. A more detailed comparison is given in Figure 5-43, where the exact quantum levels of Richens and Berry

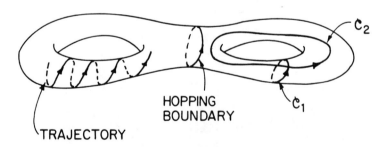

Figure 5-41
Trajectory on a "torus" of genus 2. If the trajectory only "hops" occasionally, as is the case in the $L \to 1$ limit of the truncated square torus billiard, EBK quantization can be implemented as shown. The hopping boundary is then identified with a "missing" piece of a torus, as shown heuristically in Figure 5-35. This figure arose in discussions with J. Ford relating to vague tori and hopping between vague tori. See also the discussion of reference 230.

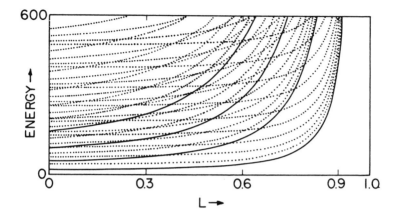

Figure 5-42
"EBK" energy levels for the truncated square torus billiard of Figure 5-22(*b*), calculated from the simple ansatz of Equation (5-26), as a function of *L*. Compare with the exact quantum results of Richens and Berry[190] in Figure 5-24.

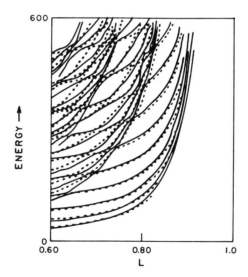

Figure 5-43
Detailed comparison of Figures 5-42 and 5-24. In this case the levels of the primitive quantization Equation (5-26) have been shifted slightly to improve visualization of the comparison. However, it is evident, in any case, that there is a one-to-one correspondence between level crossings in the EBK picture and avoided crossings in the exact results. The basic dynamics is thus almost integrable, yielding a quantum spectrum whose basic features are captured by the integrable approximation.

are superimposed on the predictions of Equation (5-26), the latter being "shifted" by about 2% in L. The comparison is then striking, showing a one-to-one correspondence of the crossings of the integrable approximation of Equation (5-26) and the avoided crossings of the exact quantum results. As the EBK quantization implicit in Equation (5-26) cannot adequately represent the necessary uniform approximation in the case that different (fragmented) tori produce degenerate quantum states, the fact that Equation (5-26) gives crossings where exact quantum mechanics shows avoided crossings is hardly a surprise, as further discussed in the next section.

Crossings and Avoided Crossings

Whenever a simplified, or approximate, solution of a quantum-mechanical problem leads to a degeneracy between states of a single symmetry, one expects that a more accurate analysis will lift the degeneracy. This is nothing but a restatement of the Wigner–von Neuman rule. Within the framework of semiclassical approximations, such degeneracies are resolved by allowance for "tunneling" or barrier penetration in the one-dimensional case, and by appropriate uniform approximations in the multidimensional case. The need for such uniform approximations has long been recognized: Noid and Marcus,[94] and Swimm and Delos[98] used an appropriately modified quantization rule to obtain EBK estimates of states for the Hénon-Heiles problem, which led to certain states having double degeneracy (those labeled A^* in Table 5-1.). Jaffé and Reinhardt[68] subsequently resolved the degeneracy by use of a uniform approximation,[235] appropriate to a threefold hindered rotor. Davis and Heller[236] have commented on the more general problem of the possibility of the occurrance of delocalized wavefunctions arising from degenerate but individually localized EBK tori.

Application of an appropriately higher-order approximation will then be expected to lift the degeneracies of the approximation of Equation (5-26) for the truncated $45°-45°-90°$ triangle, bringing this result, based on an integrable approximation to the dynamics, into even better agreement with the the exact quantum results. If this is indeed the case—as is conjectured here, but not yet shown—there would be no particular reason to associate the *robust* avoided crossings of Figure 5-24 with classical nonintegrability. Rather it would be appropriate to note only that the splittings arose via a perturbation of degenerate states arising in an integrable approximation to the dynamics.

As an example of such a higher approximation for a system *not* displaying chaos in large measure, Noid, Koszykowski, and Marcus[237] (NKM), have studied an isolated avoided crossing of levels for the Hamiltonian

$$H^{\mathrm{NKM}}(\bar{p}, \bar{q}) = \tfrac{1}{2}(\bar{p}_1^2 + \omega_1^2 q_1^2) + \tfrac{1}{2}(p_2^2 + \omega_2^2 q_2^2)$$
$$- a(q_1^3 + q_2^3) + \lambda q_1^2 q_2^2 - bq_1 q_2^3$$

with $\omega_1 = 3\omega_2 = 3$. Figure 5-44 shows a crossing of EBK quantum levels, and an avoided crossing of the exact quantum levels, as a function of λ for $b = 0.005$. NKM were ablt to use a simple semiclassical perturbation theory, based on the degnerate tori, to estimate the splitting as $\Delta E = 0.0041$, where two-state quantum (degenerate) perturbation theory gives $\Delta E = 0.0035$. The semiclassical and quantum treatments are thus in good agreement. Uzer, Noid, and Marcus have further extended the semiclassical analysis of such avoided crossings.[238] It is our conjecture that a similar analysis of the primitive approximation of Equation (5-26) would give excellent quantum levels. Analysis[239, 240] of the Hénon-Heiles problem near $\alpha = 0.088$, where an irregular spectrum was observed, gave level crossings within an analysis using

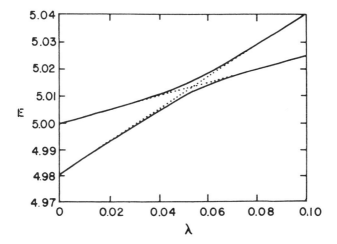

Figure 5-44
Quantum avoided crossing compared with an EBK crossing of levels. Use of an appropriate uniform approximation gives a semiclassical avoided crossing in good agreement with the quantum result. In this way, *integrable* classical dynamics can give rise to irregular quantum spectra. (*From D. W. Noid, M. L. Koszykowski, and R. A. Marcus, "Comparison of Quantal, Classical, and Semiclassical Behavior at an Isolated Avoided Crossing," Journal of Chemical Physics 78 [1983]: 4019.*)

the Birkhoff-Gustavson integrable approximation to the dynamics. An appropriate uniform approximation would lift this degeneracy, again generating an irregular spectrum (in the sense of large second differences) from integrable dynamics.

The Irregular Spectrum Revisited

Integrability and the irregular spectrum

We have argued[241] (and conjectured) that many of the features of Percival's irregular spectrum can be understood in terms of integrable approximations to the underlying dyanmics, and that conversely there is no one-to-one correspondence between classical nonintegrability and an irregular spectrum. The detail required of such integrable approximations will, of course, increase as $\hbar$ decreases, and/or local de Broglie wavelengths decrease. It is our current belief, however, that for a fixed value of $\hbar$, integrable approximations can always be generated that will give a correct global description of complex spectra. The fact that such integrable approximations will depend on $\hbar$ indicates the role of the correspondence principle.

Where are irregular spectra best studied?

The previous remarks certainly are not in conflict with existence of a quantum irregular spectrum, they just attempt to dissociate it from the necessity of an interpretation in terms of classical chaos. Where are these questions and conjectures best investigated?

Classical, quantum-mechnical, and experimental (quantum mechanical *per force!*) studies can be now conveniently carriet out for the dynamics of Rydberg electrons as function of external E and B fields as well as a discrete function of breaking the Coulomb $O(4)$ symmetry via chaning the nature of the core ion from H^+ to Li^+, and so on. The hydrogenic Stark problem is integrable both classically and quantum mechanically;[242] the equivalent Zeeman problem appears to be almost integrable quantum mechanically[196] and classically.[197] The mixed-field case is currently an open question both theoretically and experimentally. As an example of the effect of a change in atomic core from pure Coulomb[242] in the Stark case,[243] see Figure 5-45(A), where the Stark splittings of $|m_l| = 1$ states of Na Rydberg states (near $n = 15$, 16, 17) are shown. By Percival's P3 criterion this effect might be analyzed as an irregular spectrum, although it would not be so judged by the Marcus criterion of *overlapping* avoided crossings. Figure 5-45(B) shows the $|m_l| = 0$ spectrum of the same system: In the region where man-

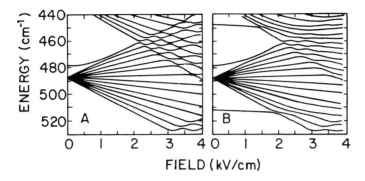

Figure 5-45
(*A*): Rydberg Stark levels for $|m| = 1$ levels in sodium. The patterns of level avoided crossings (see Figure 5-44) can be understood in terms of a separable (i.e., integrable) approximation, perturbed by the core. The spectrum is thus not very irregular, even though $P(s)$ will not be Poisson. (*B*): As in (*A*), but $|m| = 0$. In this case the interpenetrating manifolds give rise to a qualitatively new structure, suggesting the onset of a real irregular spectrum. (*From M. G. Littman, M. C. Zimmerman, T. W. Ducas, R. R. Freeman, and D. Kleppner, "Structure of Sodium Rydberg States in Weak to Strong Electric Fields," Physical Review Letters 36 [1978]: 790.*)

of states interpenetrate, no remnant of the original pattern persists and a new "regularity" sets in because all states seem to repel equally. This most irregular spectrum (evenly spaced states; see Pechukas[244]) would fail Percival's P2 and P3 criteria, but might well meet P1 and Berry's criterion for the character of an irregular wavefunction. It would be useful to examine the wavefunctions in this region; it is, of course, possible that rather then irregularity a new dynamical symmetry is present. In any case, it is certainly time to move away from the Hénon-Heiles and billiard numerical experiments and to accept the challenge of these $n = 2$ and $n = 3$ systems provided by nature.

5-5 SUMMARY

In this discursive essay many topics of current activity in the area of nonlinear dynamics have been discussed. The most important and rapidly developing themes are the role of fractal dimensionality, both in empirical analysis, and in understanding the nature of nonanalyticity associated with chaos; the use of the Painlevé property (and extensions) as a detection scheme for integrable systems and as a diagnostic of the global structure of nonanalyticity for chaotic systems; the qualitative discussion of fragmentation of tori, and the concommitant suggestion that there need be no one-to-one relationship between classical chaos and the quantum irregular spectrum. Finally, a brief plea has been made for the necessity of studying *physical* systems, rather than artificial models of approximately the same level of complexity: studies of Rydberg atoms in external E, B, and combined E and B fields would seem an ideal choice.

ACKNOWLEDGMENTS

This research has been supported, in part, by National Science Foundation Grants CHE80-11442 and PHY82-00805. The author is particularly grateful to the Molecular Spectroscopy Division of the National Bureau of Standards for the hospitality shown during his tenure as a Visiting Scientist at NBS, Washington, 1982–83. The author is most grateful for the input and stimulation of many coworkers, in particular, C. Holt, C. Jaffé, C. Cerjan, R. Shirts and D. Farrelly, and to the continuing encouragement and ruthless criticism of R. A. Marcus, M. Tabor, and J. Ford.

NOTES AND REFERENCES

1. See, for example, the expository article, R. M. May, "Simple Mathematical Models with Very Complicated Dynamics," *Nature* 261 (1976): 459–467.
2. M. Hénon and C. Heiles, "The Applicability of the Third Integral of Motion: Some Numerical Experiments," *The Astronomical Journal* 69 (1964): 73–79.
3. B. B. Mandelbrot, *The Fractal Geometry of Nature* (San Francisco: W. H. Freeman, 1983).
4. H. Goldstein, *Classical Mechanics* (Reading, Mass.: Addison-Wesley, 1951).
5. M. Born, *Mechanics of the Atom* (New York: Ungar, 1960).
6. V. I. Arnold, *Mathematical Methods of Classical Mechanics* (New York: Springer-Verlag, 1980).
7. V. I. Arnold and A. Avez, *Ergodic Problems in Classical Mechanics* (New York: Benjamin, 1968).
8. R. Abraham and J. E. Marsden, *Foundations of Mechanics*, 2nd ed., (Reading, Mass.: Benjamin/Cummings, 1978).
9. A. Einstein, "On the Quantization Condition of Sommerfeld and Epstein," *Deutsche Physikalische Gesellschaft* 19 (1917): 82–88.
10. L. Brillouin, "Remarques sur la mécanique ondulatoire," *Journal de Physique et le Radium* 7 (1926): 353–368.
11. J. B. Keller, "Corrected Bohr-Sommerfeld Quantum Conditions for Nonseparable Systems," *Annals of Physics* 4 (1958): 180–188.
12. M. Toda, "Instability of Trajectories of the Lattice with Cubic Nonlinearity," *Physics Letters* 48A (1974): 335–336.
13. P. Brumer and J. W. Duff, "A Variational Equations Approach to the Onset

of Statistical Intramolecular Energy Transfer," *Journal of Chemical Phyics* 65 (1976): 3566–3574.

14. C. Cerjan and W. P. Reinhardt, "Critical Point Analysis of Instabilities in Hamiltonian Systems: Classical Mechanics of Stochastic Intramolecular Energy Transfer," *Journal of Chemical Physics* 71 (1979): 1819–1831.

15. R. Kosloff and S. A. Rice, "Dynamical Correlations and Chaos in Classical Hamiltonian Systems," *Journal of Chemical Physics* 74 (1981): 1947–1955.

16. E. Hopf, "On Causality, Statistics and Probability," *Journal of Mathematics and Physics* 13 (1934): 51–102.

17. R. H. G. Helleman and T. Bountis, "Periodic Solutions of Arbitrary Period, Variational Methods," in *Stochastic Behavior in Classical and Quantum Hamiltonian Systems*, ed. G. Casati and J. Ford (New York: Springer-Verlag) (*Lecture Notes in Physics* 93 [1979]: 353–375).

18. M. J. Feigenbaum, "Quantitative Universality for a Class of Nonlinear Transformations," *Journal of Statistical Physics* 19 (1978): 25–52.

19. J. M. Greene, R. S. MacKay, F. Vivaldi, and M. J. Feigenbaum, "Universal Behavior in Families of Area-Preserving Maps," *Physica* 3D (1981): 468–486.

20. P. Pakavinen and R. M. Nieminen, "Period-Multiplying Bifurcations and Multifurcations in Conservative Mappings," *Journal of Physics A: Mathematical and General* 16 (1983): 2105–2119.

21. N. S. Manton and M. Nauenberg, "Universal Scaling Behaviour for Iterated Maps in the Complex Plane," *Communications in Mathematical Physics* 89 (1983): 555–570.

22. M. J. Ablowitz, A. Ramani, and H. Segur, "A Connection between Nonlinear Evolution Equations and Ordinary Differential Equations of P-type. I," *Jounral of Mathematical Physics* 21 (1980): 715–721.

23. Y. F. Chang, M. Tabor, J. Weiss, and G. Corliss, "On the Analytic Structure of the Henon-Heiles System," *Physics Letters* 85A (1981): 211–213.

24. M. Tabor, "Analytic Structure and Integrability of Dynamical Systems," *International Journal of Quantum Chemistry: Quantum Chemistry Symposium* 16 (1982): 167–181.

25. Y. F. Chang, M. Tabor, and J. Weiss, "Analytic Structure of the Henon-Heiles Hamiltonian in Integrable and Non-Integrable Regimes," *Journal of Mathematical Physics* 23 (1982): 531–538.

26. A. Ramani, B. Dorizzi and B. Grammaticos, "Painlevé Conjecture Revisited," *Physical Review Letters* 49 (1982): 1539–1541.

27. T. Bountis, H. Segur, and F. Vivaldi, "Integrable Hamiltonian Systems and the Painlevé Property," *Physical Review A* 25 (1982): 1257–1264.

28. J. Weiss, M. Tabor, and G. Carnevale, "The Painlevé Property for Partial Differential Equations," *Journal of Mathematical Physics* 24 (1983): 522–526.

29. I. C. Percival, "Semiclassical Theory of Bound States," *Advances in Chemical Physics* 36 (1977): 1–61.

30. M. V. Berry, "Regular and Irregular Motion," in *Topics in Nonlinear Dynamics, A Tribute to Sir Edward Bullard*, AIP Conference Proceedings no. 46, ed. S. Jorna (New York: American Institute of Physics, 1978), pp. 16–120.

31. A. S. Wightman, "Statistical Mechanics and Ergodic Theory: An Expository Lecture," in *Statistical Mechanics at the Turn of the Decade*, ed. E. G. D. Cohen (New York: Dekker, 1971), pp. 1–32.

32. J. Ford, "The Transition from Analytic Dynamics to Statistical Mechanics," *Advances in Chemical Physics* 24 (1973): 155–185.
33. J. Ford, "The Statistical Mechanics of Classical Analytic Dynamics," in *Fundamental Problems in Statistical Mechanics, III*, ed. E. G. D. Cohen (New York: North-Holland/Elsevier, 1975), pp. 215–255.
34. R. H. G. Helleman, 'Self-Generated Chaotic Behavior in Nonlinear Mechanics," in *Fundamental Problems in Statistical Mechanics, V*, ed. E. G. D. Cohen (Amsterdam: North-Holland, 1980), pp. 165–233. This review contains an extensive set of references.
35. M. Tabor, "The Onset of Chaotic Motion in Dynamical Systems," *Advances in Chemical Physics* 46 (1981): 73–151.
36. P. Brumer, "Intramolecular Energy Transfer: Theories for the Onset of Statistical Behavior," *Advances in Chemical Physics* 47 (1981): 201–238.
37. S. A. Rice, "Quasiperiodic and Stochastic Intramolecular Dynamics: The Nature of Intramolecular Energy Transfer," in *Quantum Dynamics of Molecules*, ed. R. G. Woolley (New York: Plenum, 1980), pp. 257–356.
38. D. W. Noid, M. L. Koszykowski, and R. A. Marcus, "Quasiperiodic and Stochastic Behavior in Molecules," *Annual Review of Physical Chemistry* 32 (1981): 267–309.
39. B. V. Chirikov, "A Universal Instability of Many-Dimensional Oscillator Systems," *Physics Reports* 52 (1979): 263–379.
40. G. M. Zaslavsky, "Stochasticity in Quantum Systems," *Physics Reports* 80 (1981): 157–250, parts I and III.
41. S. Jorna, ed., *Topics in Nonlinear Dynamics, A Tribute to Sir Edward Bullard*, AIP Conference Proceedings no. 46 (New York: American Institute of Physics, 1978).
42. G. Casati and J. Ford, eds., *Stochastic Behavior in Classical and Quantum Hamiltonian Systems* (New York: Springer-Verlag, 1979).
43. G. Casati, ed., *Chaotic Behavior in Quantum Systems* (New York: Plenum, in press).
44. V. Szebehely and B. D. Tapley, eds., *Long Time Predictions in Dynamics* (Boston: Reidel, 1976).
45. C. W. Horton, Jr., L. E. Reichl, and V. G. Szebehely, eds., *Long Time Predictions in Dynamics* (New York: John Wiley & Sons, 1983).
46. C. Bardos and D. Bessin, eds., *Bifurcation Phenomena in Mathematical Physics and Related Topics* (Berlin: Reidel, 1980).
47. G. Iooss, R. H. G. Helleman, and R. Stora, eds., Proceedings of the 1981 "Les Houches" Summer School, vol. 36, *Chaotic Behavior in Deterministic Systems* (Amsterdam: North-Holland, in press).
48. R. L. Devaney and Z. Nitecki, eds., *Classical Mechanics and Dynamical Systems* (New York: Dekker, 1981).
49. R. H. G. Helleman, ed., *Proceedings of the International Conference on Nonlinear Dynamics* (New York: New York Academy of Sciences, 1980).
50. G. D. Birkhoff, *Dynamical Systems*, Colloq. Publ. no. 9 (Providence, RI: American Mathematical Society, 1979).
51. J. Moser, *Stable and Random Motions in Dynamical Systems* (Princeton, NJ: Princeton University Press, 1973).

52. C. C. Siegel and J. Moser, *Lectures on Celestial Mechanics* (New York: Springer-Verlag, 1971).
53. I. C. Percival and D. Richards, *Introduction to Dynamics* (London, New York: Cambridge University Press, 1982).
54. R. H. Abraham and C. D. Shaw, *Dynamics—The Geometry of Behavior*, vol. 1, *Periodic Behavior* (Santa Cruz, Calif.: Aerial, 1982).
55. R. H. Abraham and C. D. Shaw, *Dynamics—The Geometry of Behavior*, vol. 2, *Chaotic Behavior* (Santa Cruz, Calif.: Aerial, 1983).
56. G. M. Zaslavsky, "Stochasticity in Quantum Systems," *Physics Reports* 80 (1981): 157–250, part II.
57. H. B. Thacker, "Exact Integrability in Quantum Field Theory and Statistical Systems," *Reviews of Modern Physics* 53 (1981): 253–285.
58. M. A. Olshanetsky and A. M. Perelomov, "Quantum Integrable Systems Related to Lie Algebras," *Physics Reports* 94 (1983): 313–404.
59. J.-P. Eckmann, "Roads to Turbulence in Dissipative Dynamical Systems," *Reviews of Modern Physics* 53 (1981): 643–654, which contains an excellent set of general references.
60. J. D. Farmer, E. Ott, and J. A. Yorke, "The Dimension of Chaotic Attractors," *Physica* 7D (1983): 153–180.
61. P. Hu, "Introduction to Real-Space Renormalization-Group Methods in Critical and Chaotic Phenomena," *Physics Reports* 91 (1982): 233–295.
62. R. H. G. Helleman and R. S. MacKay, "One Mechanism for the Onsets of Large Scale Chaos in Conservative and Dissipative Systems," in *Long Time Predictions in Dynamics*, ed. C. W. Horton, Jr., L. E. Reichl, and V. G. Szebeheley (New York: John Wiley & Sons, 1983), pp. 95–126.
63. R. S. MacKay, "Period Doubling as a Universal Route to Stochasticity," in *Long Time Predictions in Dynamics*, ed. C. W. Horton, Jr., L. E. Reichl, and V. G. Szebeheley (New York: John Wiley & Sons, 1983), pp. 127–134.
64. D. F. Escande, "Renormalization Approach to Nonintegrable Hamiltonians," in *Long Time Predictions in Dynamics*, ed. C. W. Horton, Jr., L. E. Reichl, and V. G. Szebeheley (New York: John Wiley & Sons, 1983), pp. 149–178.
65. R. B. Shirts and W. P. Reinhardt, "Formal Integrals for a Nonintegrable Dynamical System: Preliminary Report," in *Quantum Mechanics in Mathematics, Chemistry, and Physics*, ed. K. Gustafson and W. P. Reinhardt (New York: Plenum, 1981), pp. 277–287.
66. R. B. Shirts and W. P. Reinhardt, "Approximate Constants of Motion for Classicially Chaotic Vibrational Dynamic: Vague Tori, Semiclassical Quantization, and Classical Intramolecular Energy Flow," *Journal of Chemical Physics* 77 (1982): 5204–5217.
67. W. P. Reinhardt and C. Jaffé, "Problems in the Semiclassical Quantization of Integrable and Nonintegrable Dynamical Systems," in *Quantum Mechanics in Mathematics, Chemistry, and Physics*, ed. K. Gustafson and W. P. Reinhardt (New York: Plenum, 1981), pp. 167–177.
68. C. Jaffé and W. P. Reinhardt, "Uniform Semiclassical Quantization of Regular and Chaotic Classical Dynamics on the Hénon-Heiles Surface," *Journal of Chemical Physics* 77 (1982): 5191–5203.
69. W. P. Reinhardt, "Chaotic Dynamics, Semiclassical Quantization, and Mode-

Mode Energy Transfer: The Boulder View," *Journal of Physical Chemistry* 86 (1982): 2158–2165.

70. W. P. Reinhardt, "Semiclassical Quantization on Fragmented Tori," in *Chaotic Behavior in Quantum Systems*, ed. G. Casati (New York: Plenum, in press).

71. R. C. Churchill, G. Pecelli, and D. C. Rod, "A Survey of the Henon Heiles Hamiltonian with Applications to Related Examples," in *Stochastic Behavior in Classical and Quantum Hamiltonian Systems*, ed. G. Casati and J. Ford (New York: Springer-Verlag, 1979), *Lecture Notes in Physics* 93 (1979): 76–136.

72. G. Contopoulos, "On the Existence of a Third Integral of Motion," *Astronomical Journal* 68 (1963): 1–14.

73. B. Barbanis, "An Application of the Third Integral in the Velocity Space," *Zeitschrift für Astrophysik* 56 (1962): 56–67.

74. G. Contopoulos, "Resonance Cases and Small Divisors in a Third Integral of Motion.I," *Astronomical Journal* 68 (1963): 763–779.

75. G. Contopoulos and M. Moutsoulas, "Resonance Cases and Small Divisors in a Third Integral of Motion.II," *Astronomical Journal* 70 (1965): 817–835.

76. W. H. Jefferys, "Some Dynamical Systems of Two Degrees of Freedom in Celestial Mechanics," *Astronomical Journal* 71 (1966): 306–313.

77. B. Barbanis, "On the Isolating Character of the "Third" Integral in a Resonance Case," *Astronomical Journal* 71 (1966): 415–424.

78. G. Contopoulos and M. Moutsoulas, "Resonance Cases and Small Divisors in a Third Integral of Motion.III," *Astronomical Journal* 71 (1966): 687–698.

79. W. H. Jefferys and J. Moser, "Quasi-Periodic Solutions for the Three-Body Problem," *Astronomical Journal* 71 (1966): 568–578.

80. See, for example, V. Guillemin and A. Pollack, *Differential Topology* (Englewood Cliffs, N.J.: Prentice-Hall, 1974), chap. 1.

81. A. M. Ozovio de Almeida and J. H. Hannay, "Geometry of Two Dimensional Tori in Phase Space: Projections, Sections and the Wigner Function," *Annals of Physics* (New York) 138 (1982): 115–154.

82. J. Moser, "Nearly Integrable and Integrable Systems," in *Topics in Nonlinear Dynamics, A Tribute to Sir Edward Bullard*, AIP Conference Proceedings no. 46, ed. S. Jorna (New York: American Institute of Physics, 1978), pp. 1–16.

83. A. N. Kolmogorov, "The General Theory of Dynamical Systems and Classical Mechanics," Address to the 1954 International Congress of Mathematicians. An English translation is given in R. Abraham and J. E. Marsden, *Foundations of Mechanics*, 2nd ed. (Reading, Mass.: Benjamin/Cummings, 1978), pp. 741–757.

84. An excellent discussion of the idea of the theorem is in the review by M. V. Berry, "Regular and Irregular Motion," in *Topics in Nonlinear Dynamics, a Tribute to Sir Edward Bullard*, AIP Conference Proceedings no. 46, ed. S. Jorna (New York: American Institute of Physics, 1978), pp. 16–129.

85. D. W. Noid, M. L. Koszykowski, and R. A. Marcus, "A Spectral Analysis Method of Obtaining Molecular Spectra from Classical Trajectories," *Journal of Chemical Physics* 67 (1977): 404–408.

86. A. Knauf, "Das No-Vortex-System der Klassischen Mechanik," thesis, Free University of Berlin (Berlin, 1983).

87. E. A. Novikov, "Dynamics and Statistics of a System of Vortices," *Soviet Physics JETP* 41 (1975): 937–943.
88. E. A. Novikov, "Stochastization and Collapse of Vortex Systems," in *Nonlinear Dynamics*, ed. R. H. G. Helleman, *Annals of the New York Academy of Sciences* 357 (1980): 47–54.
89. H. von Koch, 'Sur une courbe continue sans tangente, obtenue par une construction géométrique élémentaire," *Arkiv for Matematik, Astronomi och Fysik* 1 (1904): 681–704.
90. For example, S. J. Shenker and L. P. Kadanoff, "Critical Behavior of a KAM Surface: I. Empirical Results," *Journal of Statistical Physics* 27 (1982): 631–656.
91. S. J. Shenker, "Scaling Behavior in a Map of a Circle onto Itself: Empirical Results," *Physica* 5D (1982): 405–411.
92. R. A. Marcus, "Semiclassical Theory for Collisions Involving Complexes (Compound State Resonances) and for Bound State Systems," *Faraday Discussions of the Chemical Society* 55 (1973): 34–44.
93. W. Eastes and R. A. Marcus, "Semiclassical Calculation of Boundstates of a Multidimensional System," *Journal of Chemical Physics* 61 (1974): 4301–4306.
94. D. W. Noid and R. A. Marcus, "Semiclassical Calculation of Bound States in a Multidimensional System for Nearly 1 : 1 Degenerate Systems," *Journal of Chemical Physics* 67 (1977): 559–567.
95. K. S. Sorbie and N. C. Handy, "Semiclassical Eigenvalues for Non-Separable Bound Systems: The Degenerate Case," *Molecular Physics* 32 (1976): 1327–1347.
96 I. C. Percival and N. Pomphrey, "Vibrational Quantization of Polyatomic Molecules," *Molecular Physics* 31 (1976): 97–114.
97. S. L. Chapman, B. C. Garrett, and W. H. Miller, "Semiclassical Eigenvalues for Nonseparable Systems: Nonperturbative Solution of the Hamilton-Jacobi Equation in Action-Angle Variables," *Journal of Chemical Physics* 64 (1976): 502–509.
98. R. T. Swimm and J. B. Delos, "Semiclassical Calculations of Vibrational Energy Levels for Nonseparable Systems Using the Birkhoff-Gustavson Normal Form," *Journal of Chemical Physics* 71 (1979): 1706–1717.
99. C. Jaffé and W. P. Reinhardt, "Time-Independent Methods in Classical Mechanics: Calculation of Invariant Tori and Semiclassical Energy Levels via Classical Van Vleck Transformations," *Journal of Chemical Physics* 71 (1979): 1862–1869.
100. M. C. Gutzwiller, "Classical Quantization Conditions for a Dynamical System with Stochastic Behavior?" in *Stochastic Behavior in Classical and Quantum Hamiltonian Systems*, ed. G. Casati and J. Ford (New York: Springer-Verlag, 1979), pp. 316–325.
101. M. C. Gutzwiller, "The Quantization of a Classically Ergodic System," *Physica* 5D (1982): 183–207. Discussion of this extraordinary work is outside the scope of the present review.
102. M. V. Berry and M. Tabor, "Calculating the Bound Spectrum by Path Summation in Action-Angle Variables," *Journal of Physics A: Mathematical and General* 10 (1977): 371–379.

103. M. V. Berry and M. Tabor, "Closed Orbits and the Regular Bound Spectrum," *Proceedings of the Royal Society of London* 349 (1976): 101–123.

104. M. V. Berry, "Semiclassical Mechanics of Regular and Irregular Motion," in *Chaotic Behavior in Deterministic Systems*, Proceedings of the 1981 "Les Houches" Summer School, vol. 36, ed. G. Iooss, R. H. G. Helleman and R. Stors (Amsterdam: North-Holland, in press).

105. R. Balian and C. Bloch, "Solution of the Schrödinger Equation in Terms of Classical Paths," *Annals of Physics* 85 (1974): 514–545.

106. A. Voros, "Semi-Classical Approximations," *Annales de l'Institut Henri Poincaré* 24 (1976): 31–90.

107. W. H. Miller, "Classical Limit Green's Function (Fixed-Energy Propagator) and Classical Quantization of Nonseparable Systems," *Journal of Chemical Physics* 56 (1972): 38–44.

108. W. H. Miller, "Semiclassical Quantization of Nonseparable Systems: A New Look at Periodic Orbit Theory," *Journal of Chemical Physics* 63 (1975): 996–999.

109. R. Kosloff and K. F. Freed, "Green's Function Semiclassical Quantization of Non-Closed Quasiperiodic Classical Trajectories," *Chemical Physics Letters* 84 (1981): 630–635.

110. N. de Leon and E. J. Heller, "Semiclassical Quantization and Extraction of Eigenfunctions Using Arbitrary Trajectories," *Journal of Chemical Physics* 78 (1983): 4005–4017.

111. M. Tabor, "A Semiclassical Quantization of Area-Preserving Maps," *Physics* 6D (1983): 195–210.

112. See, for example, the pedagogical discussion of M. V. Berry, "Regular and Irregular Motion," in *Topics in Nonlinear Dynamics, A Tribute to Sir Edward Bullard*, AIP Conference Proceedings no. 46, ed. S. Jorna (New York: American Institute of Physics, 1978), pp. 16–120, and references therein.

113. E. Borel, *Lecons sur les Fonctions Monegenes Uniformes d'une Variable Complexe* (Paris: Gauthier-Villars, 1917).

114. The possibility of the use of Borel, rather than Cauchy, analyticity was suggested to the author by A. S. Wightman, private discussions, March 1981.

115. Another is the anisotropic Kepler problem; see M. C. Gutzwiller, "The Quantization of a Classically Ergodic System," *Physics* 5D (1982): 183–207, and references therein.

116. See, for example, M. V. Berry, "Regular and Irregular Motion," in *Topics in Nonlinear Dynamics, A Tribute to Sir Edward Bullard*, AIP Conference Proceedings no. 46, ed. S. Jorna (New York: American Institute of Physics, 1978, p. 91.

117. V. I. Arnold, "Instability of Dynamical Systems with Several Degrees of Freedom," *Soviet Mathematics Doklady* 5 (1964): 581–585.

118. C. Jaffé, "Semiclassical Quantization of Nonseparable Hamiltonian Systems," Ph.D. thesis, University of Colorado (1979).

119. E. Thiele and D. J. Wilson, "Anharmonicity in Unimolecular Reactions," *Journal of Chemical Physics* 35 (1961): 1256–1263.

120. R. M. Hedges, Jr., and W. P. Reinhardt, "Classical and Quantum Dynamics of Long Lived Doubly Excited Vibrational States of Triatomic Molecules," *Journal of Chemical Physics* 78 (1983): 3964–3975.

121. P. Brumer, "Stability Concepts in the Numerical Solution of Classical Atomic and Molecular Scattering Problems," *Journal of Computational Physics* 14 (1973): 391–419.
122. M. Yamaguti and S. Ushiki, "Chaos in Numerical Analysis of Ordinary Differential Equations," *Physica* 3D (1981): 618–626.
123. J. Burnusson, *Point Mapping Stability* (New York: Pergamon, 1977).
124. A. C. Bountis, "Nonlinear Models in Hamiltonian Dynamics and Statistical Mechanics," Ph.D. thesis, University of Rochester (1978).
125. R. C. Churchill, G. Pecelli, and D. L. Rod, "A Survey of the Henon Heiles Hamiltonian with Applications to Related Examples," in *Stochastic Behavior in Classical and Quantum Hamiltonian Systems,* ed. G. Casati and J. Ford (New York: Springer, 1979). *Lecture Notes in Physics* 93 (1979): 76–136.
126. C. R. Holt, "Construction of New Integrable Hamiltonians in Two Degrees of Freedom," *Journal of Mathematical Physics* 23 (1982): 1037–1046, and references therein.
127. L. S., "A Theory of Exact and Approximate Configurational Invariants," *Physics* 8D (1983): 90–116.
128. E. Hille, *Ordinary Differential Equations in the Complex Domain* (New York: Wiley-Interscience, 1976), chap. 3.
129. S. Kovalevskaya, "Sur une propriété du system d'équations différentielles qui définit la rotation d'un solide autour d'un point fixe," *Acta Mathematica* 14 (1890): 81–93.
130. J. M. Greene and I. C. Percival, "Hamiltonian Maps in the Complex Plane," *Physica* 3D (1981): 530–548.
131. I. C. Percival, "Chaotic Boundary of a Hamiltonial Map," *Physica* 6D (1982): 67–77.
132. U. Frisch and R. Morf, "Intermittency in Nonlinear Dynamics and Singularities at Complex Times," *Physical Review A* 23 (1981): 2673–3705.
133. Another example is the Lorenz system studied by M. Tabor and J. Weiss, "Analytical Stucture of the Lorenz System," *Physical Review A* 24 (1981): 2157–2167.
134. K. S. J. Nordholm and S. A. Rice, "Quantum Ergodicity and Vibrational Relaxation in Isolated Molecules," *Journal of Chemical Physics* 61 (1974): 203–223.
135. K. S. J. Nordholm and S. A. Rice, "Quantum Ergodicity and Vibrational Relaxation in Isolated Molecules. II. λ-Independent Effects and Relaxation to the Asymptotic Limit," *Journal of Chemical Physics* 61 (1974): 768–779.
136. J. Dancz and S. A. Rice, "Large Amplitude Vibrational Motion in a One Dimensional Chain: Coherent State Representation," *Journal of Chemical Physics* 67 (1977): 1418–1426.
137. T. J. Rolfe and S. A. Rice, "Simulation Studies of the Scattering of a Solitary Wave by a Mass Impurity in a Chain of Nonlinear Oscillators," *Physica* 1D (1980): 375–382.
138. E. J. Heller, "Quantum Intramolecular Dynamics: Criteria for Stochastic and Nonstochastic Flow," *Journal of Chemical Physics* 72 (1980): 1337–1347.
139. E. J. Heller, "Potential Surface Properties and Dynamics from Molecular Spectra: A Time-Dependent Picture," in *Potential Energy Surfaces and*

Dynamics Calculations, ed. D. G. Truhlar (New York: Plenum, 1981), pp. 103–131.

140. E. L. Sibert III, W. P. Reinhardt, and J. T. Hynes, "Intramolecular Vibrational Relaxation of CH Overtones in Benzene," *Chemical Physics Letters* 92 (1982): 455–458.

141. J. S. Hutchinson, W. P. Reinhardt, and J. T. Hynes, "Nonlinear Resonances and Vibrational Energy Flow in Model Hydrocarbon Chains," *Journal of Chemical Physics* 79 (1983): 447–4260.

142. Proceedings of the 1981 American Conference on Theoretical Chemistry, *Journal of Physical Chemistry* 86 (1982): 2111–2268.

143. P. Brumer and M. Shapiro, "Intramolecular Dynamics: Time Evolution of Superposition States in the Regular and Irregular Spectrum," *Chemical Physics Letters* 72 (1980): 528–532.

144. M. J. Davis, E. B. Stechel, and E. J. Heller, "Quantum Dynamics in Classically Integrable and Non-Integrable Regions," *Chemical Physics Letters* 76 (1980): 21–26.

145. K. G. Kay, "Numerical Study of Intramolecular Vibrational Energy Transfer: Quantal, Classical, and Statistical Behavior," *Journal of Chemical Physics* 72 (1980): 5955–5975.

146. J. S. Hutchinson and R. E. Wyatt, "Quantum Ergodicity for Time-Dependent Wave-Packet Dynamics," *Physical Review A* 23 (1981): 1567–1584.

147. R. Kosloff and S. A. Rice, "The Influence of Quantization on the Onset of Chaos in Hamiltonian Systems: The Kolmogorov Entropy Interpretation," *Journal of Chemical Physics* 74 (1981): 1340–1349.

148. S. A. Rice and R. Kosloff, "Is Dynamical Chaos the Same Phenomenon in Classical and Quantum Mechanical Hamiltonian Systems?" *Journal of Physical Chemistry* 86 (1982): 2153–2157.

149. P. Pechukas, "Kolmogorov Entropy and 'Quantum Chaos,'" *Journal of Physical Chemistry* 86 (1982): 2239–2243.

150. P. Pechukas, "'Quantum chaos' in the Irregular Spectrum," *Chemical Physics Letters* 86 (1982): 553–557.

151. M. Shapiro and M. S. Child, "Quantum Stochasticity and Unimolecular Decay," *Journal of Chemical Physics* 76 (1982): 6176–6185.

152. M. Shapiro and P. Brumer, "Time Evolution with a Regular or Irregular Spectrum, Revisited," *Chemical Physics Letters* 90 (1982): 481–483.

153. E. J. Heller and E. B. Stechel, "Quantum Correspondence to Classical Chaos," *Chemical Physics Letters* 90 (1982): 484–485.

154. G. Casati, B. V. Chirikov, F. M. Izraedev, and J. Ford, "Stochastic Behavior of a Quantum Pendulum under a Periodic Perturbation," in *Stochastic Behavior in Classical and Quantum Hamiltonian Systems*, ed. G. Casati and J. Ford (New York: Springer, 1979), pp. 334–352.

155. M. L. Koszykowski, D. W. Noid, M. Tabor, and R. A. Marcus, "On Correlation Functions and the Onset of Chaotic Motion," *Journal of Chemical Physics* 74 (1981): 2530–2535.

156. J. T. Muckerman, D. W. Noid, and M. S. Child, "Local Modes and Cross Correlation Functions," *Journal of Chemical Physics* 78 (1983): 3981–3989.

157. G. Casati and I. Guarneri, "Chaos and Special Features of Quantum Systems under External Perturbations," *Physical Review Letters* 50 (1983): 640–643.

158. G. Casati, "Irreversibility and Chaos in Quantum Systems," in *Chaotic Behavior in Quantum Systems*, ed. G. Casati (New York: Plenum, in press).
159. G. Casati, F. Valz-Gris, and I. Guarneri, "On Correlation Function Methods for Detecting the Stochastic Transition," *Physica* 3D (1981): 644–648.
160. J. D. Crawford and J. R. Cary, "Decay of Correlations in a Chaotic Measure-Preserving Transformation," *Physica* 6D (1983): 223–232.
161. J. D. Meiss, J. R. Cary, C. Grebogi, J. D. Crawford, A. N. Kaufman, and H. D. I. Abarbanel, "Correlations of Periodic Area-Preserving Maps," *Physica* 6D (1983): 375–384.
162. I. C. Percival, "Regular and Irregular Spectra," *Journal of Physics B: Atomic and Molecular Physics* 6 (1973): L229–L232.
163. Percival, "Semiclassical Theory of Bound States," p. 43 (ref. 29).
164. For an application of this idea, see Noid, Koszykowski and Marcus, "Spectral Analysis Method," pp. 404–408 (ref. 85).
165. M. V. Berry, "Regular and Irregular Semiclassical Wavefunctions," *Journal of Physics A: Mathematical and General* 10 (1977): 2083–2091.
166. M. V. Berry, "Semi-Classical Mechanics in Phase Space: a Study of Wigner's Function," *Philosophical Transactions of the Royal Society of London. Series A. Mathematical and Physical Sciences* 287 (1977): 237–271.
167. See also Hutchinson and Wyatt, "Quantum Ergodicity," pp. 1567–1584 (ref. 146).
168. G. Benettin and J.-M. Strelcyn, "Numerical Experiments on the Free Motion of a Point Mass Moving in a Plane Convex Region: Stochastic Transition and Entropy," *Physical Review A* 17 (1978): 773–785.
169. Illustrations appear in, for example, D. W. Noid, M. L. Koszykowski, and R. A. Marcus, "Semiclassical Calculation of Bound States in Multidimensional Systems with Fermi Resonance," *Journal of Chemical Physics* 71 (1979): 2864–2873.
170. S. W. McDonald and A. N. Kaufman, "Spectrum and Eigenfunctions for a Hamiltonian with Stochastic Trajectories," *Physical Review Letters* 42 (1979): 1189–1191.
171. K. S. J. Nordholm and S. A. Rice, "Quantum Ergodicity and Vibrational Relaxation in Isolated Molecules," *Journal of Chemical Physics* 61 (1974): 203–223.
172. K. S. J. Nordholm and S. A. Rice, "A Quantum Ergodic Theory Approach to Unimolecular Fragmentation," *Journal of Chemical Physics* 62 (1975): 157–168.
173. R. M. Stratt, N. C. Handy, and W. H. Miller, "On the Quantum Mechanical Implications of Classical Ergodicity," *Journal of Chemical Physics* 71 (1979): 3311–3322.
174. E. J. Heller, E. B. Stechel, and M. J. Davis, "Molecular Spectra, Fermi Resonances, and Classical Motion," *Journal of Chemical Physics* 73 (1980): 4720–4735.
175. G. Hose and H. S. Taylor, "A Quantum Analog to the Classical Quasiperiodic Motion," *Journal of Chemical Physics* 76 (1981): 5356–5364.
176. M. D. Feit, J. A. Fleck, Jr., and A. Steiger, "Solution of the Schrödinger Equation by a Spectral Method," *Journal of Computational Physics* 47 (1982): 412–433.

177. M. V. Berry and M. Tabor, "Level Clustering in the Regular Spectrum," *Proceedings of the Royal Society of London A* 356 (1977): 375-394.
178. M. V. Berry, "Semiclassical Mechanics of Regular and Irregular Motion," in Proceedings of the 1981 "Les Houches" Summer School, vol. 36, *Chaotic Behavior in Deterministic Systems* (Amsterdam: North-Holland, 1983), pp. 171-272.
179. J. von Neuman and E. P. Wigner, "Über das Verhalten von Eigenwerten bei adiabatischen Prozessen," *Physikalische Zeitschrift* 30 (1929): 467-470.
180. G. M. Zaslavskii, "Statistics of Energy Levels when the Integrals of Motion are Violated," *Soviet Physics JETP* 46 (1978): 1094-1098.
181. N. Pomphrey, "Numerical Identification of Regular and Irregular Spectra," *Journal of Physics B: Atomic and Molecular Physics* 7 (1974): 1909-1915.
182. D. W. Noid, M. L. Koszykowski, M. Tabor, and R. A. Marcus, "Properties of Vibrational Energy Levels in the Quasiperiodic and Stochastic Regimes," *Journal of Chemical Physics* 72 (1980): 6169-6175.
183. A. R. Edmonds, R. A. Pullen, and I. C. Percival, "The Irregular Energy Spectrum of the Henon-Heiles System," *Chemical Physics Letters* 91 (1982): 273-276.
184. K. K. Lehmann, G. J. Scherer, and W. Klemperer, "Highly Excited HCN: The Inapplicability of Classical Dynamics," *Journal of Chemical Physics* 76 (1982): 6441-6442.
185. K. K. Lehmann, G. J. Scherer, and W. Klemperer, "Responses to 'Highly Excited States of HCN: The Probable Applicability of Classical Dynamics,' " *Journal of Chemical Physics* 78 (1983): 608-609.
186. D. Farrelly and W. P. Reinhardt, "Highly Excited States of HCN: The Probable Applicability of Classical Dynamics," *Journal of Chemical Physics* 78 (1983): 606-608.
187. R. A. Pullen and A. R. Edmonds, "Comparison of Classical and Quantal Spectra for the Hénon-Heiles Potential," *Journal of Physics A: Mathematical and General* 14 (1981): L319-L327.
188. R. A. Pullen and A. R. Edmonds, "Comparison of Classical and Quantal Spectra for a Totally Bound Potential," *Journal of Physics A: Mathematical and General* 14 (1981): L477-L484.
189. M. V. Berry, "Quantizing a Classically Ergodic System, Sinai's Billiard and the KKR Method," *Annals of Physics* 131 (1981): 163-216.
190. P. J. Richens and M. V. Berry, "Pseudointegrable Systems in Classical and Quantum Mechanics," *Physics* 2D (1981): 495-512.
191. O. Bohigas, M.-J. Giannoni, and C. Schmit, "Characterization of Fluctuations of Chaotic Quantum Spectra," in *Chaotic Behavior in Quantum Systems*, ed. G. Casati (New York: Plenum, in press).
192. See M. L. Mehta, *Random Matrices and the Statistical Theory of the Energy Levels* (New York: Academic Press, 1967).
193. R. U. Haq, A. Pandey, and O. Bohigas, "Fluctuation Properties of Nuclear Energy Levels: Do Theory and Experiment Agree?" *Physical Review Letters* 48 (1982): 1086-1089.
194. H. S. Camarda and P. D. Georgopoulos, "Statistical Behavior of Atomic Energy Levels: Agreement with Random-Matrix Theory," *Physical Review Letters* 50 (1983): 492-495.

195. J. Schaefer and R. Yaris, "Random Matrix Theory and Nuclear Magnetic Resonance Spectral Distribution," *Journal of Chemical Physics* 51 (1969): 4469–4474.
196. M. L. Zimmerman, M. M. Kash, and D. Kleppner, "Evidence of an Approximate Symmetry for Hydrogen in a Uniform Magnetic Field," *Physical Review Letters* 45 (1980): 1092–1094.
197. W. P. Reinhardt and D. Farrelly, "The Quadratic Zeeman Effect in Hydrogen: An Example of Semi-Classical Quantization of a Strongly Non-Separable but almost Integrable System," *Journal de Physique* 43 colloque supplément C2 (1982): 29–43.
198. W. P. Reinhardt, "Semiclassical Quantization on Fragmented Tori," in *Chaotic Behavior in Quantum Systems*, ed. G. Castati (New York: Plenum, in press).
199. Although the boundary line in Figure 5-26 indicates the lack of correlation of the onset of classical chaos with an irregular spectrum, it is interesting to note that this boundary gives the approximate range of validity of a semiclassical SCF quantization of the problem (D. Farrelly and W. P. Reinhardt, unpublished).
200. A. Harada and H. Hasegawa, "Correspondence between Classical and Quantum Chaos for Hydrogen in a Uniform Magnetic Field," *Journal of Physics A: Mathematical and General* 16 (1983): L259–L263.
201. C. W. Clark and K. T. Taylor, "The Quadratic Zeeman Effect in Hydrogen Rydberg Series," *Journal of Physics B: Atomic and Molecular Physics* 13 (1980): L737–L743.
202. C. W. Clark and K. T. Taylor, "Diamagnetism in Excited States of Hydrogen," *Journal de Physique* (Paris) 43 colloque supplément C2 (1982): 127–135.
203. M. Robnik, "Hydrogen Atom in a Strong Magnetic Field: On the Existence of the Third Integral of Motion," *Journal of Physics A: Mathematical and General* 14 (1981): 3195–3216.
204. R. A. Marcus, "Intramolecular Dynamics: Regular and Stochastic Vibrational States of Molecules," in *Horizons of Quantum Chemistry*, ed. K. Fukui and B. Pullman (Boston: Reidel, 1979), pp. 107–121.
205. R. A. Marcus, "On the Theory of Intramolecular Energy Transfer," *Faraday Discussions of the Chemical Society* 75 (1983): 103–115.
206. R. A. Marcus, remarks made at the Como Quantum Chaos Conference, 1983, private communication.
207. J. L. Lebowitz, "Microscopic Dynamics and Macroscopic Laws," in *Long Time Predictions in Dynamics*, ed. C. W. Horton, Jr., L. E. Reichl, and V. G. Szebehely (New York: John Wiley & Sons, 1983), pp. 3–19.
208. Percival and Pomphrey, "Vibrational Quantization" (ref. 96).
209. Chapman, Garrett, and Miller, "Semiclassical Eigenvalues for Nonseparable Systems."
210. Jaffé and Reinhardt, "Time-Independent Methods in Classical Mechanics" (ref. 99).
211. F. G. Gustavson, "On Constructing Formal Integrals of a Hamiltonian System Near an Equilibrium Point," *Astronomical Journal* 71 (1966): 670–686.
212. A. Deprit, private communication, 1983.

213. A. Deprit, "Canonical Transformations Depending on a Small Parameter," *Celestial Mechanics* 1 (1969): 12–30.
214. A. Giorgilli, "A Computer Program for Integrals of Motion," *Computer Physics Communications* 16 (1979): 331–343.
215. G. Baker, Jr., *Essentials of Padé Approximates* (New York: Academic Press, 1975).
216. G. A. Baker, Jr. and P. Graves-Morris, *Padé Approximants*, Parts I and II in *Encyclopedia of Mathematics and Its Applications*, ed. G.-C. Rota (Reading, Mass: Addison-Wesley, 1981), vols. 13 and 14.
217. G. A. Baker, Jr. *Essentials of Padé Approximates*, pp. 128–131.
218. W. P. Reinhardt, "Padé Summations for the Real and Imaginary Parts of Atomic Stark Eigenvalues," *International Journal of Quantum Chemistry* 21 (1982): 133–146.
219. R. T. Baumel, J. C. Gammel, and J. Nuttall, "Placement of Cuts in Padé-Like Approximation (1)," *Journal of Computational and Applied Mathematics* 7 (1981): 135–140.
220. M. J. Ablowitz, B. A. Funk, and A. C. Newell, "Semi-Resonant Interactions and Frequency Dividers," *Studies in Applied Mathematics* 11 (1973): 51–74.
221. S. J. Shenker and L. P. Kadanoff, "'Band to Band Hopping in One-Dimensional Maps," *Journal of Physics A: Mathematical and General* 14 (1981): L23–L26.
222. N. de Leon and B. J. Berne, "Intramolecular Rate Process: Isomerization Dynamics and the Transition to Chaos," *Journal of Chemical Physics* 75 (1981): 3495–3510.
223. M. Robnik, "Quantizing a Generic Family of Billiards with Analytic Boundaries," *Journal of Physics A: Mathematical and General* 17 (1984): 1049–1074.
224. S. J. Shenker and L. P. Kadanoff, "Critical Behavior of a KAM Surface: I. Empirical Results," *Journal of Statistical Physics* 27 (1982): 631–656.
225. S. J. Shenker, "Scaling Behavior in a Map of a Circle onto Itself: Empirical Results," *Physica* 5D (1982): 405–411.
226. J. M. Greene, "A Method for Determining a Stochastic Transition," *Journal of Mathematical Physics* 20 (1979): 1183–1201.
227. Greene and Percival, "Hamiltonian Maps" (ref. 130).
228. J. B. Keller and S. I. Rubinow, "Asymptotic Solution of Eigenvalue Problems," *Annals of Physics* 9 (1960): 24–75.
229. A. N. Zemlyakov and A. B. Katok, "Topological Transitivity of Billiards in Polygons," *Mathematical Notes of the Academy of Sciences of the USSR* 18 (1976): 760–764.
230. N. Saito, H. Hirooka, J. Ford, F. Vivaldi, and G. H. Walker, "Numerical Study of Billiard Motion in an Annulus Bounded by Non-Concentric Circles," *Physica* 5D (1982): 273–286.
231. B. Eckhardt, "Theory of Classical Billiards in a Polygon," M.S. thesis, Georgia Institute of Technology (1982).
232. B. Eckhardt, J. Ford, and F. Vivaldi, "Analytically Solvable Dynamical Systems which are not Integrable," (preprint, 1983).
233. J. Ford, "How Random Is a Coin Toss?" in *Long Time Predictions in Dynam-*

ics, ed. C. W. Horton, Jr., L. E. Reichl, and V. G. Szebehely (New York: John Wiley & Sons, 1983), pp. 79–92.

234. This has been confirmed by P. S. Richens, private communication, 1983.
235. W. H. Miller, "Semiclassical Treatment of Multiple Turning-Point Problems— Phase Shifts and Eigenvalues," *Journal of Chemical Physics* 48 (1968): 1651– 1658.
236. M. J. Davis and E. J. Heller, "Quantum Dynamical Tunneling in Bound States," *Journal of Chemical Physics* 75 (1981): 246–254.
237. D. W. Noid, M. L. Koszykowski, and R. A. Marcus, "Comparison of Quantal, Classical, and Semiclassical Behavior at an Isolated Avoided Crossing," *Journal of Chemical Physics* 78 (1983): 4018–4024.
238. T. Uzer, D. W. Noid, and R. A. Marcus, "Uniform Semiclassical Theory of Avoided Crossings," *Journal of Chemical Physics* 79 (1983): 4412–4425.
239. D. Farrelly and W. P. Reinhardt, unpublished.
240. F. Borondo and W. P. Reinhardt, unpublished.
241. See also W. P. Reinhardt, "Chaotic Dynamics" (ref. 69).
242. In the pure Coulomb case, presence of a homogeneous E field does not destroy separability [see, for example, L. D. Landau and E. M. Lifshitz, *Quantum Mechanics* (Reading, Mass: Addison-Wesley, 1965), p. 269]. The Coulomb-Stark problem is thus integrable in both classical and quantum mechanics: A quantum signature of this is exact crossings of levels of the same spin and parity as a function of field. These crossings are transformed into avoided crossings as soon as the $O(4)$ Coulomb symmetry is broken, as is always the case in nature.
243. M. G. Littman, M. C. Zimmerman, T. W. Ducas, R. R. Freeman, and D. Kleppner, "Structure of Sodium Rydberg Status in Weak to Strong Electric Fields," *Physical Review Letters* 36 (1978): 788–791.
244. P. Pechukas, " 'Quantum Chaos' in the Irregular Spectrum," *Chemical Physics Letters* 86 (1982): 553–557.

6

The Adiabatic Method in the Theory of Many-Body Systems

Vladimir Z. Kresin

Materials and Molecular Research Division
Lawrence Berkeley Laboratory

William A. Lester, Jr.

Materials and Molecular Research Division
Lawrence Berkeley Laboratory
and
Department of Chemistry
University of California, Berkeley

ABSTRACT: The adiabatic approach is presented as a method in quantum many-body theory. Recent advances in adiabatic theory for solids, molecules, and nuclei are described. The application of the adiabatic methods to systems with comparable masses is presented and applied to the problem of nuclear motion of polyatomic systems. Different types of nonadiabatic interactions (e.g., the electron-phonon interaction in metals, the pair correlation in superconducting systems, and the electron-vibrational interaction in molecules) are discussed. The continuity of electron and phonon spectra in solids and the potential energy surface crossings in molecular systems are shown to lead to large nonadiabaticity.

6-1 INTRODUCTION

An investigation of real many-body systems such as solids, molecules, or nuclei is connected with serious difficulties. The interactions between particles are not weak, and therefore the usual perturbation theory cannot be applied. It is necessary to develop different methods to approach the problem. This chapter contains a description of the so-called adiabatic method, which has been successfully applied to many systems. A many-body system is characterized by many degrees of freedom, and usually some are "fast" relative to others. With the use of the adiabatic method one can take advantage of this fact.

Born and Oppenheimer[1] developed the classical adiabatic method that is the basis of solid state theory and the theory of molecules. There has been remarkable progress since 1927, and we attempt here to describe many of the significant achievements in the field.

Section 6-2 presents a rigorous description of the electron-phonon interaction in solids, a classification of different types of excitations in molecules, and a detailed treatment of nonadiabaticity. Section 6-3 is concerned with the analysis of nuclear dynamics for polyatomic photodissociation and chemical reaction. It also contains the adiabatic approach to collective motion of nucleons.

We emphasize that this chapter cannot be considered a complete analysis of solid state theory or the theory of molecules, and nuclei, although many aspects of these areas will be discussed here. We focus on the adiabatic method, and we describe the applications of this powerful method to different fields.

Section 6-2 presents a rigorous description of the electron-phonon interaction in solids, a classification of different types of excitations in molecules, and a detailed treatment of nonadiabaticity. Section 6-3 is concerned with the analysis of nuclear dynamics for polyatomic photodissociation and chemical reation. It also contains the adiabatic approach to collective motion of nucleons.

We emphasize that this chapter cannot be considered a complete analysis of solid state theory or the theory of molecules, and nuclei, although many aspects of these areas will be discussed here. We focus on the adiabatic method, and we describe the applications of this powerful method to different fields.

6-2   THE ADIABATIC METHOD IN SOLID STATE THEORY AND THE THEORY OF MOLECULES

Solids and molecules have many common features. First, both many-body systems contain electronic and nuclear subsystems and, as will be discussed, both are characterized by electronic and vibrational manifolds. They differ in the number of particles that are typically dealt with, but this circumstance is not critical for the applicability of the adiabatic method. The important similarity is that both systems consist of heavy particles (nuclei) and light, fast particles (electrons). Hamiltonians for solids and for molecules have similar structure. In this section we describe the adiabatic approach in solid state theory and in the theory of molecules; this approach is fundamental to the theory of both types of systems. We think that the description of the general theory combined with its application to selected systems provides the most useful presentation of the adiabatic method.

Adiabatic Method

The total wavefunction of the system $\Psi(\mathbf{r}, \mathbf{R})$ is the solution of the Schrödinger equation

$$H\Psi(\mathbf{r}, \mathbf{R}) = E\Psi(\mathbf{r}, \mathbf{R}). \qquad (6\text{-}1)$$

Here $\{\mathbf{r}\}$ and $\{\mathbf{R}\}$ denote sets of electronic and nuclear coordinates, respectively.

The total Hamiltonian in (6-1) is

$$H = \hat{T}_{\mathbf{r}} + \hat{T}_{\mathbf{R}} + V(\mathbf{r}, \mathbf{R}), \qquad (6\text{-}2)$$

where $\hat{T}$ and $\hat{T}_{\mathbf{R}}$ are the kinetic energy operators

$$\hat{T}_{\mathbf{r}} = -\sum_{\mathbf{r}} \frac{\hbar^2}{2m_e} \nabla_{\mathbf{r}}^2; \; \hat{T}_{\mathbf{R}} = -\sum_{\mathbf{R}} \frac{\hbar}{2M_{\mathbf{R}}} \nabla_{\mathbf{R}}^2, \qquad (6\text{-}3)$$

and

$$V(\mathbf{r}, \mathbf{R}) = V_1(\mathbf{r}, \mathbf{r}) + V_2(\mathbf{r}, \mathbf{R}) + V_3(\mathbf{R}, \mathbf{R})$$

is the total potential energy consisting of the electron-electron repulsion term V_1, the electron-nuclear attraction term V_2, and the nuclear-nuclear repulsion term V_3.

In the usual way,[1,2] one introduces a complete orthogonal set of electronic wavefunctions that are solutions of

$$[\hat{T}_{\mathbf{r}} + V(\mathbf{r}, \mathbf{R})] \, \psi_m(\mathbf{r}, \mathbf{R}) = \epsilon_m(\mathbf{R}) \, \psi_m(\mathbf{r}, \mathbf{R}). \qquad (6\text{-}4)$$

The functions $\psi_m(\mathbf{r}, \mathbf{R})$ and the eigenvalues depend parametrically on the nuclear (ion) positions.

The total wavefunctions $\Psi_n(\mathbf{r}, \mathbf{R})$, associated with the nth electronic state, is sought in the form of an expansion in the complete orthogonal set of solutions in the electronic Schrödinger equation (6-4); that is,

$$\Psi_n(\mathbf{r}, \mathbf{R}) = \sum_m \phi_{nm}(\mathbf{R}) \, \psi_m(\mathbf{r}, \mathbf{R}). \qquad (6\text{-}5)$$

Note that the quantities $\phi_{nm}(\mathbf{R})$ appear in (6-5) as the coefficients of the expansion. Substituting the expansion (6-5) into Equation (6-1), multiplying by $\psi_\alpha^*(\mathbf{r}, \mathbf{R})$, and integrating over $\mathbf{r}$ yields the following equation for the coefficients:

$$[\hat{T}_R + \epsilon_\alpha(\mathbf{R})]\, \phi_{n\alpha}(\mathbf{R}) + \sum_m C_{\alpha m}\phi_{nm}(\mathbf{R}) = E_n \phi_{n\alpha}(\mathbf{R}), \qquad \text{(6-6)}$$

where

$$C_{\alpha m} = -\sum_j - \frac{\hbar^2}{2M_j} \sum_{i=1}^{3} \int \psi_\alpha^* \left[2\frac{\partial^2 \psi_m}{\partial X_{ji}\partial X_{ij}} + \frac{\partial^2 \psi_m}{\partial X_{ji}^2} \right] d\mathbf{r}. \qquad \text{(6-7)}$$

The summation is taken over all heavy particles j and Cartesian coordinates i.

Let us assume that one can neglect the second term of the left-hand side of Equation (6-6). Then Equation (6-6) has the usual Schrödinger equation form; that is,

$$[\hat{T}_R + \epsilon_\alpha(\mathbf{R})]\, \phi_{n\alpha}(\mathbf{R}) = E_n \phi_{n\alpha}(\mathbf{R}). \qquad \text{(6-8)}$$

Thus $\phi_{n\alpha}(\mathbf{R})$ can be treated as a nuclear (ion) wavefunction and the electronic term $\epsilon_\alpha(\mathbf{R})$ [see Eq. (6-4)] appears as the potential energy for the motion of heavy particles.

If we also neglect the off-diagonal contributions to the expansion (6-5), we arrive at the final Born-Oppenheimer (BO) expression for the total wavefunction

$$\Psi_n(\mathbf{r}, \mathbf{R}) = \psi_n(\mathbf{r}, \mathbf{R})\, \phi_{n\nu}(\mathbf{R}), \qquad \text{(6-9)}$$

where $\phi_{n\nu}(\mathbf{R}) \equiv \phi_{nn\nu}(\mathbf{R})$ and ν denotes the set of quantum numbers describing nuclear motion.

Equations (6-4), (6-8), and (6-9) define the BO approximation. Note that two approximations [neglect of the second term of the left-hand side of Eq. (6-6) and off-diagonal terms in (6-5)] were made in order to obtain this result. Therefore we see that the BO expression (6-9) is not an exact solution of the total Schrödinger equation (6-1).

Nonadiabaticity—the additional "friction" between electrons and nuclei (ions)—neglected in obtaining the BO approximation is describable by an operator H'. We now evaluate H' and estimate the accuracy of the BO adiabatic approximation.[3,4]

The operator H' can be found in the following way.[5] The total Hamiltonian H is written as a sum of two terms,

$$H = H_0 + H'. \tag{6-10}$$

First, one can find the operator H_0, because the BO function (6-9) is its eigenfunction. It is easy to see that

$$H_0 = \hat{T}_{\mathbf{r}} + V(\mathbf{r}, \mathbf{R}) + H_{\mathbf{R}}^0, \tag{6-11}$$

where the operator $H_{\mathbf{R}}^0$ is defined by the relation

$$H_{\mathbf{R}}^0 \psi_n(\mathbf{r}, \mathbf{R}) \, \phi_{n\nu}(\mathbf{R}) = \psi_n(\mathbf{r}, \mathbf{R}) \, \hat{T}_{\mathbf{R}} \phi_{n\nu}(\mathbf{R}). \tag{6-12}$$

The function (6-8) is the eigenfunction of the operator H_0. Indeed,

$$\begin{aligned} H_0 \Psi_{BO}(\mathbf{r}, \mathbf{R}) &= [\hat{T}_{\mathbf{r}} + V(\mathbf{r}, \mathbf{R})] \, \psi_n(\mathbf{r}, \mathbf{R}) \, \phi_{n\nu}(\mathbf{R}) + H_{\mathbf{R}}^0 \psi_n(\mathbf{r}, \mathbf{R}) \, \phi_{n\nu}(\mathbf{R}) \\ &= \epsilon_n(\mathbf{R}) \, \psi_n(\mathbf{r}, \mathbf{R}) \, \phi_{n\nu}(\mathbf{R}) + \psi_n(\mathbf{r}, \mathbf{R}) \, \hat{T}_{\mathbf{R}} \phi_n(\mathbf{R}) \\ &= \psi_n(\mathbf{r}, \mathbf{R}) \, [\hat{T}_{\mathbf{R}} + \epsilon_n(\mathbf{R})] \, \phi_{n\nu}(\mathbf{R}) \\ &= E_{n\nu} \Psi_{BO}(\mathbf{r}, \mathbf{R}). \end{aligned} \tag{6-13}$$

Here we used Equations (6-4), (6-8), and (6-12). Hence, $\Psi_{BO}(\mathbf{r}, \mathbf{R})$ is an eigenfunction of the operator H_0 defined by Equation (6-11).

Based on Equations (6-10), (6-11), and (6-12), one can now determine the nonadiabatic operator H' from the relation

$$H' = \hat{T}_{\mathbf{R}} - H_{\mathbf{R}}^0.$$

Hence

$$H' \Psi_{BO}(\mathbf{r}, \mathbf{R}) = \hat{T}_{\mathbf{R}} \psi_n(\mathbf{r}, \mathbf{R}) \, \phi_{n\nu}(\mathbf{R}) - \psi_n(\mathbf{r}, \mathbf{R}) \, \hat{T}_{\mathbf{R}} \, \phi_{n\nu}(\mathbf{R}) \tag{6-14}$$

or

$$\begin{aligned} \langle m\nu' | H' | n\nu \rangle = &\int \psi_m^*(\mathbf{r}, \mathbf{R}) \, \phi_{m\nu'}^*(\mathbf{R}) \, \hat{T}_{\mathbf{R}} \psi_n(\mathbf{r}, \mathbf{R}) \, \phi_{n\nu}(\mathbf{R}) \, d\mathbf{R} \, d\mathbf{r} \\ &- \delta_{mn} \int \phi_{m\nu'}^*(\mathbf{R}) \, \hat{T} \phi_{n\nu}(\mathbf{R}) \, d\mathbf{R}. \end{aligned}$$

Let us estimate the accuracy of the BO approximation. Consider a molecule or a solid. The nuclear wavefunction $\phi_{n\nu}(\mathbf{R})$ is characterized by a magnitude of order a (a is the amplitude of vibrations), whereas the characteristic "length" L of the electronic wavefunction is the length of a molecular bond or the period of a crystal lattice. On this basis we obtain

$$H'\Psi_{BO}(\mathbf{r},\ \mathbf{R}) \sim -\frac{\hbar^2}{M}\frac{\partial\psi}{\partial\mathbf{R}}\frac{\partial\phi}{\partial\mathbf{R}}$$

$$\sim -\frac{\hbar^2}{M}\frac{\psi}{L}\frac{\phi}{a} \sim \frac{\hbar^2}{Ma^2}\frac{a}{L}\ \Psi_{BO} \sim \frac{a}{L}\ \hbar\omega\Psi_{BO};$$

that is,

$$H' \sim \frac{a}{L} \ll 1.$$

Hence the nonadiabaticity H' is proportional to a small parameter $\gamma = a/L$. Here we have used the relation

$$a \simeq (\hbar/M\omega)^{1/2}. \tag{6-15}$$

The estimation Equation (6-15) may be obtained from the expression $\hbar\omega = \hbar^2/Ma^2$ (particle in a box model) or from the equation $\hbar\omega \simeq M\omega^2 a^2$ (harmonic oscillator).

Using Equation (6-15) one can also express the parameter γ in terms of a mass ratio. Since the frequency $\omega = (K_n/M)^{1/2}$, where K_n is the force constant due to the Coulomb internuclear interaction, then $a^2 \simeq \hbar/(K_nM)^{1/2}$. In a similar way, one can show that $L^2 \simeq \hbar/(K_em)^{1/2}$. The Coulomb interaction does not depend on the mass factor and, hence, $K_e \simeq K_n$. This leads to the result

$$\gamma \simeq \kappa$$

where

$$\kappa = (m/M)^{1/4}. \tag{6-16}$$

Consider again Equation (6-14), which can be written in the form[6]

$$H'\Psi_{BO}(\mathbf{r},\ \mathbf{R}) = [H_\mathbf{R},\ \psi_n]\ \phi_{n\nu}(\mathbf{R}).$$

If we assume that the motion of heavy particles is classical where $[H_\mathbf{R},\ \psi_n]$ is the commutator of $H_\mathbf{R}$ and ψ_n then the commutator $i[H_\mathbf{R},\ \psi_n] \rightarrow (\partial\psi_n/\partial t) = (\partial\psi_n/\partial R)\,v$, where v is the velocity of the heavy particles. Hence the qualitative aspect of the adiabatic approximation is connected with the small velocity of the nuclei (ions) relative to the much larger velocity of the electrons. Averaging the total potential energy $V(\mathbf{r}, \mathbf{R})$ over the fast electronic motion, we obtain the effective potential $\epsilon_\alpha(\mathbf{R})$ [see Eq. (6-8)] for nuclear motion.

The slowness of the nuclear motion results (in the time-dependent picture) in a small change of the electron-nuclear interaction during the characteristic electronic time $\tau_e \simeq (\Delta\epsilon_{el})^{-1}$. The probability of a transition from a nondegenerate electronic state to an excited state is exponentially small. Thus the quantum numbers n and ν are essentially integrals of motion.

The previously described adiabatic method is a rigorous approach that provides a basis for the theory of molecules and solids. According to Equation (6-9), the total wavefunction can be written as a product of an electronic wavefunction $\psi_n(\mathbf{r}, \mathbf{R})$ and a nuclear function $\phi_{n\nu}(\mathbf{R})$. Note that the electronic and nuclear variables $\{\mathbf{r}\}$ and $\{\mathbf{R}\}$ cannot be separated even in the zeroth-order approximation (6-9); the electronic wavefunction depends on nuclear positions. We stress that many problems are connected with deviations from the BO approximation. From the practical point of view, the approach described, despite its orderliness, has a definite shortcoming. This shortcoming is connected with the impossibility of separation of the Hamiltonian into two terms (or fields), one dependent on $\mathbf{r}$ and the other on $\mathbf{R}$.

It turns out that it is possible to develop another version of the adiabatic theory, which can be applied in many cases. Often nuclei (or ions) are located near the equilibrium geometry, which leads to simplifications of the theory—the clamped adiabatic approximation. We consider this approximation separately for solids and molecules. Although both types of systems consist of electrons and nuclei, the number of particles associated with each differs in a significant way. As a result, a solid is characterized by a continuous spectrum, whereas the energy spectra of small molecules are discrete. This difference leads to qualitatively different behavior for the two classes of systems.

Adiabatic Method in Solid-State Theory

Electron-phonon interaction

Electron-phonon interaction (EPI) plays a key role in solid-state physics. Different kinetic phenomena, such as charge transfer (electronic conductivity), energy transfer (thermal conductivity), lattice stability, and superconductivity, are connected with EPI. The EPI is one of the nonadiabatic effects that describes the "friction" between electronic and nuclear subsystems.

It is worth noticing in advance that the problem of EPI is nontrivial. At first sight, one might expect that the effects caused by EPI would be small, because nonadiabaticity [see Equation (6-16)] is characterized by a small parameter $\kappa = (m/M)^{1/4}$. Sometimes this statement corresponds to reality.

For example, normal conductivity can be described by perturbation theory (see below) in accord with the smallness of the nonadiabatic parameter. However, there are some phenomena, such as mass renormalization and superconductivity, that appear to have a very large effect. For instance, the increase of the effective mass due to EPI can be of the order of $\sim 100\%$, by comparisons with experimental data. [Note that the dispersion relation $\epsilon = \epsilon(p)$ can be often written in the effective mass approximation $\epsilon = p^2/2m^*$, where m^* is the effective mass.] The appearance of such a large nonadiabaticity has to be explained by the theory.

In this section our main goal is to describe the effect of EPI on the properties of solids. First, we should introduce EPI. It means that we should introduce two independent fields: electron and phonon fields and the term in the Hamiltonian describing EPI. In accordance with this direction, the total Hamiltonian is divided into two parts. One is the zeroth-order Hamiltonian corresponding to the independent fields and the other is the term $H' \equiv H_{\text{EPI}}$. The total wavefunction is written in the form

$$\Psi^0(\mathbf{r}, \mathbf{R}) = \psi_1(\mathbf{r})\,\psi_2(\mathbf{R}). \tag{6-17}$$

The usual adiabatic method (see previous paragraph) is not appropriate for our purpose, because the wavefunction [see Eq. (6-9)] does not have the form (6-17).

In order to solve the problem, we can take advantage of an important feature of the ion system (an ion consists of a nucleus and the electrons of filled shells). Namely, the ion system is characterized by vibrational motion; that is, the ions are located near equilibrium (in the region of the order $\sim a \ll d$, where d is the period of the lattice).

Let us choose the zeroth-order Hamiltonian in the form[7, 8]

$$H_0 = H_{0r} + H_{0R}, \tag{6-18}$$

where

$$H_{0r} = \hat{T}_\mathbf{r} + V(\mathbf{r}, R_0), \tag{6-19}$$

$$H_{0R} = \hat{T}_\mathbf{R} + \epsilon_2^{(n)}, \tag{6-20}$$

$$\epsilon_2^{(n)} = \frac{1}{2} \sum_{i\alpha, j\beta} \left. \frac{\partial^2 \epsilon_n(\mathbf{R})}{\partial X_{i\alpha} \partial X_{j\beta}} \right|_{\mathbf{R}_0} \Delta X_{i\alpha} \Delta X_{j\beta}. \tag{6-21}$$

Here $\mathbf{R}_0$ denotes the coordinates of the equilibrium position, $\Delta X_{i\alpha}$ is the deviation from equilibrium, and the operators $\hat{T}_\mathbf{r}$ and $\hat{T}_\mathbf{R}$ and the energy terms $\epsilon_n(\mathbf{R})$ are defined by Equations (6-3) and (6-4). Based on Equations

(6-4) and (6-8), one can see that the eigenfunction of the Hamiltonian H_0 is

$$\Psi^0(\mathbf{r}, \mathbf{R}) = \psi_n(\mathbf{r}, \mathbf{R}_0)\, \phi_{n\nu}(\mathbf{R}). \qquad (6\text{-}22)$$

Here $\psi_n(\mathbf{r}, \mathbf{R}_0)$ is the solution of Equation (6-4), and $\phi_{n\nu}(\mathbf{R})$ satisfies Equation (6-8) [in the harmonic approximation, see Eq. (6-21)]; that is, the function $\phi_{n\nu}(\mathbf{R})$ is a product of linear harmonic oscillator wavefunctions corresponding to different normal phonon modes. The frequencies of these normal modes can be evaluated from the adiabatic force constant matrix (6-21). Strongly speaking, the equilibrium position also depends on n, but this dependence can be neglected for large many-body systems such as solids. In other words, the shift of the equilibrium position caused by electronic transitions is negligibly small. Note that the total energy $E_{n\nu}^0$ corresponding to the operator H^0 is

$$E_{n\nu}^0 = \epsilon_n^0 + \epsilon_{\text{vib}}, \qquad (6\text{-}23)$$

where $\epsilon_n^0 \equiv \epsilon_n(\mathbf{R}_0)$ is the electronic energy, and

$$\epsilon_{\text{vib}} = \sum_{q,\lambda} \left(N_{q\lambda} + \frac{1}{2} \right) \hbar\omega_{q\lambda}$$

(the sum is over all phonon branches and phonon momenta).

The function (6-22) has the desirable form (6-17). The Hamiltonian H' decribing EPI can be found by subtraction of H_0 from the total Hamiltonian H [see Eq. (6-2)]; that is,

$$H' = H - H_0 = V(\mathbf{r}, \mathbf{R}) - V(\mathbf{r}, \mathbf{R}_0) - \epsilon_2^{(n)} \qquad (6\text{-}24)$$

or

$$H' = H_1 + \tilde{H}_2 + H_3 + \cdots, \qquad (6\text{-}25)$$

where

$$H_1 = \sum_{i\alpha} \left. \frac{\partial V(\mathbf{r}, \mathbf{R})}{\partial X_{i\alpha}} \right|_{\mathbf{R}_0} \Delta X_{i\alpha}, \qquad (6\text{-}26)$$

$$\tilde{H}_2 = H_2 - \epsilon_2^{(n)}, \qquad (6\text{-}27)$$

$$H_2 = \frac{1}{2} \sum_{\substack{i\alpha \\ j\beta}} \frac{\partial^2 V(\mathbf{r}, \mathbf{R})}{\partial X_{i\alpha}\partial X_{j\beta}} \Delta X_{i\alpha}\, \Delta X_{j\beta}, \qquad (6\text{-}28)$$

$\epsilon_2^{(n)}$ is defined by Equation (6-21), and the terms H_3, H_4, $\cdots$ contain higher-order displacements $\Delta X_{i\alpha}$.

It is worth noting that until recently EPI had been described by a semi-intuitive so-called Fröhlich model. This model is based on several assumptions. Note that in the Fröhlich model EPI is described by term (6-27) only.

Hence Equations (6-18)–(6-28) describe the electron and phonon fields, their energies, and EPI. Let us estimate the order of different terms in (6-25). We start with H_2 : $H_2 \sim \epsilon_2^{(n)} \sim \hbar\omega$. Note that each additional factor $\Delta X_{i\alpha}$ corresponds to the parameter $a/d \sim \kappa \ll 1$, where a is the amplitude of vibrations and d is the period of the lattice. Hence

$$H_1 \sim \hbar\omega/\kappa,$$

$$\tilde{H}_2 \sim \hbar\omega,$$

$$H_3 \sim \kappa\hbar\omega, \quad \text{and so on.} \tag{6-29}$$

We come to a very important finding. Only the terms H_s ($S \geq 1$) contain the small parameter κ. The term H_2 does not contain this parameter, and the leading term is proportional to a large quantity $\sim \hbar\omega/\kappa$. Hence, one can introduce the independent electron and phonon fields; however, EPI is not small and, generally speaking, cannot be described by perturbation theory. This is a main feature of EPI. We will see that, nevertheless, the corrections owing to EPI appear to be small for some phenomena but large for others.

The energy of noninteracting electron and phonon fields is described by Equation (6-23). The expression (6-25) enables us to evaluate the correction connected with EPI. It is important that the diagonal term $H_{1|nn}$ vanishes because $\partial V/\partial X_{i\alpha|nn;\mathbf{R}_0} = \partial\epsilon_n/\partial X_{i\alpha|\mathbf{R}_0} = 0$ and, hence the correction $\sim \hbar\omega/\kappa$ does not appear. The corrections ($\sim \hbar\omega$) come from the first order of $\tilde{H}_2$ and from the second order of H_1. It was proven that these terms cancel each other, and as a result the correction due to EPI appears to be small ($\sim \kappa^2\hbar\omega$).[8]

The conductivity of normal metals

In the previous section we noted that the diagonal term $H_{1|nn} = D$ is equal to zero. If we are interested in quantum transitions of electrons and phonons, it is necessary to take into account nondiagonal terms that differ from zero. We obtain

$$|\nabla_{\mathbf{R}i}V(\mathbf{r}, \mathbf{R})|_{nm;\mathbf{R}_0} = e^{i\mathbf{q}\cdot\mathbf{R}_{i0}}\mathbf{U}_{\mathbf{k},\mathbf{k}'},$$

where $q = k' - k$, and

$$U_{kk'} = - \int \psi^*(r_i) \nabla V(r_i) \, \psi_k(r_i) \, dr_i. \tag{6-30}$$

Here $r_i = r - R_{0i}$, $\nabla V(r_i) = U(r - R_{0i})$ is the Coulomb potential describing the electron-ion interaction, and $\psi_k(r) = U_k(r) e^{ik \cdot r}$ is Bloch's electron function in a crystal, and k is the electron's momentum. (We do not consider unklapp processes.)

In order to describe properties of such a many-body system as a solid, it is convenient to use second quantization formalism. Then the operator H_1, which plays the main role in the theory of conductivity, can be written in the form

$$H' = \sum_{k,k',k} \left| H_1 \right|_{k',k,q} a_{k'}^+ a_k b_q + \text{c.c.},$$

where $a_{k'}^+$ and a_k are the operators of creation and annihilation of the electron, b_q is the Bose operator describing the annihilation of the phonon, and the operator H_1 is defined by Equation (6-27) and c.c. denotes the complex conjugate.

As is well known (see, e.g., reference 9), the conductivity connected with electron-phonon scattering is described by the matrix element of the transition $k \to k'$, which is accompanied by the absorption (or the emission) of the phonon. The evaluation of this matrix element leads to the expression[4, 10]

$$\left| H_{1kk'} \right|_{N_{q\lambda}, N_{q\lambda} \pm 1} = \mp I M_{kq\lambda} (N_{q\lambda} \mp \tfrac{1}{2} + \tfrac{1}{2})^{1/2} \delta_{k', k + q} \tag{6-31}$$

Here $N_{q\lambda}$ is a number of phonons with the momentum q (λ is the index of the polarization),

$$I = \frac{\hbar \omega_{q\lambda}}{\epsilon_{k'} - \epsilon_k}. \tag{6-32}$$

ϵ_k and $\hbar \omega_{q\lambda}$ are the electron and phonon energies and

$$M_{kq\lambda} = \sqrt{\frac{\hbar N_i}{2 M \omega_{q\lambda}}} (e_{q\lambda} U_{k, k + k}) \tag{6-33}$$

is termed Bloch's matrix element [$e_{q\lambda}$ is a unit vector of the phonon's polarization and N_i is the number of ions; $U_{kk'}$ is defined by Eq. (6-30)].

The theory of conductivity deals with the scattering of electrons by real thermal phonons. (The effect of virtual phonons on the properties of solids will be considered next.) Then the energy is conserved and $I = 1$. Hence the collision integral appearing in the kinetic equation of the theory of conductivity contains the transition matrix element (6-31) with $I = 1$.

Renormalization of Fermi velocity and effective mass

Based on Equations (6-25)–(6-28), one can evaluate the adiabatic correction to the energy (see above). As was noted, this correction appears to be small ($\sim \kappa^2 \hbar\omega$) because of the cancellation of terms in H_1 and H_2.

With the use of the relation $\mathbf{V} = \partial\epsilon/\partial\mathbf{p}$, one can calculate the shift (renormalization) $\mathbf{V} = \partial(\Delta\epsilon)/\partial\mathbf{p}$ of the Fermi velocity V_F. It turns out that the cancellation does not take place in this case, and the renormalization of V_F is large.[11, 12] One can obtain the expression

$$V_F^r = V_F^0(1 + \zeta)^{-1}, \tag{6-34}$$

where

$$\zeta = M_{kq\lambda}^2 V m p_F / \omega_{q\lambda} 2\pi^2 \tag{6-35}$$

is the Fröhlich parameter. This parameter does not contain any parameters of smallness and, in principle, can be of order of unity or even greater (e.g., for Pb, $\zeta \simeq 0.7$).

The renormalization of the Fermi velocity is connected with the renormalization of the effective mass. Indeed, the electron energy ξ, referred to the Fermi level ϵ_F, is given by

$$\xi = (p^2 - p_0^2)/2m^*.$$

Here m^* is the effective mass (we use the isotropic effective mass approximation). If we consider the states near ϵ_F, we can write

$$\xi \simeq V_F(p - p_0) = p(p - p_0)/m^*.$$

One can see that the renormalization of V_F means a corresponding renormalization of the effective mass m^*. According to Equation (6-34), we obtain

$$m^* = m_b(1 + \zeta),$$

where m_b is the "band" value of the effective mass obtained by neglecting the EPI.

The value of the effective mass can be measured experimentally, for example, by the investigation of the cyclotron effect;[13-15] and the large effect of renormalization previously described has been observed. This increase of the effective mass is caused by a large nonadiabatic contribution.

Such a large value of the nonadiabatic contribution arises for two reasons. First, as was noted, the term H_1 [Eq. (6-27)] is not small ($\sim \hbar\omega/\kappa$). On the other hand, the renormalization is caused not by EPI with real thermal phonons as in the theory of conductivity, but by EPI with virtual phonons (that is why this effect does not disappear at $T = 0°$). If we consider virtual processes, the conservation of energy does not hold. If we consider electron transitions that correspond to the condition $\epsilon_{k'} - \epsilon_k < \hbar\omega_D$ (ω_D is the Debye frequency), then we obtain a noticeable increase of the amplitude of the transition [see Eqs. (6-32) and (6-33)]. Hence the domain of the width $\sim \omega_D$ near ϵ_F plays an important role, because EPI with virtual phonons accompanied by electron transitions in this region is characterized by a very large intensity.

This strong nonadiabatic behavior in the region $\Delta\epsilon < \omega_D$ near the Fermi surface is connected with the previously mentioned characteristic of solids, that is, that the electron and phonon spectra are continuous. In molecular systems such a situation can appear for special nuclear configurations only, for example, near potential crossings.

The described renormalization effect takes place at $T = 0°$. If $T \neq 0°$, then an additional factor owing to the appearance of thermal phonons should be taken into account. When this is done, the electron effective mass becomes a temperature-dependent quantity.[14, 16, 17] This effect has been observed experimentally.[18, 19] The relevant phenomenon is the deviation of the electronic heat capacity from linear behavior.[17, 20-23] The analysis of the dependence of $C_{el}(T)$ enables one to obtain detailed information about EPI.[17] For example, it was established that for Va the EPI with the transverse phonon branches considerably exceeds the interaction with the longitudinal phonons.

Renormalization of the phonon frequency. Stability of the lattice

We have considered the shift (renormalization) of the Fermi velocity and the effective mass. Let us now turn to evaluation of the shift of the phonon frequency. The problem of the influence of EPI on the phonon frequency is not new and is characterized by an interesting development. According to the Fröhlich model, which until recently was considered the starting point for studies of EPI, the phonon frequency is given by $\omega(q) = \omega_0(q)$

$\sqrt{1 - 2\zeta}$, where ζ is the Fröhlich parameter [see Eq. (6.35)] and $\omega_0(q)$ is the zeroth-order frequency corresponding to the free phonon field. One can easily see that if $\zeta > \frac{1}{2}$, the frequency $\omega(q)$ becomes imaginary. This result means that the lattice can be stable if it does not exceed the critical value $\zeta_c = \frac{1}{2}$. The greater values result in imaginary frequencies that lead to the lattice instability caused by the EPI. It was a valuable conclusion, particularly in the theory of superconductivity. (The superconducting state will be discussed next.) The critical temperature T_c for the transition to the superconducting state depends directly on ζ, and the condition $\zeta < \frac{1}{2}$ was considered a reason for the limitation of T_c. However, this conclusion is inconsistent with experimental data.

The Fröhlich model is based on two assumptions. First, it is assumed that EPI describes the Hamiltonian H_1 [Eq. (6-27)], and second, it is assumed that $\omega_0(q)|_{q \to 0} = uq$ (u is the velocity of the sound); that is, the bare phonon frequency (in the absence of H_1) is described by the usual acoustic law. However, according to rigorous adiabatic theory, these two assumptions cannot be combined. The zeroth-order phonon frequency can be chosen to have the form $\omega(q)|_{q \to 0} = uq$ [see Eq. (6-21)], but then the zeroth-order Hamiltonian is not given by H_1 but by H' [see Eq. (6-25)]. On the other hand, if one chooses H_1 as a zeroth-order Hamiltonian, then the force matrix will be different from the adiabatic matrix (6-25) and the dispersion relation will be different. In particular, the limiting value $\omega|_{q \to 0}$ will be equal to $\omega_0 = \text{const}$, where ω_0 is the plasma frequency. This result has been obtained in the bare ions model.[24-27] The EPI results in the transformation of the plasmon frequency into an acoustic law, but not in the instability of the lattice.

To evaluate the shift of the adiabatic frequency caused by the EPI, one should calculate the variational derivative $\delta \Delta E'/\delta N_{q\lambda}$, where E' is the correction to the energy of the system owing to EPI and $N_{q\lambda}$ is the number of phonons. This calculation[7, 10] leads to a small shift $\Delta \omega_{q\lambda} \sim \kappa^2 \omega_{q\lambda}$. Hence the expression obtained in the Fröhlich model and the conclusion about the instability of the lattice, if $\zeta > \frac{1}{2}$, are incorrect.

The conclusion that the EPI leads to a small renormalization of the adiabatic frequency does not mean that the lattice is stable for any strength of the EPI. A more detailed analysis of the EPI (see, e.g., reference 28) shows that some value of the phonon frequency, which does not belong to the region $q \to 0$, vanishes, and it is accompanied by a structural transition. The corresponding value of ζ should be very large ($\zeta \gg 1$). The total solution of this problem has not been obtained.

We have discussed the problem of lattice stability for usual three-dimensional systems. It turns out that the situation for one-dimensional or two-dimensional systems is entirely different. The EPI results in lattice instability, and it is possible to observe a specific type of structural transition.

Consider a one-dimensional crystal. An electron moving in the field formed by the one-dimensional lattice is characterized by the dispersion relation $E(p)$, where p is the quasi-momentum. The correction to the total energy owing to the EPI is (see, e.g., reference 29)

$$\Delta E = \sum \frac{|V_\kappa|^2}{E(p) - E}. \tag{6-36}$$

In the region where the denominator of Equation (6-36) is small, the correction is not applicable. One should use instead perturbation theory for degenerate levels. As a result of this approach, one can get two branches

$$E = \frac{1}{2}(E_1 + E_2) \pm \left[\frac{(E_1 - E_2)^2}{2} + V_\kappa^2 \right]^{1/2}$$

separated by the energy gap.

Assume that the one-dimensional crystal contains N ions and N free electrons. Then only half of the energy band will be filled and, according to band theory, the system will display the metallic properties. An interesting situation appears if we consider a small shift of each odd ion. This shift leads to the change of a lattice period, which then becomes $2a$. Correspondingly, the period of the inverse lattice decreases: $g_{2a} = \pi/2a$. The correction to the total energy caused by the EPI can be evaluated from Equation (6-36); it is only necessary to make replacement $g_a \to g_{2a}$. Then the EPI results in the appearance of the energy gap and consequently a metal-insulator transition. Hence, a one-dimensional metallic state is unstable with respect to the shift described (a doubling of the period). The described transition is called a Peierl's instability and represents a giant nonadiabatic phenomenon.[29]

An analogous type of instability appears in three-dimensional systems if the Fermi surface is characterized by the presence of plane parts. Then we deal with the same type of the degeneracy as in the case of Peierl's instability. However, a detailed description of structural transitions caused by the EPI is beyond the purpose of this paper.

Exchange of virtual phonons. Superconducting state.

In this section we consider a nonadiabatic effect that appears to be very large. It is well known that the Coulomb interaction can be described as an exchange of phonons between electrons. In a similar way one can consider the exchange of phonons to consist of the emission of a phonon by one electron followed by the absorption of this phonon by another electron. As a result, an additional effective interaction between electrons is obtainable.

Let us estimate the strength of this interaction.[7,8] Note that the matrix elements describing this interaction [see Equation (6-37)] contain nondiagonal terms of the operator H_1 [see Eq. (6-27)]. These nondiagonal terms are nonzero and, therefore, the term H_1 plays a principal role.

The matrix elements of interest $M_{p_1, p_2}^{p_3 p_4}$, where p_1, p_2 are initial momenta, and p_3, p_4 are final momenta) can be written in the form

$$M_{p_1, p_2; p_3, p_4} = \frac{\left| H_1 \right|_{p_1, p_2; N_{q\lambda}=0}^{p_3, p_2; N_{q\lambda}=1} \left| H_1 \right|_{p_3, p_2 1; N_{q\lambda}=1}^{p_3, p_4; N_{q\lambda}=0}}{\epsilon_{p_1} - \epsilon_{p_3} - \omega_{q\lambda}}$$

$$+ \frac{\left| H_1 \right|_{p_1, p_2; N_{-q\lambda}=0}^{p_1, p_4; N_{-q\lambda}=1} \left| H \right|_{p_1, p_4 1; N_{-q\lambda}=1}^{p_3, p_4; N_{-q\lambda}=0}}{\epsilon_{p_2} - \epsilon_{p_4} - \epsilon_{q\lambda}}. \qquad (6\text{-}37)$$

There are the following conservation rules:

$$p_3 - p_1 = p_2 - p_4 = q$$

where q is the phonon momenta and

$$\epsilon_{p_1} + \epsilon_{p_2} = \epsilon_{p_3} + \epsilon_{p_4}.$$

We consider the exchange of the virtual phonons so that conservation of energy for the elementary act of the electron-phonon interaction does not hold. The main contribution to the exchange interaction comes from the nonadiabatic region

$$\epsilon_{p_1} - \epsilon_{p_3} < \omega_D,$$

that is, from the region that is characterized by a small change of the electron energy. (The existence of such a region is a consequence of the continuous nature of the electron and phonon spectra.) Indeed, let us estimate the value of the matrix element (6-37) in this region. Using Equation (6-29), we obtain

$$M \sim \frac{(\hbar\omega/\kappa)^2}{\omega} = \frac{\hbar\omega}{2} = \frac{\hbar\omega}{(\hbar\omega/\epsilon_F)} = \epsilon_F.$$

Hence the strength of the exchange interaction is very large ($\sim \epsilon_F$). This large value is caused by the large contribution of H_2 [Eq. (6-29)]; the main contribution comes from the nonadiabatic domain $\Delta\epsilon_{el} < \omega_D$.

The large value of the exchange interaction shows that considerations based on Equation (6-37) can only lead to a qualitative approach. A rigorous analysis of such a large effect cannot be based on perturbation theory.

Note that the exchange interaction described forms the basis of the theory of superconductivity.[30] In this connection, one should note that the theory of superconductivity is based on the methods different from the perturbation theory. The diagrammatic method is a most powerful method that has been applied to the theory of superconductivity. The description of this method falls outside of the framework of this paper and we give only selected references.[12,31] For completeness we write out the main equation of the theory of superconductivity Eliashberg's equation,[32]

$$\Delta(p, \omega_n)$$

$$= \frac{\pi T}{Z} \zeta \sum_{\omega_{n'}} \int dq \, \frac{\omega^2(\mathbf{q})}{\omega^2(\mathbf{q}) + (\omega_n - \omega_{n'})^2} \frac{\Delta(\mathbf{p'}, \omega_{n'})}{\omega_{n'}^2 + \xi^2(\mathbf{p}) + \Delta^2(\mathbf{p'}, \omega_{n'})}.$$

$$(6\text{-}38)$$

Here $\omega_n = (2n + 1)\pi T$, $\Delta(\mathbf{p}, \omega_n)$ is the so-called order parameter [if $T > T_c$, the order parameter $\Delta(\mathbf{p}, \omega_n) = 0$], ζ is the Fröhlich parameter, $\omega^2(\mathbf{q})[\omega^2(\mathbf{q}) + (\omega_n - \omega_{n'})^2]^{-1}$ is the phonon Green function, $\omega(\mathbf{q})$ is the dispersion relation for phonons, $\xi(\mathbf{p}) = \epsilon(\mathbf{p}) - \epsilon_F$ is the electron energy referred to the Fermi energy, and Z is the renormalization factor (we will not write the expression for Z).

Equation (6-38) is the nonlinear equation that enables evaluation of the main parameters describing the superconducting state. This state is caused by the interaction of electrons with virtual excitations of the lattice that appear to be a huge nonadiabatic phenomenon.

Adiabatic Method in the Theory of Molecules

A molecule as well as a solid contains electron and nuclear subsystems. Everything that has been written in the section on the adiabatic method is also applicable to molecules, so that their properties can be described by Equations (6-4), (6-8), and (6-9) (BO approximation). However, one should note some distinctive molecular properties.

The number of particles in molecules and solids and, correspondingly, the size of the systems, differ in a striking way. This difference is shown in discrete structure of the energy spectrum of a molecule, whereas the spectrum of solids is continuous; note that sometimes the spectrum of molecular systems is partly continuous.

Molecules are characterized not only by electron and nuclear internal motion, but also by rotational motion. Correspondingly, it is necessary to take into account the rotational manifold of the energy levels.

The lattice is an example of a bound state of a nuclear subsystem analogous to a stable molecule. Along with investigation of bound states of molecules, there is another important direction related to scattering problems, namely, the study of unbounded states.

The properties of giant molecules are similar to the properties of solids; we focus on descriptions of small molecules that are characterized by the indicated differences.

"Crude" approach

Molecular properties can be described by Equations (6-4), (6-8), and (6-9). But often it is convenient to use the "clamped" adiabatic approach (see the earlier discussion of the adiabatic method). Sometimes this approach is called the "crude" approximation.[33,34]

Consider Equation (6-4) and denote some fixed nuclear configuration by R_0. If we study a bound state R_0 corresponds to some equilibrium configurations, then the electronic wavefunctions $\psi_m(r, R_0)$, which are the solutions of the equation

$$[\hat{T}_r + V(r, R_0)] \, \psi_m(r, R_0) = \epsilon_m(R_0) \, \psi_m(r, R_0), \qquad (6\text{-}39)$$

can be used as a basis set for the expansion of the total wavefunction; that is [cf. Eq. (6-5)],

$$\Psi_n(r, R) = \sum_m \tilde{\phi}_{nm}(R) \psi_m(r, R_0). \qquad (6\text{-}40)$$

Substituting (6-40) into Equation (6-1), one can obtain, after simple manipulations, the equation for the nuclear wavefunction $\tilde{\phi}_n \equiv \phi_{nn}$. The total wavefunction $\Psi(r, R)$ in zeroth-order approximation is [cf. Eq. (6-6)]

$$\Psi_n(r, R) = \psi_n(r, R_0) \tilde{\phi}_{n\nu}(R_0) \qquad (6\text{-}41)$$

The total Hamiltonian H can be written in the form

$$H = \hat{T}_\mathbf{r} + V(\mathbf{r}, \mathbf{R}_0) + \Delta H, \qquad (6\text{-}42)$$

where

$$\Delta H = \hat{T}_\mathbf{R} + V(\mathbf{r}, \mathbf{R}) - V(\mathbf{r}, \mathbf{R}_0)$$

or

$$\Delta H = \hat{T}_\mathbf{R} + H_1 + H_2 + \cdots, \qquad (6\text{-}43)$$

where

$$H_1 = \sum_{i,\alpha} \frac{\partial V}{\partial X_{i\alpha}|_0} \Delta X_{i\alpha} \qquad (6\text{-}44)$$

and

$$H_2 = \frac{1}{2} \sum_{i\alpha, i\beta} \frac{\partial^2 V}{\partial X_{i\alpha} \partial X_{i\beta}|_0} \Delta X_{i\alpha} \, \Delta X_{i\beta} \qquad (6\text{-}45)$$

There is an interesting question about the relation between the usual BO approximation [see Eqs. (6-4), (6-8), (6-9)], and the "crude" approach. First of all, one should emphasize that both basis sets—$\psi_n(\mathbf{r}, \mathbf{R})$ [see Eq. (6-4)] and $\psi_n(\mathbf{r}, \mathbf{R}_0)$ [see Eq. (6-39)]—are complete and can be used in order to evaluate the total wavefunction $\Psi(\mathbf{r}, \mathbf{R})$. The operator describing the perturbation in the "crude" approach [see Eqs. (6-43)–(6-45)] is simpler than expression (6-14), and it makes the "crude" approximation more convenient in practical calculations.

Nuclear wavefunctions ϕ_n [see Eq. (6-8)] and $\tilde{\phi}_n$ are different because of the difference between nuclear Hamiltonians. However, one can prove[7] that if we restrict consideration to the second-order terms of H_1 [see Eq. (6-44)], and to the harmonic approximation in Equation (6-8), the nuclear Hamiltonians will coincide. Indeed, the nuclear Schrödinger equation in the "crude" approximation is

$$\left[|\Delta H|_{nn} + \sum_{m \neq n} |\Delta H_{nm}| \, |\Delta H_{mn}| \, (E_n^0 - E_m^0)^{-1} + \cdots \right] \tilde{\phi}_n$$

$$= (E_n - E_n^0) \tilde{\phi}_n; \qquad E_n^0 \equiv \epsilon_n(\mathbf{R}_0)$$

(the matrix elements are taken with respect to electronic wavefunctions) or

$$\left\{\hat{T}_{\mathbf{R}} + \frac{1}{2} \sum_{i\alpha,j\beta}\left[\left|\frac{\partial^2 V}{\partial X_{i\alpha}\,\partial X_{j\beta}}\right|_{nn}\right|_{\mathbf{R}_{0n}} + \sum_m \left|\frac{\partial H}{\partial X_{i\alpha}}\right|_{nm} \left|\frac{\partial H}{\partial X_{j\beta}}\right|_{mn\mathbf{R}} = \mathbf{R}_{0n}\right.$$

$$\left. \cdot\, (E_n^0 - E_m^0)^{-1}\right]\Delta x_{i\alpha}\,\Delta X_{j\beta}\bigg\}\phi_{n\nu} = E_{n\nu}\phi_{n\nu}; \quad E_{n\nu} = E_n - E_n^0.$$

The expression in square brackets[8] is equal to $\partial^2\epsilon_n(\mathbf{R})/\partial X_{i\alpha}\,\partial X_{j\beta}$, and we obtain

$$\left[\hat{T}_{\mathbf{R}} + \frac{1}{2}\sum_{i\alpha,j\beta}\frac{\partial^2\epsilon_n(\mathbf{R})}{\partial X_{i\alpha}\,\partial X_{j\beta}}\,\Delta X_{i\alpha}\,\Delta X_{j\beta}\right]\tilde{\phi}_{n\nu} = E_{n\nu}\tilde{\phi}_{n\nu}.$$

This equation coincides with Equation (6-8) in the harmonic approximation.

The usual BO method [see Eqs. (6-4), (6-5), (6-8), (6-9)] as well as the "crude" approach [Eqs. (6-39)–(6-42)] are characterized by a common shortcoming. Namely, it is difficult to evaluate higher nonadiabatic corrections based on these versions of adiabatic method. Neither of them is adaptable to the usual perturbation theory development. In this connection, the development of a method that would enable the application of perturbation theory is of interest. Such a method has been determined by Geilikman in solid state theory (see the earlier discussion on the adiabatic method in solid state theory). An analogous method can be developed for molecules.[38]

The main idea of the method can be described in the following way. Instead of seeking the solution in the form of (6-5) or (6-40), one can start from an alternative choice of zeroth-order Hamiltonian. Let us choose this Hamiltonian H_0 in the form

$$H_0 = \hat{T}_{\mathbf{r}} + \hat{T}_{\mathbf{R}} + V(\mathbf{r}, \mathbf{R}_0) + \tilde{\epsilon}_n(\mathbf{R}), \tag{6-46}$$

where

$$\tilde{\epsilon}_n(\mathbf{R}) = \epsilon_n(\mathbf{R}) - \epsilon_n(\mathbf{R}_{0n}).$$

The term $\epsilon_n(R)$ is defined by Equation (6-4). It is easy to see that the function

$$\Psi_n(\mathbf{r}, \mathbf{R}) = \psi_n(\mathbf{r}, \mathbf{R}_{0n})\phi_{n\nu}(\mathbf{R}) \tag{6-47}$$

is an eigenfunction of the operator H_0. [Note that $\psi_n(\mathbf{r}, \mathbf{R}_0)$ is a solution of Eq. (6-4), and $\phi_{n\nu}(\mathbf{R})$ is a solution of Eq. (6-8).] Hence, according to the definition of H_0 [see Eq. (6-46), the functions on the right-hand side of Equation (6-47) are taken from different versions of the adiabatic theory.

The perturbation H' is

$$H' = V(\mathbf{r}, \mathbf{R}) - V(\mathbf{r}, \mathbf{R}_{0n}) - \epsilon_n(\mathbf{R}). \qquad (6\text{-}48)$$

Unlike the solid-state theory it is necessary here to explicitly take into account the dependence of $\mathbf{R}_0$ on the energy level and the anharmonicity. The expression (6-48) can be expanded near $\mathbf{R}_{0n}$. The operator H' and the functions (6-47) can be used in order to evaluate higher corrections on the basis of the usual perturbation theory (or the usual diagrammatic technique).

Rotation of molecules

Along with vibrational motion, a molecule is characterized by the rotational manifold of energy levels. The Hamiltonian H, describing the rotational motion, can be written in the form (see, e.g., reference 39)

$$H = \frac{\hbar^2}{2} \frac{\hat{J}_\xi^2}{I_A} + \frac{\hat{J}_\eta^2}{I_B} + \frac{\hat{J}_\zeta^2}{I_C}. \qquad (6\text{-}49)$$

One can estimate the spacing of rotational levels. According to (6-49), we obtain $\Delta E_{\text{rot}} \sim (ML^2)^{-1}$, where L is the length of the bond, or

$$\Delta E_{\text{rot}} \sim (Ma^2)^{-1} (a^2/L^2) \sim (a^2/L^2)\hbar\omega \sim \kappa^2 \hbar\omega, \qquad (6\text{-}50)$$

where a is the amplitude of vibrations and the parameter κ is defined by Equation (6-16). Hence the spacing between rotational levels is considerably smaller ($\sim \kappa^2$) than the vibrational level spacing. Therefore,

$$\Delta E_{\text{rot}} \ll \Delta E_{\text{vib}} \ll \Delta E_{\text{el}}.$$

The motion of a molecule is characterized by several steps of adiabaticity. The fast electronic motion is accompanied by slow vibrational motion, and the latter motion is faster than the rotational motion of the molecule. The total wavefunction can be written as a product

$$\Psi(\mathbf{r}, \mathbf{R}) = \psi_{\text{el}}(\mathbf{r}, \mathbf{R})\phi_{\text{vib}}(\mathbf{R})\varphi_{\text{rot}}(Q). \qquad (6\text{-}51)$$

If the "crude" approach is used, then the electronic wavefunction is taken at some configuration $\mathbf{R}_0$.

One should emphasize that the present description is valid for nonde-generate electronic states. The case of degeneracy (or near degeneracy) will be discussed.

Optical transitions

The adiabatic approach just described underlies the theory of molecular spectroscopy. Recall that the interaction of a molecule with radiation is usu-ally describable by the dipole matrix element

$$ d_{\mathrm{fi}} = \int \Psi_{\mathrm{f}}^* \, \mathbf{d} \Psi_{\mathrm{i}} \, d\tau, \qquad (6\text{-}52) $$

where Ψ_{i} and Ψ_{f} are the total wavefunctions describing the initial and final states of the molecule and $\mathbf{d}$ is the dipole operator. Each wavefunction can be taken in the adiabatic approximation [see Eq. (6-51)] which leads to

$$ d_{\mathrm{fi}} = \int \psi_{n'}^* \, \phi_{n'\nu'}^* \, \varphi_{n'\nu'j'}^* \, \mathbf{d}\psi_n \phi_{n\nu} \varphi_{n\nu j} \, d\tau. \qquad (6\text{-}53) $$

This matrix element contains integrations over many different degrees of freedom. First, we integrate over fast electronic motion, then over vibra-tional motion and, finally, over the slowest rotational motion. In accordance with this ordering, one could initially neglect the rotational motion, which means that the integration over electronic and vibrational coordinates should be carried out in a rotating system of coordinates. To do so, it is necessary to express the components of the dipole moment in the moving system, which leads to

$$ (d_\lambda)_{\mathrm{fi}} = \int dO \, \varphi_{n'\nu'j'} \, \varphi_{n\nu j} \sum_\alpha C_{\lambda\alpha}(O) \, (d_\alpha)_{\mathrm{fi}}, \qquad (6\text{-}54) $$

where

$$ (d_\alpha)_{\mathrm{fi}} \equiv (d_\alpha)_{n'\nu';n\nu} = \int \psi_n^*(\mathbf{r}, \boldsymbol{\rho}) \phi_{n'\nu'}^*(\boldsymbol{\rho}) d_\alpha(\mathbf{r}, \boldsymbol{\rho}) \, \psi_n(\mathbf{r}, \boldsymbol{\rho}) \, \phi_{n\nu}(\boldsymbol{\rho}) \, d\mathbf{r} \, d\boldsymbol{\rho}. $$

$$ (6\text{-}55) $$

Here $\mathbf{r}$ and $\boldsymbol{\rho}$ are the sets of electronic and vibrational coordinates; the appearance of the sum in Equation (6-54) is owing to the transformation of the dipole moment; d_λ and d_α are the components of the dipole moment in the rotating and in the space-fixed systems of coordinates, respectively.

Consider the matrix element $(d_\alpha)_{\mathrm{fi}}$ describing electronic and vibrational transitions. In accord with the adiabatic method, one integrates over electronic motion and obtains

$$(\tilde{d}_\alpha)_{\mathrm{fi}} = \int d\boldsymbol{\rho}\, \phi^*_{n'v'}(\boldsymbol{\rho})\, \phi_{nv}(\boldsymbol{\rho})\, l_{nn'}(\boldsymbol{\rho}),$$

where

$$l_{nn'}(\;) = \int \psi^*_n(\mathbf{r}, \boldsymbol{\rho})\, d_\alpha(\mathbf{r}, \boldsymbol{\rho})\, \psi_n(\mathbf{r}, \boldsymbol{\rho})\, d\mathbf{r}.$$

The factor $l_{nn'}(\boldsymbol{\rho})$ that appears as a result of the integration over the fast electronic motion depends on vibrational coordinates $\boldsymbol{\rho}$.

The vibrational motion can be conveniently described in terms of normal modes Q_i. In the harmonic approximation, the vibrational functions $\phi_{n'v'}$ and ϕ_{nv} can be written as products; that is,

$$\phi_{nv} = \prod_i \phi_{v^i}(Q_i); \qquad \phi_{v'v'} = \prod_k \phi_{v^k}(Q_k). \tag{6-56}$$

Here $\phi_{v^i}(Q_i)$ is a linear harmonic oscillator wavefunction, and v^i and v^k are sets of vibrational quantum numbers.

The factor $l_{nn'}(\;)$ can be expanded in a series of normal coordinates near equilibrium ρ_0

$$l_{nn'}(\;) = l_0 + \sum_i l_{1,i} Q_i + \sum_{i,k} l_{1;ik}\, Q_i Q_k + \cdots, \tag{6-57}$$

where

$$l_0 = l_{nn'}(\boldsymbol{\rho}_0);$$

$$l_{1,i} = (\partial l_{nn'}(\boldsymbol{\rho})/\partial Q_i)_0;$$

$$l_{1;ik} = \tfrac{1}{2}\, [\partial^2 l_{nn'}(\boldsymbol{\rho})/\partial Q_i\, \partial Q_k]_0.$$

The characteristic scale of the quantity $l_{nn'}(\boldsymbol{\rho})$ is the length of the bond L. Hence, the series (6-57) is an expansion in powers of the small adiabatic parameter $a/L = \kappa$ [see Eq. (6-16)].

Consider an optical transition between two different electronic states $(n \neq n')$. With the use of the Equations (6-52) and (6-55), we obtain (in zeroth-order approximation)

$$(\tilde{d}_\alpha)_{fi} = l_0 \int d\rho \, \phi^*_{n'\nu'}(\rho) \, \phi_{n\nu}(\rho).$$

This expression is the well-known Frank-Condon factor. The amplitude of the transition depends upon the degree of overlap of the vibrational wavefunctions. It is important to note that because the functions $\phi_{n\nu}(\rho)$ and $\phi_{n'\nu'}(\rho)$ belong to different electronic states, they are not orthogonal. Note also that the dipole moment d is a sum of two terms: $d = d_{el}(\nu) + d_{nuc}(\rho)$. The second term d_{nuc} makes no contribution in Equation (6-53) because of orthogonality of the electronic wavefunctions ψ_n and $\psi_{n'}$. Hence the electronic-vibrational transitions are caused by the electronic dipole moment only.

For vibrational transitions with no change of electronic state $(n = n')$, the Frank-Condon factor vanishes because of the orthogonality of the vibrational wavefunctions belonging to the same electronic state. Hence for this case it is necessary to take into account the next term in the series (6-57). We do so and obtain

$$(d_\alpha)_{fi} = \sum_i l_{1i} \int d\rho \, \phi^*_{n\nu'}(\rho) \, \phi_{n\nu}(\rho) \, Q_i. \qquad (6\text{-}58)$$

With the use of the expression (6-56), one can evaluate the matrix element (6-58). This matrix element describes the vibronic transitions between two adjacent vibrational levels $(\nu^i \rightarrow \nu^i + 1)$. Transitions between other vibrational levels are also possible, but in order to describe transitions, it is necessary to include successively higher terms of the expansion (6-57). These higher-order transitions are characterized by decreasing intensity, and this feature is directly related to the property that the series (6-57) is an expansion in powers of the small adiabatic parameter κ.

By analogy with the above discussion, one analyzes rotational transitions (see, e.g., reference 40). We emphasize that the adiabatic method forms the basis for the investigation of the various types of transitions occurring in molecules.

Degenerate states

The method described here is applicable in the absence of degenerate electronic states. If a molecule configuration is degenerate, then the criteria for the applicability of perturbation theory is, of course, not satisfied, and the nonadiabatic terms on the left-hand side of Equation (6-6) cannot be considered a small correction. This situation is sometimes called "the breakdown of the BO approximation." Fortunately, it does not mean the total renunciation of the scheme developed above. Equation (6-4) remains a starting point. But the degeneracy has to be taken into account in accord with the theory of degenerate states.

Returning to Equation (6-6), the sum on the left-hand side of Equation (6-6) cannot be neglected in the presence of degenerate states. One can see from Equation (6-7) that different electronic states are mixed by it. Moreover, it is impossible to separate electronic and nuclear motion and so one obtains a set of complicated electronic-nuclear states. It is necessary to consider the coupled equations of (6-6). If one considers the case of two degenerate electronic states, then the set contains two equations. From the theory of degenerate states, the solution in the zeroth-order approximation should be taken as a linear combination of functions of type (6-9).

The Jahn-Teller theorem[34,41-43] describes the behavior of degenerate electronic terms. According to this theorem, the adiabatic potential $\epsilon(\mathbf{R})$, which is characterized by the intersection of several branches $\epsilon_\alpha(\mathbf{R})$, at some configuration $\mathbf{R}_0$, does not have a minimum $\mathbf{R}_0$. There are the nuclear displacements Q_ν for which the relation $(\partial\epsilon/\partial Q_\nu)_{\mathbf{R}_0} \neq 0$ holds. The general evidence for this theorem, based on group theory, was carried out by Ruch and Schonhofer[44] (see, e.g., references 39 and 45). An analogous situation arises when the electronic states are not strictly degenerate, but the terms are close enough (quasi-degeneracy). Then perturbation theory again cannot be applied and one encounters the pseudo Jahn-Teller effect.[42]

The Jahn-Teller theorem is applicable only to nonlinear molecules. Linear molecules represent a separate case. Degenerate states of a linear molecule differ in the sign of the angular momenta. The matrix element of a nuclear displacement for a linear molecule vanishes, whereas the matrix element is nonzero for some displacements of nonlinear molecules (Jahn-Teller theorem). However, it turns out that higher terms result in a situation (Renner effect) that is analogous to the Jahn-Teller effect.[34,42,46,47]

The investigation of a system in the presence of degeneracy consists of two steps. First, one obtains the expression describing the behavior of the electronic terms, and then one solves the coupled equations (6-6). The usual case corresponds to the orbital degeneracy with the multiplicity $f = 2$ (E term) or $f = 3$ (T term). The detailed description of calculations is beyond the scope of this review. The interested reader is be referred to the literature.[42,48,49] An interesting approach to the multistate coupling problem has been developed by Cederbaum and colleagues.[50-52]

The nature of the Jahn-Teller and Renner effects is analogous to the Peierls instability (or, more generally, to the appearance of charge density waves) in solids. These phenomena are strong nonadiabatic effects owing to the degeneracy of the electronic states. Both of them appear as a consequence of the loss of symmetry that characterized the degeneracy.

Raman scattering

Raman scattering of molecules is another example of a phenomenon connected with nonadiabatic interactions. From a macroscopic point of view, this scattering is caused by the modulation of the induced dipole moment by the molecular vibrations. As a result, one observes scattering with a shift of the frequency.

From the quantum-mechanical point of view, Raman scattering of molecules can be considered as a phenomenon that proceeds through several intermediate virtual steps. These steps are

1. The absorption of a photon $\hbar\omega_0$ and the transition of the molecule to some electronic-vibrational state E_n.
2. The absorption of the vibrational quantum $\hbar\omega$, that is, the virtual transition between two adjacent vibrational states $n \to k$, so that $E_k = E_n + \hbar\omega$.
3. The emission of a photon $\hbar\omega' = \hbar\omega_0 - \hbar\omega$.

As a result, one obtains a frequency shift of the radiation and vibrational excitation of the molecule. For this process, one has conservation of energy for the process as a whole, but not for each virtual transition. Each intermediate step is described by a corresponding matrix element. We now focus on the second step. This step represents a radiationless virtual transition caused by the electron-vibrational interaction (6-44), that is, by nonadiabaticity. According to perturbation theory, the contribution of the nonadiabatic term is of order

$$\frac{H_1}{\Delta E_{el}} \sim \frac{\hbar\omega/\kappa}{\Delta E} = \kappa.$$

[Here we have used Eqs. (6-16) and (6-29).] As was just noted, Raman scattering is due to the modulation of the electric dipole moment by the vibrational motion. Hence this phenomenon is caused by the deviation from the BO adiabatic approximation, and it is natural that the probability of the scattering contains the small nonadiabatic parameter.

Radiationless transitions

In this section we discuss quantum transitions that are accompanied by the absorption or the emission of the radiation. Nonadiabaticity—that is, the interaction between the electron and nuclear subsystems—is the main mechanism of radiationless transitions. Note that radiationless transitions also play an important role in solids. Indeed, such transitions are the basis of the kinetics of solids (see, e.g., references 29, 9, and 53). For example, let us consider the conductivity of metals or semiconductors. An applied constant external electric field changes the energy distribution of electrons. This distribution becomes nonequilibrium. In the absence of the lattice it would result in nonstationary charge transfer. But the effect of the external field is balanced by the electron-ion collisions (electron-phonon interaction). The electron-phonon interaction, which is a nonadiabatic phenomenon, causes energy (or momentum) transfer from the electronic subsystem to the lattice. This relaxation process results in the possibility of a stationary state with electric current. This balance is described by the equation

$$-\frac{\partial f}{\partial \mathbf{p}} e\mathbf{E} = I_{\text{coll}}. \tag{6-59}$$

The expression on the left-hand side describes the change of the distribution function f. Indeed, $(-\partial f/\partial t)_{\text{ext}} = -(\partial f/\partial \mathbf{p})\dot{\mathbf{p}} = (\partial f/\partial p)e\mathbf{E}$. The "collision" integral (I_{coll}) describes the radiationless transition in the electron-phonon system, and it is given by

$$I_{\text{coll}} = |H'|^2 [f'(1 - f)(N_q + 1)$$
$$- f(1 - f')N_q]\delta(E' - E - \hbar\omega_{\bar{q}})$$
$$+ [1 - f)f' N_{\bar{q}} - f(1 - f')(N_{\bar{q}} + 1)]\delta(E' - E + \hbar\omega_{\bar{q}})\, d\bar{q}.$$

Here $|H'|^2$ describes the electron-phonon interaction and N_q is the number of phonons. The "collision" integral can be obtained from the Hamiltonian that describes the electron-phonon interaction

$$H' = \sum_{\bar{k},\bar{k}',\bar{q}} H'_{\bar{k}',\bar{k}\bar{q}} a_{\bar{k}'}^+ a_{\bar{k}} b_{\bar{q}} + \text{c.c.},$$

where $a_{\bar{k}'}^+$ is a Fermi operator for the creation of electron with momentum $\bar{k}'$, $b_{\bar{q}}$ is the Bose operator for the annihilation of the phonon, and so on. The "collision" integral can be obtained from the Golden Rule and describes different radiationless transitions in a metal. For example, an electron makes a transition $E' \rightarrow E_{|E=E'-\hbar\omega}$, which is accompanied by the creation of the phonon $\hbar\omega$. As was mentioned, this exchange of energy (or momentum) between electrons and the lattice balances the effect of the external field. Note that there are other relaxation mechanisms in solids (e.g., the scattering of electrons by impurities), and the electron-phonon interaction; that is, nonadiabaticity plays an important role, particularly in pure samples.

The distribution function f can be found from the kinetic equation (6-59), and then, using the expression

$$\mathbf{j} = \int ev f \, d\mathbf{p},$$

where $\mathbf{j}$ is the electric current and v is the velocity, one can evaluate the electric conductivity. Hence, radiationless transitions described by the "collision" integral play an important role in solid-state theory.

There are many properties of molecules, particularly large molecules, which are also connected with radiationless transitions (see, e.g., references 54–59). The origin of these transitions are the same as in solids. These transitions are caused by a nonadiabaticity, for example, by the electron-vibrational interaction. According to Kasha's rule,[54,55,58,60,61] radiation will be emitted only from the lowest excited electronic state of a given multiplicity. Hence, one can usually observe two luminescence spectra: fluorescence from the lowest excited singlet states and phosphorescence from the lowest triplet state. Therefore, after initial excitation, the molecule undergoes a radiationless transition to the lowest excited singlet and triplet states.

A radiationless transition can also occur between two degenerate states. Of course, these states cannot be exact eigenstates of the total Hamiltonian. These states are BO states, and it is important that the BO description is not exact. The deviation from the BO approximation, that is, the nondiabaticity, is a part of the total Hamiltonian providing radiationless transitions.

The absorption of radiation results in electronic excitation. If there is an isoenergetic manifold of vibronic levels belonging to lower electronic states, including the ground state, then the probability of a radiationless transition is nonzero. The number of such degenerate (or near degenerate) states increases noticeably with the size of the molecule. Many-phonon radiationless transitions that occur in large molecules with torsional motion have been considered by Gelbart, Freed, and Rice.[62]

Note that indirect polyatomic photodissociation is another example of a radiationless transition. This phenomenon will be discussed next.

Nonadiabatic scattering and nonadiabatic reactions.

The study of the dynamics of chemical reactions and nonreactive scattering processes is based on the adiabatic method. Nuclei are considered as particles moving on an adiabatic potential energy surface (PES).

For present purposes, we distinguish two types of chemical reactions: (1) adiabatic chemical reactions that occur on a single PES, and (2) nonadiabatic reactions characterized by a change of PES.

The same distinctions are useful for nonreactive scattering. If potential energy surfaces cross or they are very close in some region, nonadiabatic collisions are possible. Then the transition is provided by a nonadiabatic interaction (e.g., by spin-orbit interaction). Note that if a perturbation V leads to the described transition, then the intersection disappears if V is included in the Hamiltonian of the system (see, e.g., reference 39, p. 348).

Landau and Zener[63] have evaluated the probability of a nonadiabatic transition for a transition characterized by a single parameter. The nuclear motion has been considered on the basis of the semiclassical method. The wavefunction is a linear combination of the two functions corresponding to the two different electronic terms. The probability is given by the expression

$$W = 2 \exp\left(-\frac{2\pi V^2}{hv|F_2 - F_1|}\right)\left[1 - \exp\left(-\frac{2\pi V^2}{hv|F_2 - F_1|}\right)\right],$$

where V is the perturbation, $F_i = -\partial E_i/\partial \mathbf{R}$, and v is the nuclear velocity. The probability is small in both limiting cases: $V \gg hv|F_2 - F_1|$, $V \ll hv|F_2 - F_1|$ (for a detailed analysis, see reference 39).

A very powerful method for the calculation of collision cross sections is connected with the introduction of diabatic states (see, e.g., references 63–66). These states are coupled by nondiagonal terms of the potential energy matrix rather than matrix elements of the nuclear kinetic energy. This method has been used to evaluate cross sections including nonadiabatic effects arising in nonreactive collisions of $F(^2P) + H_2(^1\Sigma_g^+)$.

The simplest model describing reactive scattering is the collinear model for the reaction[67]

$$A + BC \longrightarrow AB + C.$$

This model was applied in order to consider the nonadiabatic transition for the reaction

$$Ba + N_2O \longrightarrow BaO + N_2.$$

The BaO molecules are produced in the ground $X^1\Sigma$ state or in the excited $a^3\pi$ state. In the Kupperman approach,[68] in order to treat the nonadiabatic transition, it was necessary to solve the following coupled equations:

$$\left[-\frac{\hbar^2}{2\mu}\left(\frac{\partial^2}{\partial R_\alpha^2} + \frac{\partial^2}{\partial r_\alpha^2}\right) + V_1(R_\alpha, r_\alpha) - E\right]$$
$$\times \psi_1(R_\alpha, r_\alpha) = -V_{12}(R_\alpha, r_\alpha)\,\psi_2(R_\alpha, r_\alpha),$$

$$\left[-\left(\frac{\hbar^2}{2\mu}\frac{\partial^2}{\partial R_\alpha^2} + \frac{\partial^2}{\partial r_\alpha^2}\right) + V_2(R_\alpha, r_\alpha) - E\right]\psi_2(R_\alpha, r_\alpha)$$
$$= -V_{21}(R_\alpha, r_\alpha)\psi_1(R_\alpha, r_\alpha).$$

Here $V_{12} = V_{21}$ is the coupling, R_α and r_α are scaled coordinates defined by

$$R_\alpha = \alpha_\alpha R'_\alpha, \qquad r_\alpha = \alpha_\alpha^{-1} r'_\alpha,$$

where $\alpha_\alpha = (\mu_{A,BC}/\mu_{BC})$, R'_α is the distance between the atom A and the center of mass of the diatom BC, and $r_\alpha \equiv r_{BC}$. The solution that satisfies the asymptotic conditions was obtained using the trajectory method.[68] The coupling V_{12} is the spin-orbit interaction.

A general analysis of the diabatic and adiabatic representation for the problem of atomic collisions has been carried out by Smith,[69] and reviews for molecular systems have appeared by Tully,[70] Rebentrost,[71] and Garrett and Truhlar.[72] An interesting result has been obtained by Mead and Truhlar.[73] According to their analysis a diabatic basis set, strictly speaking, does not exist. That is, it is impossible, in general, to transform to a basis set that is characterized by the total absence of nonadiabatic coupling. The one exception noted is the case of diatomic states of the same symmetry. However, for polyatomic systems one can obtain approximate diabatic states that can be used for practical applications with high accuracy.

6-3   NUCLEAR DYNAMICS

In the previous section we described the adiabatic approach for systems containing heavy and light particles (nuclei and electrons). But the adiabatic method is not restricted by such relation of the masses. Even if the many-body system contains particles that are comparable in mass, the adiabatic method can be successfully applied. The point is that the many-body system is characterized by many degrees of freedom, and some of them can be fast with respect to others. A typical example is rotational motion relative to vibrational motion of molecules (see the previous section). In this section we focus on systems with "slow" and "fast" degrees of freedom. The application of the adiabatic method to nuclear dynamics of polyatomic molecules and to some problems of nuclear physics will be described.

A polyatomic molecule contains electronic and nuclear subsystems. Based on the BO approximation (see Section 6-1), one can separate the electronic and nuclear motion. The latter motion is described by the Schrödinger equation

$$\left[\hat{T}_R + V(\bar{R}) \right] \phi(\bar{R}) = E\omega(\bar{R}). \tag{6-60}$$

For the dynamics of a chemical reaction or polyatomic photodissociation, the nuclear wavefunction describes the vibrational motion of the reactants (or products) or the photofragments as well as their relative translational and rotational motion. Equation (6-60) can be easily solved for the bound states in the harmonic approximation, where it is possible to separate variables and to introduce normal modes. (The anharmonic case presents no limitations.) But if one is concerned with a bound and continuous state corresponding to a scattering event in which one or both of the particles has internal vibrational structure such as the state of photofragments, then the problem becomes much more complicated. One can develop an adiabatic method in the framework of the nuclear dynamics,[5, 74, 75] which can be used in the theory of polyatomic photodissociation and in the theory of chemical reactions. For concreteness we focus on photodissociation and, more specifically, on the description of the dissociative state ϕ^D, of the photofragments.

Polyatomic Photodissociation

Dissociative state

We consider here a typical case of photodissociation into two fragments A and B by the breaking of one bond. The nuclear Hamiltonian $H^N = \hat{T}_R + \epsilon_n(\mathbf{R})$ [see Eq. (6-8)] can be written in the following form:

$$H^N = \sum_{v=A,B} \hat{T}_v + V(\mathbf{R}), \qquad (6\text{-}61)$$

where T_A and T_B are the kinetic energy operators of fragments A and B; that is,

$$\hat{T}_v = \sum_i^v \frac{\hbar^2}{2M_i} \frac{\partial^2}{\partial \mathbf{R}_i^2},$$

where the summation is over all nuclei of fragment v, and $V(\mathbf{R}) \equiv \epsilon_n(\mathbf{R})$ is the potential energy of nuclear motion.

The nuclear Hamiltonian can be written in the following form:

$$H^N = -\sum_{i,k}^A \frac{\hbar^2}{2\tilde{M}_{ik}} \frac{\partial^2}{\partial q_i\, \partial q_k} - \sum_{l,m}^B \frac{\hbar^2}{2\tilde{M}_{lm}} \frac{\partial^2}{\partial q_l\, \partial q_m}$$

$$- \sum \frac{\hbar^2}{2\mu} \frac{\partial^2}{\partial \rho^2} + V(\rho, q_i^A, q_l^B).$$

Here $\rho = \mathbf{R}_B - \mathbf{R}_A$ ($\mathbf{R}_A$ and $\mathbf{R}_B$ correspond to the centers of mass of the separate fragments), $\mu = M^A M^B/(M_A + M_B)$ is the reduced mass, $\tilde{M}_{ik}$ and

$\tilde{M}_{1m}$ are effective masses,[40] and q_i^A, and q_l^B are the internal coordinates of fragments A and B. The center of mass of the total system is at rest. The potential energy is conveniently expressed in terms of ρ, q_i^A, q_l^B. The internal coordinates are the aggregates

$$q_f^{\nu} = \{L_f^{\nu}, \alpha_f^{\nu}, \chi_f^{\nu}\},$$

where L_f^{ν} are the bond lengths of the fragments, α_f^{ν} are the bond angles, and the sets of angles χ_f^{ν} describe the relative orientation of the fragments A and B.

If, for instance, we consider the simplest polyatomic system, a triatomic molecule PQR, the aggregate of coordinates q_f^{ν} does not contain the set α_f^{ν}. For this system, $q_f^{\nu} = q_f^{PQ} = \{L_f^{PQ}, \chi_f^{PQ}\}$, where $L_f^{PQ} = q$ is the bond length of the diatom PQ and $\chi_f^{PQ} = \theta$, the angle formed by the diatom PQ and the vector from the atom R to the center of mass of PQ.

The simplest model of polyatomic photodissociation corresponds to collinear photodissociation of a triatomic system. Then $q_f^{\nu} \equiv q$ and

$$H^N = -\frac{\hbar^2}{2\mu}\frac{\partial^2}{\partial\rho^2} - \frac{\hbar^2}{2\mu_F}\frac{\partial^2}{\partial q^2} + V(\rho, q), \qquad \textbf{(6-62)}$$

where

$$\mu = \frac{M_A(M_B + M_c)}{M}, \qquad \mu_F = \frac{M_B M_C}{M_B + M_C}.$$

To obtain the expression for the nuclear wavefunction $\phi^D(\rho, q_i^A, q_l^B)$, we must solve the Schrödinger equation (6-60) with Hamiltonian given by (6-61).

The method of obtaining a valid expression for $\phi^D(\rho, q_i^A, q_l^B)$, is not obvious because of necessity to take into account the interfragment interaction. Unlike a bound state, the potential energy $V(\rho, q^{Ai}, q_l^B)$ cannot be expanded in a series of the deviations of all variables from equilibrium because ρ corresponds to translational motion. Hence the variables are not separable; this is the main difficulty for solving the equation

$$H^N\phi^D(\rho, q_i^A, q_l^m) = E\phi^D. \qquad \textbf{(6-63)}$$

Because of the impossibility of separating variables, the nuclear wavefunction $\phi^D(\rho, q_i^A, q_l^B)$ generally cannot be well approximated as a product $\phi_1^D(\rho)\,\phi_2^D(q_i^A, q_l^B)$. In the asymptotic region the separable variable approximation (SVA) is valid because the potential energy no longer depends on ρ. However, the main contribution to the coupling matrix element H'_{DQ} is not connected with the asymptotic region. Because of the Frank-Condon principle, this main contribution comes from a region $\rho \sim \rho_0$, where ρ_0 corresponds to the equilibrium configuration of the precursor molecule. Hence it is necessary to take into account the interfragment interaction, that is, the dependence of the potential energy on ρ. In the region of short distances, the interfragment interaction is not small and, generally speaking, the deviation from the SVA is not governed by a small parameter.

We consider a polyatomic system that is characterized by several types of motion. We first consider the case of relative translational motion of photofragments slow compared to fragment internal vibrational motion. We show below how it is possible to develop a specific adiabatic method for the determination of ϕ^D for this case. One should not confuse this method with the Born-Oppenheimer adiabatic approximation. It is an adiabatic approach in the framework of the nuclear problem only.

We consider next the opposite case of relative translational motion fast with respect to internal motion. Afterward we describe the approach for the intermediate case. Note also that because it is often a good approximation to separate rotational from translational and vibrational motion. We do so here and consider first the latter motions of the photofragments.

Slow translational motion

Consider the case of the vibrational motion of the fragments accompanied by slow relative motion. The potential function $U(\rho, q_i^A \, q_l^B)$ can be expanded

$$V(\rho, q_i^A, q_l^B) = V(\rho, q_{io}^A (\rho), q_{lo}^B (\rho))$$

$$+ \sum_{i,j}^{A} k_{ij}^A(\rho) \, \tau_i^A \, \tau_j^A + \sum_{l,m}^{B} k_{lm}^B(\rho)\tau_l^B\tau_m^B, \qquad (6\text{-}64)$$

where $\tau_i^A = q_i^A - q_{io}^A(\rho)$, $\tau^B = q_l^B - q_{lo}^B(\rho)$ are the deviations of the internal coordinates from equilibrium, and $k_{ij}^A(\rho)$ and $k_{lm}^B(\rho)$ are the force constants. The coordinate ρ does not appear as an explicit expansion variable in Equation (6-64). One can neglect terms of the type $\tau_i^A \, \tau_l^B$ because a coordinate of fragment A is not adjacent to a coordinate of fragment B (see, e.g., reference 76).

Hence, according to Equations (6-61) and (6-62), the nuclear Hamiltonian H^N can be written as a sum

$$H^N = H_{tr} + H^A + H^B, \qquad (6\text{-}65)$$

where

$$H^{\text{tr}} = \frac{\hbar^2}{2\mu} \frac{\partial^2}{\partial \rho^2} + V(\rho, q_{i0}^A(\rho), q_{i0}^B(\rho))$$

$$H^Z = - \sum_{i,j}^Z \left(\frac{\hbar^2}{2\tilde{M}_{ij}} \frac{\partial^2}{\partial q_i \, \partial q_j} - K_{ij}^Z(\rho)\tau_i^Z\tau_j^Z \right); \quad Z \equiv A, B.$$

The force constants $K_{ij}^Z(\rho)$ and the quantities $q_{i0}^Z(\rho)$ can be determined from the conditions

$$\left. \frac{\partial V}{\partial q_i^Z} \right|_{q_i = q_{i0(\rho)}} = 0$$

and

$$K_{ij}^Z(\rho) = \frac{\partial^2 V}{\partial q_i^Z \, \partial q_j^Z} \left. \right|_{\substack{q_i = q_{i0}(\rho) \\ q_j = q_{j0}(\rho)}}.$$

Our goal is to evaluate the nuclear wavefunction ϕ^D, which is an eigenfunction of Hamiltonian H^N [see Eq. (6-63)]. First, we consider the simplest case of a collinear triatomic system; afterward we evaluate ϕ^D for the general case.

For the collinear triatomic system, the Hamiltonian H^N contains two terms H_{tr} and H^A and $q_i^A \equiv q$ (q is the length of the bond). The system is characterized by slow relative motion, and the photofragment is a one-dimensional oscillator.

The relative motion, which is the slow degree of freedom, is analogous to nuclear motion in the Born-Oppenheimer approach, whereas the fast vibrational motion is analogous to electronic motion. The nuclear wavefunction $\phi^D(\rho, q)$ can be written as a product:

$$\phi^D(\rho, q) = \phi_{p\nu}^{\text{tr}}(\rho)\phi_\nu^{\text{vib}}(\rho, q), \tag{6-66}$$

where $\phi_\nu^{\text{vib}}(\rho, q)$ is the solution of the equation

$$\left[-\frac{\hbar^2}{2\mu_F} \frac{\partial^2}{\partial q^2} + K(\rho)\,\tau^2 \right] \phi_\nu^{\text{vib}}(\rho, q) = \epsilon^{\text{vib}}(\rho)\,\phi_\nu^{\text{vib}}(\rho, q)$$

and represents a harmonic oscillator wavefunction whose frequency and the equilibrium position depends on the distance ρ; that is,

$$\phi_v^{\text{vib}}(\rho, q) = \left(\frac{\mu_F \omega(\rho)}{\pi\hbar}\right)^{1/4} \frac{1}{\sqrt{2^v v!}} - \frac{\mu_F \omega(\rho)}{2\hbar} \tau^2 H_v\left(\sum \sqrt{\frac{\mu_F \omega(\rho)}{\hbar}}\right), \quad \text{(6-67)}$$

where

$$\tau = q - q_0(\rho) \quad \text{and} \quad \omega(\rho) = [\mu_F^{-1} K(\rho)]^{1/2}. \quad \text{(6-68)}$$

The translational part of $\phi_{pv}^{\text{tr}}(\rho)$ satisfies the following equation:

$$\left[-\frac{\hbar^2}{2\mu}\frac{\partial^2}{\partial\rho^2} + V^{\text{eff}}(\rho)\right]\phi_{pv}^{\text{tr}}(\rho) = E\phi^{\text{tr}}(\rho), \quad \text{(6-69)}$$

where

$$V^{\text{eff}} = V[\rho, q_0(\rho)] + \epsilon^{\text{vib}}(\rho) \quad \text{(6-70)}$$

and

$$\epsilon^{\text{vib}}(\rho) = (v + \tfrac{1}{2})\hbar\omega(\rho). \quad \text{(6-71)}$$

The subscript pv means that the function ϕ_{pv}^{tr} describes translational motion with momentum p for the fragment in a specific vibrational level v.

The solution (6-66) is not exact. We have neglected the term [cf. Eqs. (6-6) and (6-7)]

$$H_1^N \phi^D \sim \frac{1}{\mu}\frac{\partial\phi^{\text{tr}}}{\partial\rho}\frac{2\phi^{\text{vib}}}{2\rho}. \quad \text{(6-72)}$$

(See references 74 and 75 for details.) We shall estimate this neglected term in the following.

Equation (6-69) has a solution in the semiclassical approximation (except the region near classical turning point α_t) of the form

$$\phi_{pv}^{\text{tr}} = \frac{2\mu}{\pi\hbar p}\cos[\sigma(\rho) + \delta], \quad \text{(6-73)}$$

where

$$\sigma(\rho) = \hbar^{-1}\int_{a_t}^{\rho} p(\rho)\,d\rho \quad \text{(6-74)}$$

and

$$p = \{2\mu[E - V^{\text{eff}}(\rho)]\}^{1/2}. \qquad (6\text{-}75)$$

The function V^{eff} is defined by Equations (6-70) and (6-71) and ϕ_{pv}^{tr} is normalized to a delta function of energy.

The solution (6-73) corresponds to the semiclassical approximation. Hence, we assume that (see, e.g., reference 31)

$$\frac{\mu}{p^3} \frac{\partial V^{\text{eff}}}{\partial \rho} \ll 1$$

or

$$pL/\hbar \gg 1, \qquad (6\text{-}76)$$

where L is the characteristic distance for variation of the field V^{eff}.

The semiclassical approximation is not applicable in the region of the turning point a_t. As is well known, one can write the solution of Equation (6-69) which also valid in the region of a_t in terms of an Airy function (see, e.g., reference 77), that is,

$$\phi^{\text{tr}}(\rho) = \frac{(2\mu)^{1/3}}{\pi^{1/2}F_0^{1/6}\hbar^{2/3}} \, \text{Ai}\left[- \left(\frac{2\mu F_0}{\hbar^2}\right)^{1/3} (\rho - a_t) \right], \qquad (6\text{-}77)$$

where

$$F_0 = \left. \frac{\partial V^{\text{eff}}}{\partial \rho} \right|_{a_t},$$

and we have introduced the expansion

$$V^{\text{eff}}(\rho) = V^{\text{eff}}(a_t) - F_0(\rho - a_t).$$

The function (6-77) asymptotically has the form of (6-73). Hence the increase of $s \equiv \rho - a_t$ results in the gradual transformation of (6-77) into (6-73). Based on this correspondence, one can write out the expression for $\phi^{\text{tr}}(\rho)$:

$$\phi^{\text{tr}}(\rho) = \left(\frac{2\mu}{\pi\hbar p}\right)^{1/2} \sigma^{1/4} \, \text{Ai}\left[- \left(\frac{3}{2}\right)^{2/3} \sigma(\rho) \right].$$

Now let us estimate the accuracy of (6-66). As was mentioned, we neglected the effect of the term (6-72) in writing (6-66). Using Equations (6-72) and (6-73), we obtain

$$
H_1^N \phi^D \sim \frac{\hbar}{\mu} p \phi^{tr} \frac{\partial \phi^{vib}}{\partial \omega} \frac{\partial \omega}{\partial \rho}
$$

$$
\sim \frac{\hbar}{\mu} p \frac{\phi^{vib}}{\omega} \frac{\partial \omega}{\partial \rho} \phi^{tr} = \frac{p^4}{\mu^2 \omega} \gamma \phi^{vib} \phi^{tr} ,
\tag{6-78}
$$

where we introduced the parameter

$$
\gamma \equiv \frac{\mu \hbar^2}{p^3} \frac{\partial \omega}{\partial p} \ll 1 .
\tag{6-79}
$$

The first term on the left-hand side of Equation (6-69) is $\sim p^2/\mu$. Hence the neglected term (6-72) is small, as may be seen from

$$
\left(\frac{p^4}{\mu^2 \omega} \gamma \right) \left(\frac{p^2}{\mu} \right)^{-1} \sim \frac{p^2}{\mu \omega} \cdot \gamma \ll 1
\tag{6-80}
$$

relative to the remaining term in Equation (6-69).

The magnitude of the semiclassical parameter γ can be estimated,

$$
\gamma \sim p^{-2} \left(\frac{dp}{d\rho} \right) \sim \left(\frac{p}{\hbar} L \right)^{-1} .
\tag{6-81}
$$

Hence our adiabatic approach is applicable, and Equation (6-66) is valid if

$$
B = \frac{p^2}{\mu \omega} \cdot \frac{\hbar}{pL} \ll 1 .
\tag{6-82}
$$

The ratio $(p^2/\mu\omega)$ has the same order of the magnitude as the parameter $\alpha \equiv (\Delta \epsilon^{tr}/\Delta \epsilon^{vib})$. Hence the validity of the solution (6-66) is a consequence not only of the smallness of the recoil energy relative to the internal energy, but the smallness of the inverse semiclassical parameter defined by Equation (6-76).

The condition (6-82) has a clear physical meaning. Indeed, this condition can be reduced to the form

$$
(v/\omega L) \ll 1
\tag{6-83}
$$

or

$$T^{\text{vib}} \ll \tau^{\text{tr}},$$

where $\tau^{\text{tr}} = L/v$, that is, the inverse frequency of vibrations is smaller than the characteristic time for the translational motion. Hence, (6-82) is the condition for adiabaticity of the vibrational motion with respect to slow relative motion.

The condition (6-82) [or (6-83)] can also be written in the following form:

$$pL(M\omega a^2)^{-1} \cdot (a^2/L^2) \ll 1,$$

where a is the amplitude of the vibrations. Because $M\omega a^2 \sim \hbar$, we arrive at the relation

$$\gamma(a^2/L^2) \ll 1; \quad \gamma = pL/\hbar.$$

The previous considerations can be generalized and one can evalute ϕ^{D} for any polyatomic molecule that dissociates into two fragments.

The nuclear Schrödinger equation can be reduced to the form

$$H^{\text{N}}\phi^{\text{D}}(\rho, \theta_i^{\text{A}}, \theta_l^m) = E_{n\nu}\phi(\rho, \theta_i^{\text{A}}, \theta_l^{\text{B}}), \tag{6-84}$$

where H^{N} is defined by Equation (6-65), and

$$H^{\text{A}} = -\sum_i^{\text{A}} \left(\frac{\hbar^2 \partial^2}{\partial \theta_{i\text{A}}^2} - K_i^{\text{A}}(\rho)\, \theta_{i\text{A}}^2 \right),$$

$$H^{\text{B}} = -\sum_l^{\text{B}} \left(\frac{\hbar^2 \partial^2}{\partial \theta_{l\text{B}}^2} - K_l^{\text{B}}(\rho)\, \theta_{l\text{B}}^2 \right).$$

Here we have introduced the normal coordinates θ_i of each fragment. A mass factor has been included in θ_i. We seek the total solution of Equation (6-84) in the form

$$\phi_N^{\text{D}}(\theta_i^{\text{A}}, \theta_l^{\text{B}}; \rho) = \sum_K \phi_{NK}(\rho)\, \tilde{\phi}_K(\theta_i^{\text{A}}, \theta_l^{\text{B}}; \rho), \tag{6-85}$$

where $\tilde{\phi}_K = \tilde{\phi}_{\nu\text{A}}(\theta_i^{\text{A}}, \rho)\, \tilde{\phi}_{\nu\text{B}}(\theta_l^{\text{B}}; \rho)$; the functions $\tilde{\phi}_{\nu\text{B}}$ and $\tilde{\phi}_{\nu\text{B}}$ are solutions of

$$H^Z \phi_{\nu Z}(\theta_i^Z; \rho) = \epsilon_{\nu Z}(\rho)\, \phi_{\nu Z}(\theta_i^Z; \rho); \quad Z = \{\text{A}, \text{B}\}.$$

Hence in the harmonic approximation we obtain

$$\tilde{\phi}_{\nu Z}(\theta_i^Z, \rho) = \prod_i \phi_{n_i}^{\text{vib}}(\theta_i^Z, \rho), \tag{6-86}$$

where $\phi_{n_i}^{\text{vib}}$ are harmonic oscillator wavefunctions and the total vibrational energy is given by

$$\epsilon_N(\rho) = \epsilon^A(\rho) + \epsilon^B(\rho) \tag{6-87}$$

$$= \sum_i^A \left(n_i + \frac{1}{2}\right) \hbar\omega_i^A(\rho) + \sum_l^B \left(n_l + \frac{1}{2}\right) \hbar\omega^B(\rho).$$

Here $N = \{\nu^A, \nu^B\}$ is the set of fragment vibrational quantum numbers.

If we neglect the off-diagonal part of the sum (6-85), we obtain the following expression for θ^D:

$$\phi_{NP}^D(\theta_i^A, \theta_l^B, \rho) = \phi_{NP}(\boldsymbol{\rho}) \, \phi_{\nu^A}(\theta_i^A; \rho) \, \phi_{\nu^B}(\theta_l^B; \rho). \tag{6-88}$$

The functions ϕ_{ν^A} and ϕ_{ν^B} are given by Equation (6-86), and $\phi_{NP}(\boldsymbol{\rho}) \equiv \phi_{NNP}(\boldsymbol{\rho})$ is a solution of

$$\left[-\frac{\hbar^2}{2\mu} \frac{\partial^2}{\partial\boldsymbol{\rho}^2} + V^{\text{eff}}(\rho) \right] \phi_{NP}(\boldsymbol{\rho}) = E_{nNP} \phi_{NP}(\boldsymbol{\rho}). \tag{6-89}$$

Here

$$V^{\text{eff}}(\rho) = V(\rho; q_{i0}^A(\rho), q_{l0}^B(\rho)) + \epsilon_N(\rho).$$

Equation (6-89) can be solved in the semiclassical approximation. The expression for the radial part of $\phi_{NP}(\rho) = \rho\chi_{NP}(\rho)$ can be written by analogy with (6-73). It is necessary, of course, to take into account the centrifugal energy.

The estimation of the accuracy of the solution can be carried out by analogy with (6-78)–(6-82), and we obtain

$$\beta = \left(\frac{p^2}{\mu\tilde{\omega}}\right)\left(\frac{\hbar}{pL}\right) \ll 1,$$

where

$$\tilde{\omega}^{-1} = \sum_i \omega_i^{-1}.$$

Note that the discussion following Equation (6-82) is also applicable to the general case of polyatomic systems.

We have considered the case in which the vibrational motion of the photofragments is accompanied by slow relative motion. We developed the adiabatic approach to evaluate the nuclear wavefunction ϕ^D and arrived at Equations (6-66) and (6-88). Here, instead of a system of electrons and nuclei (Born-Oppenheimer approximation), we considered only nuclear motion of a polyatomic system with several degrees of freedom, one of which was "fast" relative to the others.

We did not separate the part of Hamiltonian describing the interfragment interaction. This interaction is included in the zeroth-order approximation (6-89).

We emphasize that the interaction between fragments is taken into account in two ways. First, the vibrational frequency and the equilibrium position depend on the distance between the fragments [see Eqs. (6-67) and (6-68)]. Second, the effective potential energy V_{eff}, describing the relative motion, contains the vibrational energy [see (6-70) and (6-87)].

Note that the frequencies $\omega_i^Z(\rho) \neq \omega_{i,\rho \to \infty}^Z \equiv \omega_F^Z(\omega_F^Z$ are the frequencies of free fragments). It is important to note that the main contribution to the matrix element

$$H_{\text{fi}}' = \int \phi^{D*} H' \phi^Q \, d\tau$$

comes from the region of overlap of the functions ϕ^Q and ϕ^D, that is, from short intermolecular distances. (The function ϕ^Q describes the initial bound state.)

Slow internal motion

Let us now consider the opposite case, in which the recoil energy is considerably larger than the vibrational energy. We can again apply the adiabatic approach by first neglecting the kinetic energy of internal motion. We obtain the following expression for the nucleon wavefunction:

$$\phi_{pv}^D = \phi_p(\boldsymbol{\rho}, q) \, \phi_{pv}(q).$$

The function $\phi_p(\boldsymbol{\rho}, q)$ may be evaluated at the equilibrium geometry $q = q_0$. This approximation is similar to the crude adiabatic approach (see Section 6-2). Hence, the function ϕ_{pv}^D can be written as a product

$$\phi_{pv}^D = \phi_p(\boldsymbol{\rho}, q_0) \, \phi_{pv}(q). \tag{6-90}$$

This approach is analogous to the "crude" adiabatic approximation (see above). The function $\phi_p(\boldsymbol{\rho}, q_0)$ satisfies the equation

$$\left[-\frac{\hbar^2}{2\mu_F} \frac{\partial^2}{\partial \boldsymbol{\rho}^2} + V(\boldsymbol{\rho}, q_0) \right] \phi_p(\boldsymbol{\rho}, q_0) = E(q_0)\, \phi_p(\boldsymbol{\rho}, q_0).$$

The angular dependence can be separated in the usual way. Using the semiclassical approximation, we arrive at the following expression for the radial part of the wavefunction:

$$\phi_p(\rho, q_0) = \rho_t^{-1} \cos\left[\sigma(\rho, q_0) + \delta\right]. \tag{6-91}$$

Here

$$\sigma(\rho, q_0) = \int_{a_t}^{t} p(\rho, q_0)\, d\rho,$$

where

$$p = \{2\mu[E(q_0) - U(\rho, q_0)]\}^{1/2} \tag{6-92}$$

or

$$p = \{2\mu[E - U(\rho, q_0) - (v + \tfrac{1}{2})\hbar\omega]\}^{1/2}. \tag{6-93}$$

The vibrational function ϕ^{vib} is given by

$$\phi^{\text{vib}} = \left(\frac{\mu_F\omega}{\pi}\right)^{1/4} \frac{1}{(2^v v!)^{1/2}} \exp\left(-\frac{\mu_F\omega}{2\hbar} \tau^2\right) H_v\left[\tau\left(\frac{\mu_F\omega}{\hbar}\right)^{1/2}\right]. \tag{6-94}$$

Note that the function ϕ_{pv}^{D} pf (6-90) has the form of the zeroth-order approximation of separated variables (SVA). Indeed, the ϕ_{pv}^{D} is a product of two functions. One of them depends on ρ and the other on q. Hence we arrive at the conclusion that SVA is valid, if internal motion is slower than relative motion.

If the translational motion is slower than the internal motion, the dissociative D state is described by Equations (6-67) and (6-73)–(6-75). The wavefunction describing the opposite case is given by Equations (6-91)–(6.94). Note that we can carry out the direct comparison of the two limiting cases. Let us also note the common features of both cases. (Recall, for comparison, that the basic principles of the band theory of solids were obtained by the consideration of two limits: strong and weak coupling.) In both cases the wavefunction ϕ^D can be written as a product of two functions. Moreover, the structure of the corresponding functions is similar. As a matter of fact, the limiting cases differ by the behavior of the vibrational frequency and the equilibrium position. If we consider slow translational motion, then the frequency and the equilibrium position depend on ρ, while in the opposite case the quantities ω_c and q_{0c} are constant. Hence we come to the conclusion that the gradual transition from the case of slow relative motion to the opposite one is accompanied by a gradual change of the dependence $\omega(\rho)$ and $q_0(\rho)$. (For a more detailed analysis, see references 74 and 75.)

The adiabatic method just described can also be applied in the theory of chemical reaction.[78] Polyatomic photodissociation is the transition from a bound to a bound-continuous state, whereas a chemical reaction is a transition from a bound-continuous state of reactants to a bound-continuous state of the products. Each of these states can be described by a function of the type (6-66). This adiabatic approach facilitates the development of a state-to-state description of the dynamics of chemical reaction.

Vibrational Motion of Molecules

The adiabatic method can also be used to study the properties of bound molecular systems. The analysis of rotational motion is an example of such an approach. Here we consider the application of the adiabatic method to the problem of molecular vibrations.[78] Based on the usual method (see, e.g., references 76 and 79), one can determine normal frequencies and the vibrational wavefunctions. Nevertheless, if some frequency is noticeably less than others (e.g., some bending frequencies, or an additional frequency appearing as a result of the joining of an additional complex with weak coupling), one can use the adiabatic approach. In order to illustrate the method, consider the triatomic system ABA. We assume that the bending motion is characterized by small force constant and a bending angle 2α that is near π (more exact criteria will be obtained next). The Lagrangian of the molecule can be written (see reference 43)

$$L = \frac{m_A}{2}\dot{Q}_a^2 + \left(1 + \frac{2m_A}{m_B}\sin^2\alpha\right)\frac{K_1}{2}Q_a^2 + A_1(\Delta\dot{q}_{S1})^2$$
$$+ B_1(\Delta\dot{q}_{S2})^2 - A_2(\Delta q_{S1})^2 - B_2(\Delta q_{S2})^2 + C_1\Delta q_{S1}\Delta q_{S2}. \quad \textbf{(6-95)}$$

Here $Q_a = \Delta x_1 + \Delta x_3$, $q_{S1} = x_1 - x_3$; $q_{S2} = y_1 + y_3$. (x_1, x_3, y_1, y_3 are the Cartesian coordinates at the outer atoms A.)

$$A_1 = \frac{m_A}{2L}; \qquad B_1 = \frac{m_A M}{2m_B L};$$

$$A_2 = (K_1 \sin^2 \alpha + 2K_2 \cos^2 \alpha)/2L; \qquad B_2 = \frac{M^2(K_1 \cos^2 \alpha + 2K_2 \sin^2 \alpha)}{2m_B^2 L}$$

$$C_1 = M(2K_2 - K_1) \sin \alpha \cos \alpha / M_B L,$$

where

$$M = 2m_A + m_B; \qquad L = (2m_A/M_B) + \sin^{-2} \alpha.$$

Based on Equation (6-95), one can evaluate the Hamiltonian and write the Schrödinger equation

$$H\phi = W\phi,$$

where

$$H = T_x + T_y + T_z + \tfrac{1}{2}\delta x^2 + \tfrac{1}{2}\beta y^2 - \tfrac{1}{2}\gamma xy + \tfrac{1}{2}rz^2$$

$$T_i = - \left(\frac{1}{2}\right) \frac{\partial^2}{\partial x_i^2}, \qquad\qquad (6\text{-}96)$$

and

$$x = \sqrt{2A_1}\, \Delta q_{S1}; \qquad y = \sqrt{2B_1}\, \Delta q_{S2}; \qquad z = \sqrt{m_A}\, Q_a.$$

The parameters δ, β, γ, r are given by

$$\delta = (K_1 \sin^2 \alpha + 2K_2 \cos^2 \alpha)/m_A;$$

$$\beta = M(K_1 \cos^2 \alpha + 2K_2 \sin^2 \alpha)/m_A m_B;$$

$$\gamma = 2(M/m_B)^{1/2} (2K_2 - K_1) \sin \alpha \cos \alpha / m_A;$$

$$r = K_1(1 + 2m_A \sin^2 \alpha/m_B)/m_A. \qquad\qquad (6\text{-}97)$$

As one can see from Equation (6-96) the variable z can be separated so that the wavefunction ϕ can be written in the form

$$\phi(x, y, z) = \phi(x, y)\, \varphi(z), \qquad\qquad (6\text{-}98)$$

where $\varphi(z)$ is an eigenfunction of the Hamiltonian $H_z = T_z + \tfrac{1}{2} rz^2$ and the function $\phi(x, y)$ is a solution of the equation

$$[T_x + T_y + \tfrac{1}{2}\delta x^2 + \tfrac{1}{2}\beta y^2 - \tfrac{1}{2}\gamma xy]\phi(x, y) = E\phi(x, y). \qquad \textbf{(6-99)}$$

Assuming that the bending degree of freedom y is slow relative to the symmetric mode x, we seek the solution of (6-99) in the form

$$\phi(x, y) = f(x, y)\, g(y),$$

where $f(x, y)$ and $g(y)$ are solutions of the usual adiabatic equations

$$[T_x + \tfrac{1}{2}\delta x^2 + \beta y^2 + \tfrac{1}{2}\gamma xy]\, f(x, y) = \epsilon(y) f(x, y), \qquad \textbf{(6-100)}$$

$$[T_y + \epsilon(y)]\, g(y) = Eg(y). \qquad \textbf{(6-101)}$$

The system (6-100) and (6-101) can be easily solved. The "potential" $\epsilon(y)$ given by

$$\eta(y) = (n + \tfrac{1}{2})\, \hbar\omega_2 + \tfrac{1}{2}\lambda y^2,$$

where

$$\lambda = \beta - \gamma^2/4\alpha; \qquad \omega_2 = \sqrt{\delta}. \qquad \textbf{(6-102)}$$

After some manipulations we arrive at the following expression:

$$\phi(\theta_1, \theta_2, \theta_3) = \prod_i \phi(\theta_i).$$

Here

$$\phi(\theta_i) = \left(\frac{\omega_i}{\pi}\right)^{1/4} \frac{1}{(2^{n_i} n_i!)^{1/2}} \exp\left(-\frac{m_i\omega_i}{2}\theta_i^2\right) H_{n_i}(\theta_i \sqrt{m_i\omega_i}) \qquad \textbf{(6-103)}$$

$$\theta_1 = \Delta x_1 + \Delta x_3; \qquad \theta_2 = \Delta x_1 - \Delta x_3$$

$$-\frac{1}{2}\left(\frac{\gamma}{\alpha}\right)\sqrt{\frac{B_1}{A_1}}\, (\Delta y_1 + \Delta y_3);$$

$$\theta_3 = \Delta y_1 + \Delta y_3, \qquad \textbf{(6-104)}$$

$$\omega_1 = \sqrt{r}; \qquad \omega_2 = \sqrt{\delta} \quad \text{and} \quad \omega_3 = \sqrt{\lambda}; \qquad \textbf{(6-105)}$$

the quantities r, α, and λ are defined by Equations (6-98) and (6-102).
    The energy spectrum is given by

$$E = \sum_i (n_i + \tfrac{1}{2}) \, \hbar\omega_i. \qquad \textbf{(6-106)}$$

The expressions (6-103)–(6-106) are valid if $\omega_3 \ll \omega_2$, or, more exactly, if

$$\left(\frac{k_2}{k_1}\right) \left(\frac{2M}{M_B}\right) \left[\sin^2 \alpha + 2\left(\frac{k_2}{k_1}\right) \cos^2 \alpha \right]^{-2} \ll 1. \qquad \textbf{(6-107)}$$

Equation (6-107) is the condition for the applicability of the adiabatic method described here. One can see that the inequality requires α to be close to $\pi/2$ and the constant k_2 to be relatively small.

Nuclear Collective Motion

    The atomic nucleus is another example of a many-body system, and the adiabatic method has been successfully applied to the theory of atomic nuclei.
    A most interesting application is the one concerned with the collective motion of nucleons. Along with one-particle excitations (excitations of independent nucleons moving in some self-consistent field), there are so-called collective excitations. These excitations describe rotational and vibrational motions of the nuclear substance. Their description is the subject of the unified model.[80-83]
    Let us consider a deformed nucleus with an axial symmetry. Our goal is to describe the rotational state of this nucleus.
    The wavefunction Ψ of the nucleus can be found on the basis of the adiabatic approach. Assume that the rotational motion is much slower than the internal motion. Then the function Ψ can be written as a product [cf. Eq. (6-51)]:

$$\Psi = \phi(r) \, \phi_{rot}(\eta).$$

Here $\phi_{rot}(\eta)$ describes the rotational motion (η is a set of Euler's angles).
    The total Hamiltonian can be written in the form

$$H = H_0 + H_{rot} + H',$$

where

$$H_0 = H(r) + \frac{\hbar^2}{2I}\mathbf{j}^2$$

is the part describing the internal motion (I is the moment of inertia, j is the angular momentum for the internal degrees of freedom, and r is the set of internal coordinates), and

$$H_{rot} = \frac{\hbar^2}{2I}\mathbf{J}^2,$$

$$H' = -\frac{\hbar^2}{2I}\mathbf{J}\cdot\mathbf{j}. \tag{6-108}$$

The term H_{rot} describes the rotational motion of the nucleus and H' is the interaction between the internal and rotational motions.

The rotational wavefunction $\phi_{rot}(\eta)$ is the eigenfunction of the operator H_r (see, e.g., reference 43, pp. 58, 103). Therefore the adiabatic approach allows one to write the total wavefunction in the form

$$\Psi = \left(\frac{2J+1}{8\pi^2}\right)^{1/2} D_{KM}^J(\eta)\,\phi_i(r).$$

Here D is the finite-rotation matrix, and K, M are the components of the angular momentum along the axis of the "top" and along the Z-axis fixed in space. $\phi_i(r)$ is an eigenfunction of the operator H_0.

If, for example, $K = 0$, then

$$\Psi = \frac{1}{2\pi} Y_{IM}(\eta)\,\phi_i(v)$$

The energy spectrum is described by the expression[81]

$$E = E^0 + \frac{\hbar^2}{2I} J(J+1) + (-1)^{J-1/2}\, \mathit{\beta}\left(J + \frac{1}{2}\right).$$

Here E^0 is the internal energy, the second term describes the rotational manifold, and the last term is owing to the term H' in the Hamiltonian [see Eq. (6-108)]; $\mathit{\beta}$ is the so-called coupling constant.

Another type of collective motion is the collective vibrational motion that is the vibration of the surface with respect to equilibrium. One should notice that the low-lying excited states in spherical even-even nuclei are connected with these vibrations. One introduces the distance $R(\eta, \phi)$ from the center of the nucleus to the surface (η and ϕ are angles in the spherical coordinate system) and expand this quantity in a series of spherical functions;

$$R(\eta, \phi) = \left\{ 1 + \sum_{\lambda\mu} \alpha_{\lambda\mu} Y_{\lambda\mu}(\eta, \phi) \right\} \tilde{R},$$

where $\tilde{R}$ is the radius of the equivalent sphere. The set of parameters $\alpha_{\lambda\mu}$ describes the shape of the deformed nucleus. If the deviation from the spherical shape is not large, then $\alpha_{\lambda\mu}$ are small quantities.

The energy of surface vibrations can be written in the form

$$E = \tfrac{1}{2} \sum_{\lambda} \left\{ B_\lambda |\dot\alpha_{\lambda\mu}|^2 + C_\lambda |\alpha_{\lambda\mu}|^2 \right\}.$$

The quantities $\alpha_{\lambda\mu}$ and $\dot\alpha_{\lambda\mu}$ can be considered as generalized coordinates and generalized velocities, respectively. (B_λ and C_λ are some coefficients.) Introducing the generalized momenta $\pi_{\lambda\mu} = \partial E/\partial\dot\alpha_{\lambda\mu} = B_\lambda \dot\alpha_{\lambda\mu}$, one can express the Hamiltonian as

$$H = \tfrac{1}{2} \sum_{\lambda\mu} \left\{ B_\lambda^{-1} |\hat\pi_{\lambda\mu}|^2 + C_\lambda |\alpha_{\lambda\mu}|^2 \right\}$$

and then find the spectrum

$$E = \sum_{\lambda\mu} (n_{\lambda\mu} + \tfrac{1}{2}) \hbar\omega_\lambda$$

where $\omega_\lambda = (C_\lambda/B_\lambda)^{1/2}$.

These collective excitations are called "phonons," by analogy with the vibrational quasi-particles in solids.

Vibrational and rotational levels in molecules have been described in Section 6-1. It turns out that these manifolds can be separated. Vibrational-rotational interaction in molecules is proportional to the adiabatic parameter κ^2 [see Eq. (6-50)] and, therefore, is small. The situation in nuclei is different. Generally speaking, the vibrational-rotational interaction does not contain any small parameter and collective motion often represents a complex combination of rotational and vibrational motions. One can introduce a parameter of nonadiabaticity μ describing the coupling of these two motions.[84] Some nuclei are characterized by a small value of μ (e.g., $^{156}_{64}$Cd, $^{230}_{90}$Th, $^{238}_{92}$U) and the vibrational manifold can be separated out with high accuracy. However, there are nuclei (e.g., $^{56}_{26}$Fe, $^{198}_{80}$Hg, $^{72}_{32}$Ge) with strong vibrational-rotational coupling.

The described adiabatic approach allows us to separate collective levels from one-particle excitations. Such separation means that the spacing of collective, rotational and vibrational, excitations is much less than the distance between the levels of one-particle excitations. According to experimental data, this assumption is particularly correct for even-even nuclei. This feature of nuclei containing even number of nucleons is due to the pair correlation of the particles. The amount of energy ΔE, which is required in order to create a one-particle excitation, should exceed the coupling energy. Usually $\Delta E \simeq 1-2$ MeV, whereas the manifold of collective levels is located much lower. For example, the low-lying rotational levels of the $^{238}_{94}$Pu are $\Delta E_1^{rot} = 44.2$ keV, $\Delta E_2^{rot} = 145.8$ keV, and $\Delta E_3^{rot} = 303.4$ keV. The adiabatic approach is the basis of the unified theory.

Thus the adiabatic approach is a powerful method in the theory of solids, molecules, and nuclei. A rigorous quantum-mechanical treatment based on the adiabatic approximation allows us to obtain interesting information about various many-body systems.

ACKNOWLEDGMENTS

The authors are indebted to C. E. Dateo for his assistance during the preparation of this manuscript.

This work was supported for V. Z. Kresin by the U.S. Office of Naval Research under Contract N00014-83-F-0109 and for W. A. Lester, Jr. by the Director, Office of Energy Research, Office of Basic Energy Sciences, Chemical Sciences, Division of the U.S. Department of Energy under Contract No. DE-AC03-76SF00098.

REFERENCES

1. M. Born and R. Oppenheimer, "Zur Quantentheorie der Molkeln," *Annalen der Physik* 84 (1927): 457–484.
2. M. Born and K. Huang, *Dynamical Theory of Crystal Lattices* (Oxford: Clarendon Press, 1954).
3. R. Kubo, "Thermal Ionization of Trapped Electrons," *Physical Review* 86 (1952): 929–937.
4. J. Ziman, "The Electron-Phonon Interaction, According to the Adiabatic Approximation," *Proceedings of the Cambridge Philosophical Society* 51 (1955): 707–712.
5. V. Z. Kresin and W. A. Lester, Jr., "A New Adiabatic Approach to the Photodissociation of Polyatomic Molecules," in *International Journal of Quantum Chemistry: Quantum Chemistry Symposium*, vol. 51, (New York: Wiley-Interscience, 1981), pp. 703–714.

6. A. Migdal, *Qualitative Methods in Quantum Theory* (New York: W. Benjamin, 1977).
7. B. Geilikman, "The Adiabatic Approximation and Fröhlich Model in the Theory of Metals," *Journal of Low Temperature Physics* 4 (1971): 189–208.
8. B. Geilikman, "Adiabatic Perturbation Theory for Metals and the Problem of Lattice Stability," *Soviet Physics-Uspekhi* 18 (1975): 190–202.
9. J. Ziman, *Electrons and Phonons* (Oxford: Clarendon Press, 1960).
10. E. Brovman and Y. Kagan, "The Phonon Spectrum of Metals," *Soviet Physics JETP* 25 (1967): 365–376.
11. A. Migdal, "Interaction between Electrons and Lattice Vibrations in a Normal Metal," *Soviet Physics JETP* 7 (1958): 996–1001.
12. A. Abrikosov, L. Gor'kor, and I. Dzyaloshinskii, *Quantum Field Theoretical Methods in Statistical Physics* (New York: Pergamon Press, 1965).
13. N. Ashcroft and J. Wilkins, "Low Temperature Electronic Specific Heat of Simple Metals," *Physics Letters* 14 (1965): 285–287.
14. P. Allen and M. L. Cohen, "Pseudopotential Calibration of the Mass Enhancement and Superconducting Transition Temperature of Simple Metals," *Physical Review* 187 (1969): 525–538.
15. G. Grimvall, "The Electron-Phonon Interaction in Normal Metals," *Physica Scripta* 14 (1976): 63–78.
16. G. Grimvall, "Temperature Dependent Effective Masses of Conduction Electrons," *Journal of Physics and Chemistry of Solids* 29 (1968): 1221–1225.
17. V. Z. Kresin and G. Zaitsev, "Temperature Dependence of the Electron Specific Heat and Effective Mass," *Soviet Physics-JEPT* 47 (1978): 983–989.
18. J. Sabo, Jr., "Temperature Dependence of the Cyclotron Effective Mass in Zinc," *Physics Review* B1 (1969): 1325–1328.
19. Y. Krasnopolin and M. Khaikin, "Electron-Phonon Interaction and Cyclotron Resonance in Lead," *Soviet Physics JETP* 37 (1975): 883–889.
20. G. Eliashberg, "The Low Temperature Specific Heat of Metals," *Soviet Physics-JETP* 16 (1963): 780–781.
21. A. Miller and B. Brockhouse, "Crystal Dynamics and Electronic Specific Heats of Palladium and Copper," *Canadian Journal of Physics* 49 (1971): 704–723.
22. G. Knapp and R. Jones, "Determination of the Electron-Phonon Enhancement Factor from Specific-Heat Data," *Physics Review* B6 (1972): 1761–1767.
23. R. Prange and L. Kadanoff, "Transport Theory for Electron-Phonon Interactions in Metals," *Physical Review* A134 (1954): 566–580.
24. D. Bohm and T. Staver, "Application of Collective Treatment of Electron and Ion Vibrations to Theories of Conductivity and Superconductivity," *Physical Review* 84 (1952): 836–837.
25. J. Bardeen and D. Pines, "Electron-Phonon Interaction in Metals," *Physical Review* 99 (1955): 1140–1150.
26. A. Rajagopal and M. Cohen, "Dynamic Theory of Crystal Lattices from a Microscopic Standpoint," *Collective Phenomena* 1 (1972): 9–27.
27. D. Pines, *Elementary Excitations in Solids* (New York: W. Benjamin, 1963).
28. I. Gomersal and B. Gyorffy, "Variation of T_c with Electron-per-Atom Ratio in Superconducting Transition Metals and Their Alloys," *Physical Review Letters* 33 (1974): 1286–1290.

296 Vladimir Z. Kresin and William A. Lester, Jr.

29. R. Peierls, *Quantum Theory of Solids* (Oxford: Clarendon Press, 1955).
30. J. Bardeen, L. Cooper, and J. Schrieffer, "Theory of Superconductivity," *Physical Review* 108 (1957): 1175–1206.
31. J. Schrieffer, *Superconductivity* (New York: W. Benjamin, 1964).
32. G. Eliashberg, "Temperature Green's Function for Electrons in a Superconductor," *Soviet-Physics-JETP* 12 (1961): 1000–1002.
33. G. Herzberg and E. Teller, "Schwingungsstruktur der Elektronenübergänge bei Mehratomigen Molekulen," *Zeitschrift für physikalische Chemie* B21 (1933): 410–446.
34. H. Longuet-Higgins, "Some Recent Developments in the Theory of Molecular Energy Levels," in *Advances in Spectroscopy*, vol. 2, ed. H. W. Thompson (New York: Wiley-Interscience, 1961), pp. 429–472.
35. K. Freed and W. Gelbart, "On the Born-Oppenheimer Separation and the Calculation of Nonradiative Transition Rates," *Chemical Physics Letters* 10 (1971): 187–192.
36. A. Nitzan and J. Jortner, "What is the Nature of Intramolecular Coupling Responsible for Internal Conversion in Large Molecules?" *Chemical Physics Letters* 11 (1971): 458–463.
37. M. Kemper, J. Van Dijk, and H. Buck, "On the Comparison between Crude and Adiabatic Born-Oppenheimer Coupling Elements," *Chemical Physics Letters* 48 (1977): 590–592.
38. V. Z. Kresin and W. A. Lester, Jr., "The Perturbative Approach in Adiabatic Theory," to be published.
39. L. Landau and E. Lifshitz, *Quantum Mechanics* (New York: Pergamon Press, 1975).
40. G. Herzberg, *Electronic Spectra and Electronic Structure of Polyatomic Molecules* (Princeton: Van Nostrand, 1966).
41. H. A. Jahn and E. Teller, "Stability of Polyatomic Molecules in Degenerate Electronic States. I-Orbital Degeneracy," *Proceedings of the Royal Society* A161 (1937): 220–235.
42. K. Englman, *The Jahn-Teller Effect in Molecules and Crystals* (New York: Wiley-Interscience, 1972).
43. L. Landau and E. Lifshitz, *Mechanics* (Oxford: Pergamon Press, 1976).
44. E. Ruch and A. Schonhofer, "Ein Beweis des Jahn-Teller-Theorems mit Hilfe eines Satzes uber die Induktion von Darstellungen endlicher Gruppen," *Theoretica Chimica Acta* 3 (1965): 291–304.
45. S. Aronowitz, "Adiabatic Approximation and the Jahn-Teller Theorem," *Physical Review* A14 (1976): 1319–1325.
46. R. Renner, "Zur Theorie der Wechselwirkung Zwischen Elektronen-und Kernbewegung bei Dreiatomigen, Stabförmigen Molekülen," *Zeitschrift für Physik* 92 (1934): 172–193.
47. H. Sponer and E. Teller, "Electronic Spectra of Polyatomic Molecules," *Reviews of Modern Physics* 13 (1941): 75–170.
48. J. Van-Vleck, "On the Magnetic Behavior of Vanadium, Titanium, and Chrome Alum," *Journal of Chemical Physics* 7 (1939): 61–71.
49. R. Coffman, "Operator Form for the Linear and Quadratic Jahn-Teller Coupling in Octahedrally Coordinated Eg Electronic States," *Journal of Chemical Physics* 44, (1966): 2305–2306.

50. L. Cederbaum, H. Koppel, and W. Domcke, "Multimode Vibronic Coupling Effects in Molecules," *International Journal of Quantum Chemistry Symposium* 15 (1981): 251–267.

51. L. Cederbaum and H. Koppel, "What Happens When Several Closely Lying Electronic States Interact through Nuclear Motion?" *Chemical Physics Letters* 87 (1982): 14–17.

52. L. Cederbaum, "The Multistate Vibronic Coupling Problem," *Journal of Chemical Physics* 78 (1983): 5714–5728.

53. F. Blatt, *Theory of Mobility of Electrons in Solids* (New York: Academic Press, 1957).

54. W. Siebrand, "Nonradiative Processes in Molecular Systems," in *Modern Theoretical Chemistry,* vol. 1A, ed. W. H. Miller, Dynamics of Molecular Collisions (New York: Plenum Press, 1976), pp. 249–302.

55. J. Jortner, S. Rice, and R. Hochstrasser, "Radiationless Transitions in Photochemistry," in *Advances in Photochemistry,* vol. 7, J. Pitts, Jr., G. Hammond, and W. Noyes, Jr. (New York: Wiley-Interscience, 1969), pp. 149–309.

56. K. Freed, "Energy Dependence of Electronic Relaxation Processes in Polyatomic Molecules," in *Topics in Applied Physics,* vol. 15, *Radiationless Processes in Molecular and Condensed Phases,* ed. F. Fong (New York: Springer-Verlag, 1976), pp. 23–168.

57. F. Fong, "Introduction," in *Topics and Applied Physics,* vol. 15, *Radiationless Processes in Molecular and Condensed Phases,* ed. F. Fong (New York: Springer-Verlag, 1976), pp. 1–21.

58. F. Fong, *Theory of Molecular Relaxation: Applications in Chemistry and Biology* (New York: Wiley-Interscience, 1975).

59. M. Kasha, "Characterization of Electronic Transitions in Complex Molecules," *Discussions, Faraday Society* 9 (1950): 14–19.

60. B. Henry and M. Kasha, "Radiationless Molecular Electronic Transition," *Annual Review of Physical Chemistry* 19 (1968): 161–192.

61. B. Henry and W. Siebrand, "Radiationless Transitions," in *Organic Molecular Photophysics,* vol. 1, ed. J. Birks (New York: Wiley-Interscience, 1973), pp. 153–237. (See also the postscript in vol. 2, pp. 303–311).

62. W. Gelbart, K. Freed, and S. Rice, "Internal Rotations and the Breakdown of the Adiabatic Approximation: Many-Phonon Radiationless Transitions," *Journal of Chemical Physics* 52 (1970): 2460–2473.

63. F. Rebentrost and W. A. Lester, Jr., "Nonadiabatic Effects in the Collision of $F(^2P)$ with $H_2(^1\Sigma_g^+)$. I," *Journal of Chemical Physics* 63 (1975): 3737–3740.

64. F. Rebentrost and W. A. Lester, Jr., "Nonadiabatic Effects in the Collision of $F(^2P)$ with $H_2(^1\Sigma_g^+)$. II," *Journal of Chemical Physics* 64 (1976): 3879–3884.

65. F. Rebentrost and W. A. Lester, Jr., "Resonant Electronic-to-Rotational Energy Transfer: Quenching of $F(^2P_{1/2})$ by $H_2(j=0)$," *Journal of Chemical Physics* 64 (1976): 4223–4224.

66. F. Rebentrost and W. A. Lester, Jr., "Nonadiabatic Effects in the Collision of $F(^2P)$ with $H_2(^1\Sigma_g^+)$. III," *Journal of Chemical Physics* 67 (1977): 3367–3375.

67. Z. H. Top and M. Baer, "Incorporation of Electronically Nonadiabatic Effects into Bimolecular Reactive Systems. I. Theory," *Journal of Chemical Physics* 66 (1977): 1363–1371.

68. K. Bowman, S. Leasure, and A. Kuppermann, "Large Quantum Effects in a

Model Electronically Nonadiabatic Reaction: Ba + $N_2O \rightarrow$ BaO* + N_2^+,'' *Chemical Physics Letters* 43 (1976): 374–376.

69. F. Smith, "Diabatic and Adiabatic Representations for Atomic Collision Problems," *Physical Review* 179 (1969): 111–123.

70. J. C. Tully, "Nonadiabatic Processes in Molecular Collisions," in *Modern Theoretical Chemistry*, vol. 1B, ed. W. H. Miller, *Dynamics of Molecular Collisions* (New York: Plenum Press, 1976), pp. 217–267.

71. F. Rebentrost, "Nonadiabatic Molecular Collisions," in *Theoretical Chemistry*, vol. 6B, ed. D. Henderson (New York: Academic Press, 1981), pp. 1–77.

72. B. C. Garrett and D. G. Truhlar, "The Coupling of Electronically Adiabatic States in Atomic and Molecular Collisions," in *Theoretical Chemistry, Part A*, ed. D. Henderson (New York: Academic Press, 1981).

73. C. Mead and D. Truhlar, "Conditions for the Definition of a Strictly Diabatic Electronic Basis for Molecular Systems," *Journal of Chemical Physics* 77 (1982): 6090–6098.

74. V. Z. Kresin and W. A. Lester, Jr., "Inverse Vibrational Distributions of Photofragments," *Chemical Physics Letters* 87 (1982): 392–396.

75. V. Z. Kresin and W. A. Lester, Jr., "Theory of Polyatomic Photodissociation Adiabatic Description of the Dissociative State and the Translation-Vibration Interaction," *Journal of Physical Chemistry* 86 (1982): 2182–2187.

76. G. Herzberg, *Infrared and Raman Spectra of Polyatomic Molecules* (Princeton: Van Nostrand, 1965).

77. Y. Band and K. Freed, "Dissociation Processes of Polyatomic Molecules," *Journal of Chemical Physics* 63 (1975): 3382–3397.

78. V. Z. Kresin and W. A. Lester, Jr., "Reaction Hamiltonian and the Adiabatic Approach to the Dynamics of Chemical Reactions," *Journal of Chemical Physics* 90 (1984): 335; V. Z. Kresin, W. A. Lester, Jr., M. Dupuis, and C. Dateo, "Chemical Reaction as a Quantum Transition," in *International Journal of Quantum Chemistry: Quantum Chemistry Symposium*, vol. 18, (New York: Wiley-Interscience), in press.

79. F. Wilson, J. Decius, and R. Cross, *Molecular Vibrations* (New York: Dover Publications, 1983).

80. A. Bohr, "The Coupling of Nuclear Surface Oscillations to the Motion of Individual Nucleons," *Danske Videnskabernes Selskab-Matem-Fysiske Meddelelser* 26 (1952): 1–40.

81. A. Bohr and B. Mottelson, "Collective and Individual-Particle Aspects of Nuclear Structure," *Danske Videnskabernes Selskab-Matem-Fysiske Meddeleser* 27 (1953): 1–174.

82. A. Lane, *Nuclear Theory* (New York: W. Benjamin, 1964).

83. D. Rowe, *Nuclear Collective Motion* (London: Methuen, 1970).

84. A. Davydov and A. Chaban, "Rotation-Vibration Interaction in Non-Axial Even Nuclei," *Nuclear Physics* 20 (1960): 499–508.

7

Catastrophe Theory: What It Is; Why It Exists; How It Works

Robert Gilmore

Drexel University

ABSTRACT: Catastrophe theory is a program. The thrust of the program is to describe and classify the qualitative properties of the solutions of equations, and in particular to determine how these qualitative properties change as the parameters that appear in the equations change. Elementary catastrophe theory is the third in a sequence of developments in elementary calculus, whose two preceding stages include the implicit function theorem (for linearization) and the Morse lemma (for quadratic forms). We shall survey mathematics to find the niche occupied by elementary catastrophe theory.

A family of functions may depend on control parameters as well as state variables. As the control parameters are changed, the qualitative properties of a critical point may change. These interesting critical points can be transformed to canonical form by using two distinct mathematical tools: the degrees of freedom by means of which control parameters can annihilate the leading terms in a Taylor series expansion; and the degrees of freedom inherent in a nonlinear coordinate transformation, which can be used to transform away the tail of the Taylor series expansion. What is left over, in the middle, is the catastrophe function that represents the interesting critical point.

When a physical system is represented by a mathematical function (e.g., potential or equation of state) that contains a catastrophe, many characteristic qualitative features are observed. The eight features that are often observed are described. The applications of catastrophe theory to physical systems, and in particular the manifestations of these eight features in physical systems, are illustrated with respect to six systems.

Catastrophe Theory: What It Is

7-1   THE PROGRAM OF CATASTROPHE THEORY

Catastrophe theory[1-6] is a program. The object of this program is to determine how the solutions to families of equations change as the parameters that appear in these equations change.[4]

In general, a small change in parameter values (Reynolds number, interaction strength) has only a small quantitative effect on the solutions of these equations. However, under certain conditions a small change in the value of some parameter has a very large quantitative effect on the solutions of these equations. Large quantitative changes in solutions describe qualitative changes in the behavior of the system modeled.

Catastrophe theory is therefore concerned with determining the parameter values at which there occur qualitative changes in the solutions of families of equations described by parameters.

This program has had few successes because of its great generality. Even for systems of equations of the form

$$F(x, \dot{x}, t; c) = 0, \qquad (7\text{-}1)$$

where

x is an n-vector: $x = (x_1, x_2, \ldots, x_n) \in \mathbf{R}^n$, called a state vector; c is a k-vector: $c = (c_1, c_2, \ldots, c_k) \in \mathbf{R}^k$, called the control parameters; and the dot indicates differentiation with respect to t, little can be said. Even when the form of Equations (7-1) is restricted to dynamical systems form

$$\dot{x}_i = f_i(x, t; c),$$

almost nothing is known. For time-independent, or autonomous, dynamical systems of the form

$$\dot{x}_i = f_i(x; c), \qquad (7\text{-}2)$$

a few results are available. For $n = 2$, the qualitative behavior of these systems is well understood for $k = 1$ and not well understood for $k > 1$. For $n = 3$, the behavior of these systems is provocative for any value of k. For $n = 4$, the qualitative behavior of these systems seems sometimes almost within reach. For larger values of n, new methods of study are required.

As in physics, so also in dynamical systems theory: Marvelous simplifications occur when the force is derivable from a potential. For autonomous dynamical systems of the form (7-2), with

$$f_i = -\frac{\partial V(x; c)}{\partial x_i} \tag{7-3}$$

(gradient dynamical systems), a great many statements can be made about the qualitative nature of the solutions and how the qualitative nature depends on the control parameters c.

These qualitative properties can be ferretted out of the potential $V(x; c)$ fairly easily. Following standard procedure in dynamical systems theory, the equilibria ($\dot{x}_i = 0$) of the system are located. The qualitative nature of the flow in the neighborhood of each equilibrium is then determined. The flows around each equilibrium are then joined in a smooth way (intuition often is the best method).

If a small change in the control-parameter values produces only a small displacement of the equilibria, then the system undergoes no qualitative change. Qualitative changes occur when small changes in control-parameter values cause two or more equilibria either to coalesce and annihilate each other or to be created and disperse.

Elementary catastrophe theory is the study of how the equilibria of $V(x; c)$ move about, coalesce and annihilate, or bifurcate and disperse, as the control parameters c are varied.

7-2 THREE THEOREMS OF ELEMENTARY CALCULUS

Elementary catastrophe theory exists at the intersection of two lines of mathematical development. One is the program of catastrophe theory, which attempts to study the qualitative properties of solutions of equations. The other is a series of results in elementary calculus dealing with the canonical forms for functions. The first two landmarks in this sequence of results are the implicit function theorem and the Morse lemma.[7] The implicit function theorem deals with functions that have a good linear approximation. The Morse lemma deals with functions that can adequately be approximated by a quadratic form. The third in this sequence of developments is the Thom theorem. This theorem provides canonical forms for functions in neighborhoods where neither the linear approximation nor the quadratic approximation is adequate.

Implicit Function Theorem

The implicit function theorem tells us that if the slope of a function is nonzero at a point, then that function can be represented locally by a linear function in some appropriate coordinate system.

This result is well known and generally covered in classes in elementary calculus. It is also intuitively obvious: Ask any physicist if the linear approximation is adequate when it is nonzero and the answer is likely to be "probably, at least locally."

Implicit function theorem

Let $f(x) = f(x_1, x_2, \ldots, x_n)$ be a function with nonzero gradient at a point x^0:

$$\nabla f|_{x^0} \neq 0.$$

Then it is possible to find a new coordinate system $y = y(x)$ such that

$$f \doteq y_1.$$

That is, f is equal, after a smooth change of coordinates, to y_1.

Morse Lemma

If $\nabla f = 0$ at a point, then physicists are likely to say "the implicit function theorem fails." Of course, theorems do not fail; rather, the conditions for the theorem fail to be satisfied.

When the implicit function theorem is not applicable, is there a followup theorem that can be used to provide a canonical form for f where $\nabla f = 0$? Once again, the vaunted intuition of physicists has successfully proposed a solution (canonical quadratic form) that was subsequently given a rigorous mathematical formulation, the Morse lemma.[19] This lemma states that if $\nabla f = 0$ but the determinant of the matrix of mixed second partial derivatives is nonzero,

$$\det \left[\frac{\partial^2 f}{\partial x_i \partial x_j} \right] \neq 0,$$

then $f(x)$ has a canonical quadratic form.

Morse lemma

Let $f(x) = f(x_1, x_2, \ldots, x_n)$ be a function with the property

$$\nabla f = 0,$$

$$\det \left[\frac{\partial^2 f}{\partial x_i \partial x_j} \right] \neq 0,$$

at a point. Then there is a smooth change of variables $x' = x'(x)$ such that

$$f \doteq \sum_{i=1}^{n} \lambda_i (x_i')^2, \qquad (7\text{-}4)$$

where λ_i are the eigenvalues of the stability matrix $[\partial^2 f/\partial x_i \partial x_j]$.

By absorbing the nonzero eigenvalues into the length scale according to

$$y_i = |\lambda_i|^{1/2} \, X_i',$$

the quadratic form (7-4) is reduced to the Morse canonical form

$$f \doteq M_i^n, \qquad (7\text{-}5)$$

$$M_i^n(y) = -y_1^2 - \cdots - y_i^2 + y_{i+1}^2 + \cdots + y_n^2. \qquad (7\text{-}6)$$

The forms (7-6) are called Morse saddles. The Morse saddle M_0^n has a minimum at $y = 0$. The only stable Morse saddle is M_0^n.

Thom Theorem

If, at an equilibrium (defined by $\nabla f = 0$), the stability matrix is singular (det $[\partial^2 f/\partial x_i \partial x_j] = 0$), we might be inclined to say that the Morse lemma fails. Once again, lemmas do not fail; rather, the conditions for the lemma fail to be satisfied.

We can now ask for a theorem or a canonical form to apply when the Morse lemma is inapplicable, just as the Morse lemma (second derivatives) was used to follow up the implicit function theorem (first derivatives). It is here that the vaunted intuition of physicists has failed miserably, and success in the search for canonical forms has only come through the proof of rigorous mathematical theorems first proposed by Thom.

These results work generally as follows.[10, 11] The function $f(x)$ is expanded up to second order in Taylor series expansion. The stability matrix is brought to diagonal form by a linear transformation. A new coordinate system $x' = x'(x)$ is determined in which the first l coordinates, $x'_i, \ldots, x'_l$, are tangent to the eigenvectors with zero eigenvalue. The remaining $n - l$ coordinates $x'_{l+1}, \ldots, x'_n$ are tangent to the remaining eigenvectors with nonzero eigenvalue. This effects a decomposition of the form

$$f(x) = f_{NM}(x'_1, \ldots, x'_l) + f_M(x'_{l+1}, \ldots, x'_n).$$

The function f_M is a Morse function (i.e., satisfies the Morse lemma) and can be put into Morse canonical form (7-5). The function f_{NM} does not satisfy the Morse lemma. It is only this function of l variables, rather than the original function f of n variables, that must be put into a non-Morse canonical form.

Thom theorem (splitting lemma)

Let $f(x)$ be a function with the properties

$$\nabla f = 0,$$

$$\det \left[\frac{\partial^2 f}{\partial x_i \partial x_j} \right] = 0,$$

at a point. If the stability matrix $[\partial^2 f/\partial x_i \partial x_j]$ has l zero eigenvalues and i negative eigenvalues, then

$$f(x) \doteq f_{NM}(x'_1, \ldots, x'_l) + M_i^{n-l}(x'_{l+1}, \ldots, x'_n).$$

This result does not classify the possible forms of the non-Morse functions $f_{NM}(x')$. The classification results are consequences of another theorem proposed originally by Thom. This theorem involves explicit reference to control parameters.

For a typical function $f(x)$ at a critical point $\nabla f = 0$, the eigenvalues of the stability matrix will generally be nonzero. If the function depends on one or more control parameters, $f = f(x; c)$, then the eigenvalues of the stability matrix, $\lambda_i(c)$, will also depend on the control parameters. As a result, one or more eigenvalues can naturally vanish for certain control-parameter values. Thom's classification theorem provides a list of possible canonical forms of non-Morse functions, depending on the number of control parameters, k, and on the number of vanishing eigenvalues, l.

Thom theorem (classification theorem)

Let $f_{NM}(x'; c)$ be a non-Morse function, derived from the splitting lemma, depending on k control parameters and l state variables. Then

$$f_{NM}(x'; c) \doteq \text{Cat}(l, k),$$

$$\text{Cat}(l, k) \equiv \text{CG}(l) + \text{Pert}(l, k).$$

Here $\text{Cat}(l, k)$, the catastrophe function, is a function of l canonical state variables $y_1, \ldots, y_l$ and k canonical control parameters $a_1, \ldots, a_k$. The catastrophe function $\text{Cat}(l, k)$ has a further decomposition (another splitting) into two parts, $\text{CG}(l)$ and $\text{Pert}(l, k)$. The catastrophe germ, $\text{CG}(l)$, depends on only the l state variables. All of its mixed second partial derivatives vanish at the critical point. The perturbation, $\text{Pert}(l, k)$, depends on the l state variables and on the k canonical control parameters. The dependence on the canonical control parameters is linear.

We reemphasize, in closing this section that the three results, the Implicit function theorem, the Morse lemma, and the Thom classification theorem, depending on first, second, and higher derivatives, respectively, are all local in nature.

7-3 THE LIST OF ELEMENTARY CATASTROPHES

The canonical catastrophe functions for $k < 6$, and therefore $l < 3$, are summarized in Table 7-1.[12-15]

Table 7-1
The Canonical Catastrophe Functions Cat $(l, k) = \text{CG}(l) + \text{Pert}$ (l, k) Listed for Fewer Than Six Control Parameters ($k < 6$) and Therefore One or Two State Variables ($k = 1, 2$).

Name	k	l	CG(l)	Pert(l,k)
A_2	1	1	x^3	$a_1 x$
$A_{\pm 3}$	2	1	$\pm x^4$	$a_1 x + a_2 x^2$
A_4	3	1	x^5	$a_1 x + a_2 x^2 + a_3 x^3$
$A_{\pm 5}$	4	1	$\pm x^6$	$a_1 x + a_2 x^2 + a_3 x^3 + a_4 x^4$
A_6	5	1	x^7	$a_1 x + a_2 x^2 + a_3 x^3 + a_4 x^4 + a_5 x^5$
D_{-4}	3	2	$x^2 y - y^3$	$a_1 x + a_2 y + a_3 y^2$
D_{+4}	3	2	$x^2 y + y^3$	$a_1 x + a_2 y + a_3 y^2$
D_5	4	2	$x^2 y + y^4$	$a_1 x + a_2 y + a_3 x^2 + a_4 y^2$
D_{-6}	5	2	$x^2 y - y^5$	$a_1 x + a_2 y + a_3 x^2 + a_4 y^2 + a_5 y^3$
D_{+6}	5	2	$x^2 y + y^5$	$a_1 x + a_2 y + a_3 x^2 + a_4 y^2 + a_5 y^3$
$E_{\pm 6}$	5	2	$x^3 \pm y^4$	$a_1 x + a_2 y + a_3 xy + a_4 y^2 + a_5 xy^2$

The catastrophe functions listed in Table 7-1 are elementary in the sense that the catastrophe germ depends on no free parameters. That is, any coefficient in the germ can be renormalized to a specific numerical value, such as ± 1 or 0. For example, if the germ for A_4 were λx^5 ($\lambda \neq 0$), the value of λ could be renormalized to ± 1 by a simple scale transformation: $x' = \pm |\lambda|^{1/5} x$.

Table 7-1 does not list all the elementary catastrophe functions. There are two infinite series of elementary catastrophes and one finite series of elementary catastrophes. These are summarized in Table 7-2. One infinite series of elementary catastrophes, A_n, depends on a single state variable; the other, D_n, depends on two state variables. The finite series of catastrophes, called exceptional, contains only three members: E_6, E_7, and E_8.[16-18] The classification of the elementary catastrophes follows the classification of simple Lie algebras, all of whose nonzero roots have the same length.)

The reason that the list in Table 7-1 terminates at $k = 5$ is the following. For $k = 6$, it is possible that the six control parameters can be used to annihilate all six second-degree coefficients in the Taylor series of a function of three state variables. If the third-degree terms are then brought to canonical form, one of the ten cubic coefficients cannot be given a canonical

Table 7-2
Listing of All Elementary Catastrophes. [There are two infinite series of catastrophes, the cuspoid catastrophes (A_k), depending on one state variable ($l = 1$) and the umbilics (D_k), depending on two. There is in addition one finite chain, the exceptional catastrophes E_k, with $l = 2$. The subscript (Milnor number, 8 for E_8) is the maximum number of isolated critical points into which the degenerate critical point splits under perturbation. The Milnor number exceeds the dimension of the perturbation by one.]

Name	Catastrophe Germ	Perturbation
$A_{\pm k}$[a]	$\pm x^{k \pm 1}$	$\sum\limits_{j=1}^{k-1} a_j x^j$
	$x^2 y \pm y^{k-l}$, k even	
$D_{\pm k}$		$\sum\limits_{j=1}^{k-3} a_j y^j \pm \sum\limits_{j=k-2}^{k-1} a_j x^{j-(k-3)}$[b]
	$\pm(x^2 y \pm y^{k-1})$, k odd	
$E_{\pm 6}$	$\pm(x^3 + y^4)$	$\sum\limits_{j=1}^{2} a_j y^j + \sum\limits_{j=3}^{5} a_j xy^{j-3}$
E_7	$x^3 + xy^3$	$\sum\limits_{j=1}^{4} a_j y^j + \sum\limits_{j=5}^{6} a_j xy^{j-5}$[b]
E_8	$x^3 + y^5$	$\sum\limits_{j=1}^{3} a_j y^j + \sum\limits_{j=4}^{7} a_j xy^{j-4}$

[a] $A_{+k} = A_{-k}$ if k is even.
[b] The expression of the perturbation in terms of the monomials $x_p y^q$ is not unique.

value. The resulting catastrophe germ is then not elementary. This calculation is carried out in Section 7-9. For $k \geqslant 6$, catastrophes exist that are both elementary and not elementary. For $k < 6$, only elementary catastrophes exist.[19]

7-4   WHY A LIST OF PERTURBATIONS IS REQUIRED

The implicit function theorem and the Morse lemma are statements about transformation to canonical forms. So also is the Thom classification theorem. However, the latter comes equipped not only with a list of canonical forms, called catastrophe germs, but also with a list of perturbations, one for each germ. Why this difference between the implicit function theorem and the Morse lemma on the one hand, and the Thom classification theorem on the other?

The answer is as follows. If the implicit function theorem is applicable to a function at a point, it will be applicable to a perturbation of that function at that point. Similarly, if a function can be approximated by a nondegenerate quadratic form at a point, a perturbation of the function can be so approximated. In short, perturbation of a function for which a linear or quadratic approximation is valid at a point does not change the qualitative properties of the function at the point.[20] This is not the case of a catastrophe germ. The multiplicity of possible consequences of the perturbation of a catastrophe germ are completely encapsulated in the germ's canonical perturbation: Pert(l, k).

These ideas are summarized in a series of three figures. Figure 7-1 shows a function, $f(x)$, that satisfies the conditions for the implicit function theorem at x_0 [i.e., $f'(x)_{x_0} \neq 0$]. The perturbed function, $F(x) = f(x) + p(x)$, also obeys the implicit function theorem at x_0. The qualitative properties of the function in the neighborhood of x_0 are not changed by perturbation.

Figure 7-2 shows a function, $f(x)$, that obeys the Morse lemma at $x = 0$. Under perturbation, the minimum shifts from 0 to δx and the value of the perturbed function at the critical point shifts from 0 to δF. The slope of the perturbed function at $x = 0$, though small, is nonzero, so it may be argued that the implicit function theorem is applicable to the perturbed function at $x = 0$. This point of view is not fruitful. A more useful viewpoint is that perturbation moves the critical point but does not change its type. That is, $f''(x = 0)$ and $F''(x = \delta x)$ have the same sign.

More generally, it is not difficult to show (Section 7-6) that if a function $f(x_1, \ldots, x_n)$ has a Morse critical point of index i, M_i^n, at a point p, then any perturbation of f will have a Morse critical point of the same type in the neighborhood of p.

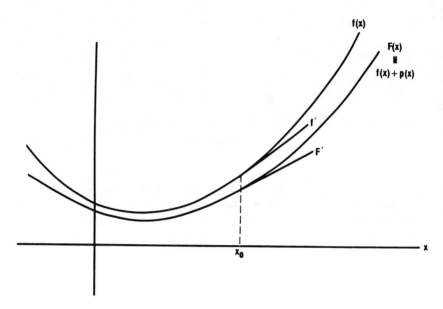

Figure 7-1
If a function, $f(x)$, satisfies the conditions of the implicit function theorem at x_0, then any sufficiently small perturbation of that function, $F(x) = f(x) + p(x)$, will also satisfy the conditions of the implicit function theorem at x_0. If $f' \neq 0$ at x_0, then $F' \neq 0$ at x_0. (*From R. Gilmore, Catastrophe Theory for Scientists and Engineers* [*New York: John Wiley & Sons, 1981*], *p. 35; copyright © 1981 by John Wiley & Sons, Inc.*)

Figure 7-3 shows the simplest of the elementary catastrophes,

$$A_2: f(x) = x^3 + a_1 x, \qquad (7\text{-}7)$$

plotted for three values of the canonical control parameter, a_1. The catastrophe germ, x^3, corresponds to the curve labeled $a_1 = 0$ in Figure 7-3. This function has a critical point at $x = 0$, since

$$\left. \frac{d}{dx} (x^3) \right|_0 = 0.$$

However, this critical point is degenerate, since

$$\left. \frac{d^2}{dx^2} (x^3) \right|_0 = 0.$$

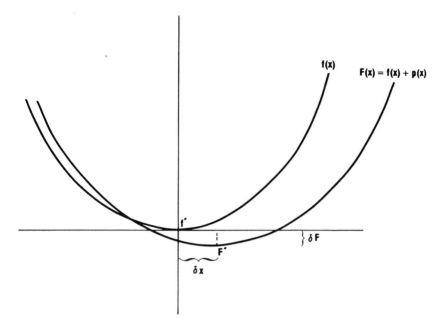

Figure 7-2
If a function, $f(x)$, satisfies the conditions of the Morse lemma at a point, then any sufficiently small perturbation of that function, $F(x) = f(x) + p(x)$, will also satisfy the conditions of the Morse lemma at a point in the neighborhood of the original point. If $f' = 0$ and $f'' \neq 0$ at x_0, then $F'(x) = 0$ and $F''(x) \neq 0$ near x_0. *(From R. Gilmore, Catastrophe Theory for Scientists and Engineers [New York: Wiley & Sons, 1981], p. 40; copyright © 1981 by John Wiley & Sons, Inc.)*

Under perturbation, two qualitatively distinct things can happen to the function x^3. The slope of the perturbed function at the origin may be positive ($a_1 > 0$), in which case there are no (real) critical points at all. Or the slope of the perturbed function at the origin may be negative ($a_1 < 0$), in which case there are two isolated critical points, a local minimum at x_+, and a local maximum at x_-.

Under perturbation, only two qualitatively distinct things can happen to the function x^3, which has a doubly degenerate critical point at the origin. Either the two critical points split up and move apart to form isolated critical points, or else they move off the real axis into the imaginary plane, and the perturbed function has no (real) critical points. This full spectrum of possibilities is completely contained in the canonical perturbation of x^3:

$$\text{Pert}\,(l = 1, k = 1) = a_1 x.$$

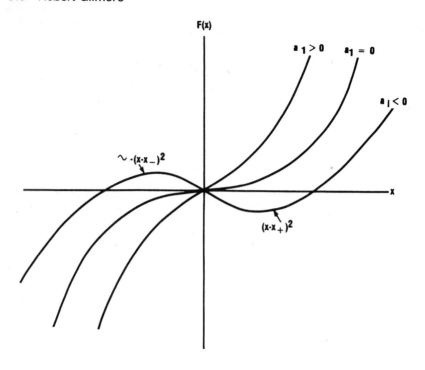

Figure 7-3

If a function, $f(x)$, does not satisfy the conditions either of the implicit function theorem or of the Morse lemma at a point, then different perturbations of that function, $F(x) = f(x) + p(x)$, will be qualitatively different. For $f(x) = x^3$ and $p(x) = a_1 x$, the perturbed function has two isolated critical points for $a_1 < 0$ and no critical points at all if $a_1 > 0$. (*From R. Gilmore, Catastrophe Theory for Scientists and Engineers [New York: Wiley & Sons, 1981], p. 45; copyright © 1981 by John Wiley & Sons, Inc.*)

The behavior of the critical points of the catastrophe function A_2 can be followed as a function of changing control parameter a_1. For a_1 negative, there are two isolated critical points of type M_1^1 and M_0^1. As a_1 is increased toward 0, the critical points approach each other. For $a_1 = 0$, the two critical points coalesce at the origin to form a doubly degenerate critical point. For $a_1 = 0$, the two critical points have annihilated each other. Sweeping a_1 in the opposite direction, we appear to have the spontaneous creation of a doubly degenerate critical point at the origin (as the two complex critical points approach the real axis) followed by bifurcation to two isolated critical points. (A field theory of interacting Morse critical points is amusing to work out.)

It is now apparent why Thom's list of canonical catastrophe germs must include also a list of perturbations, whereas a list of perturbations is not required either for the implicit function theorem or for the Morse lemma. The perturbation function, Pert(l, k), for each catastrophe germ, CG(l), contains the entire spectrum of qualitatively distinct consequences of applying an arbitrary perturbation of CG(l). Furthermore, it is the function of smallest dimension (number of independent, or control parameters, like a_1) that will do the job.

Finally, the independent control parameters all'occur linearly in the canonical perturbation.

7-5 GEOMETRY OF TWO ELEMENTARY CATASTROPHES

The question may nag: Why invest all this time and effort to learn about elementary catastrophes? The answer is the same for the catastrophes as it is whenever there are canonical mathematical structures available. The properties of these structures need be studied only once. Then, if the mathematical description of a system can be transformed to a canonical mathematical form, all of its properties are immediately known, having been previously determined for the canonical mathematical structure.

In this section we determine the properties of the two simplest catastrophes. These canonical properties are most simply displayed visually. For these reasons we shall discuss the geometry of the A_2 (fold) and the A_3 (cusp) catastrophes. For each of these functions, we shall determine four properties:

1. representative graphs of the catastrophe functions;
2. locations of the critical points;
3. values of the function at the critical points;
4. curvature of the function at the critical points.

All properties are determined as a function of the catastrophe's control parameters.

Geometry of the Fold Catastrophe (A_2)

The fold catastrophe is defined by

$$A_2: f(x; a) = \tfrac{1}{3} x^3 + ax.$$

We have renormalized the coefficient of the germ x^3 for convenience. The canonical properties of this function are summarized below and shown in Figure 7-4.[14]

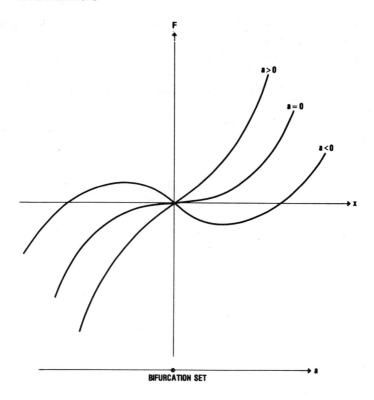

Figure 7-4(a)
The canonical geometrical properties of the fold (A_2) catastrophe $f(x; a) = \frac{1}{3}x^3 + ax$ are shown. Representative graphs of $f(x; a)$ are shown for three values of the control parameter a (top). The value $a = 0$ separates graphs with qualitatively distinct properties (two critical points versus no critical points) and is called the bifurcation set (bottom). (*From R. Gilmore, Catastrophe Theory for Scientists and Engineers* [*New York: Wiley & Sons, 1981*], *pp. 95–96; copyright* © *1981 by John Wiley & Sons, Inc.*)

Representative graphs of $f(x; a)$ are shown in Figure 7-4(a) for three values of a: $a > 0$, $a = 0$, $a < 0$. In the lower part of Figure 7-4(a) the control-parameter space is indicated. This space is a line. Each point on this line parameterizes a function in the family $f(x; a)$. Values of $a > 0$ parameterize functions with no critical points. Values of $a < 0$ parameterize functions with two isolated critical points. The point $a = 0$ separates these two qualitatively distinct types of functions. Points in the control-parameter space (such as $a = 0$ here) separating qualitatively different types of functions are called bifurcation sets. The bifurcation set of the fold catastrophe (one-parameter family of functions) is the point $a = 0$.

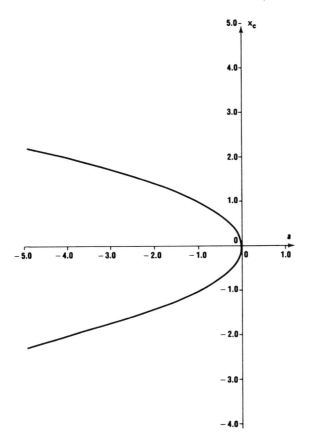

Figure 7-4(b)
The canonical geometrical properties of the fold (A_2) catastrophe $f(x; a) = \frac{1}{3}x^3 + ax$ are shown. The location of the two critical points has a canonical half-power-law dependence on the control parameter a. (*From R. Gilmore, Catastrophe Theory for Scientists and Engineers* [*New York: Wiley & Sons, 1981*], *pp. 95–96; copyright © 1981 by John Wiley & Sons, Inc.*)

The *critical points* are determined from

$$\frac{d}{dx}\left(\frac{1}{3}x^3 + ax\right) = x^2 + a = 0.$$

Two nondegenerate critical points exist only for $a < 0$. Their distances from the origin have the canonical $(-a)^{1/2}$ dependence, as shown in Figure 7-4(b).

The *value of the function* $f(x; a)$ *at the critical points* has the canonical

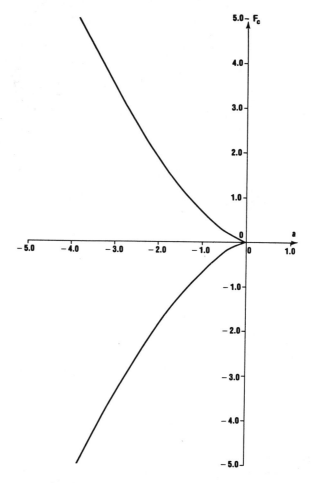

Figure 7-4(c)
The canonical geometrical properties of the fold (A_2) catastrophe $f(x; a) = \frac{1}{3}x^3 + ax$ are shown. The values of the function $f(x; a)$ at the critical points (critical values) have a canonical $\frac{3}{2}$-power-law dependence on the control parameter a. (*From R. Gilmore, Catastrophe Theory for Scientists and Engineers [New York: Wiley & Sons, 1981], pp. 95–96; copyright © 1981 by John Wiley & Sons, Inc.*)

$a^{3/2}$ power-law dependence:

$$f(x(a); a) = \pm \tfrac{2}{3}(-a)^{3/2}.$$

This dependence is illustrated in Figure 7-4(c).

The *curvature of the function $f(x; a)$ at the critical points* also possesses

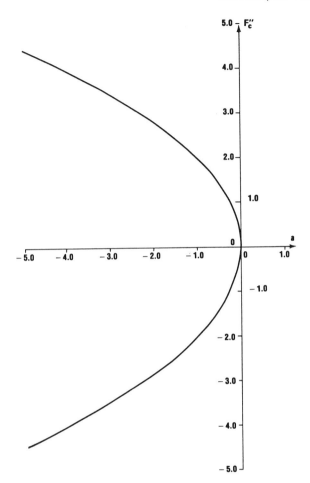

Figure 7-4(*d*)
The canonical geometrical properties of the fold (A_2) catastrophe $f(x; a) = \frac{1}{3}x^3 + ax$ are shown. The curvature of the function $f(x; a)$ at the critical points (critical curvature) has a canonical half-power-law dependence on the control parameter *a*. *(From R. Gilmore, Catastrophe Theory for Scientists and Engineers [New York: Wiley & Sons, 1981], pp. 95–96; copyright © 1981 by John Wiley & Sons, Inc.)*

a canonical $|a|^{1/2}$ power-law dependence:

$$f''(x(a); a) = \pm 2(-a)^{1/2}.$$

The curvature naturally goes to zero as $a \to 0$. This canonical behavior is shown in Figure 7-4(*d*).

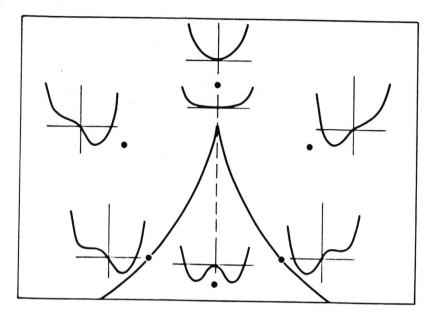

Figure 7-5(a)
The canonical geometrical properties of the cusp (A_{+3}) catastrophe $f(x; a; b) = \frac{1}{4}x^4 + \frac{1}{2}ax^2 + bx$ are shown. Every choice of control parameter values (a, b) or point in the control plane $\mathbf{R}^2$, describes a function in this family. Points within the cusp-shaped region describe functions with three isolated critical points; those outside describe functions with one isolated critical point. Points on the cusp-shaped fold lines parameterize functions either with a doubly degenerate critical point or with a triply degenerate critical point (cusp point). (*From R. Gilmore, Catastrophe Theory for Scientists and Engineers* [*New York: Wiley & Sons, 1981*], *pp. 98–100; copyright © 1981 by John Wiley & Sons, Inc.*)

Geometry of the Cusp Catastrophe (A_{+3})

The cusp catastrophe is defined by

$$A_{+3}\colon f(x; a, b) = +\tfrac{1}{4}x^4 + \tfrac{1}{2}ax^2 + bx. \tag{7-8}$$

We have again renormalized the coefficient of the germ x^4 for convenience. The canonical properties of this function are summarized next and shown in Figure 7.5.[21-23]

Representative graphs of the cusp function (7-8) are shown in Figure 7-5(a) for seven values of the control parameters (a, b). Each choice of control-parameter values, or point in the control-parameter plane $\mathbf{R}^2$, describes a single function. Functions parameterized by points within the cusp-shaped region have two local minima separated by a local maximum. On the left-

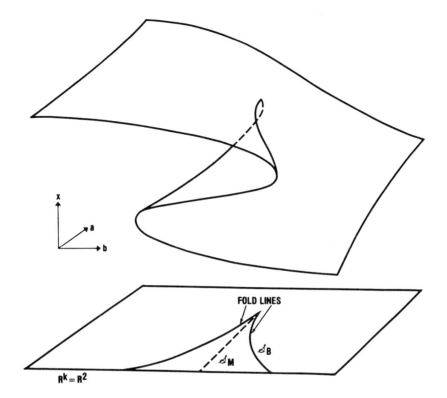

Figure 7-5(b)
The canonical properties of the cusp (A_{+3}) catastrophe $f(x; a; b) = \frac{1}{4}x^4 + \frac{1}{2}ax^2 + bx$ are shown. The locus of critical points $x^3 + ax + b = 0$ is a smooth, pleated manifold in the state variable-control parameter space $\mathbf{R}^l \otimes \mathbf{R}^k \to \mathbf{R}^1 \otimes \mathbf{R}^2 \to (x; a; b)$. The projection of the fold into the control parameter plane produces the cusp-shaped fold lines. The surface in (b) is a manifold, but the surfaces in (c) and (d) are not. (*From R. Gilmore, Catastrophe Theory for Scientists and Engineers [New York: Wiley & Sons, 1981], pp. 98–100; copyright © 1981 by John Wiley & Sons, Inc.*)

hand portion of the cusp-shaped curve, the lefthand minimum and the local maximum coalesce to form a doubly degenerate critical point. Outside the cusp-shaped region, the function has a single isolated minimum. At the cusp point $(a, b) = (0, 0)$, the function $\frac{1}{4}x^4$ is very flat. The cusp-shaped curve is a bifurcation set, separating functions of one qualitative type (two local minima, one local maximum) from those of another type (one local minimum). To get from functions of one type to those of another, it is necessary to pass through the cusp-shaped line somewhere, encountering a function with a non-Morse critical point.

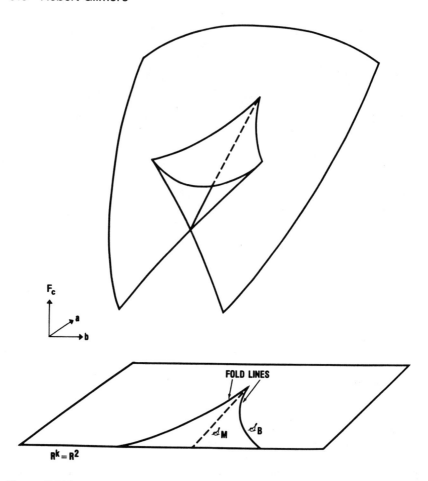

Figure 7-5(c)
The canonical geometrical properties of the cusp (A_{+3}) catastrophe $f(x; a; b) = \frac{1}{4}x^4 + \frac{1}{2}ax^2 + bx$ are shown. The critical value surface is shown in the space (F_c; a, b), where F_c is the value of $f(x; a; b)$ at a critical point. (*From R. Gilmore, Catastrophe Theory for Scientists and Engineers* [New York: Wiley & Sons, 1981], *pp. 98–100; copyright © 1981 by John Wiley & Sons, Inc.*)

The *critical points* are determined from

$$\frac{d}{dx}\left(\frac{1}{4}x^4 + \frac{1}{2}ax^2 + bx\right) = x^3 + ax + b = 0. \qquad (7\text{-}9)$$

Equation (7-9) represents a smooth manifold in the three-dimensional x-a-b space. This manifold is shown in Figure 7-5(b). Where the manifold "folds over" on itself, critical points are coalescing to form a degenerate critical

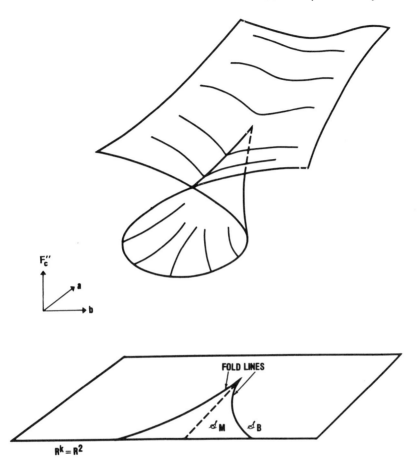

Figure 7-5(d)
The canonical geometrical properties of the cusp (A_{+3}) catastrophe $f(x; a; b) = \frac{1}{4}x^4 + \frac{1}{2}ax^2 + bx$ are shown. The critical curvature surface is shown in the space (F_c''; a, b), where F_c'' is the value of $f''(x; a; b)$ at a critical point. (*From R. Gilmore, Catastrophe Theory for Scientists and Engineers [New York: Wiley & Sons, 1981], pp. 98–100; copyright © 1981 by John Wiley & Sons, Inc.*)

point. The projection onto the control plane of these folded-over regions forms the cusp-shaped curve separating functions with two distinct types of qualitative properties.

The *value(s) of the cusp catastrophe function at its critical points* are shown in Figure 7-5(c).

The *curvature of the function at each of its critical points* is shown in Figure 7-5(d). The curvature vanishes at the folded-over part of this self-intersecting surface.

7-6 PERTURBATION OF GRADIENT DYNAMICAL SYSTEMS

In Section 7-1 we saw how the qualitative properties of a gradient dynamical system are determined by the type and distribution of its critical points. If all its critical points are isolated, then under perturbation each critical point may move slightly, but retains its Morse type, and none are created or destroyed. Therefore, under perturbation, the qualitative type of a gradient dynamical system with isolated critical points remains unchanged.

To illustrate the effect of perturbations on a gradient dynamical system, let $V(x; c)$ be a family of potentials. Assume that for control-parameter value c^0, x^0 is one of the critical points. We wish to determine the effect of changing the control-parameter value from c^0 to $c^0 + \delta c$.

This determination can be done by making a Taylor series expansion in the infinitesimals $\delta x_i = (x_i - x_i^0)$ and $\delta c_\alpha = (c_\alpha - c_\alpha^0)$:

$$V(x^0 + \delta x; c^0 + \delta c) = \quad V(x^0; c^0)$$
$$+ V_i \delta x_i + V_\alpha \delta c_\alpha$$
$$+ \tfrac{1}{2} V_{ij} \delta x_i \delta x_j + V_{i\alpha} \delta x_i \delta c_\alpha + \tfrac{1}{2} V_{\alpha\beta} \delta c_\alpha \delta c_\beta + \cdots. \tag{7-10}$$

The location of the critical point after perturbation is determined by setting $dV(x; c)/dx = 0$:

$$\frac{\partial}{\partial x_i} V(x; c) = V_{ij} \delta x_j + V_{i\alpha} \delta c_\alpha = 0. \tag{7-11}$$

The coefficient $V_i = \partial V(x^0; c^0)/\partial x_i$ is zero, since x^0 is assumed to be a critical point. If the stability matrix $V_{ij} = \partial^2 V(x^0; c^0)/\partial x_i \partial x_j$ is nonsingular, then (7-11) can be solved uniquely to give

$$\delta x_i = -(V^{-1})_{ij} V_{j\alpha} \delta c_\alpha, \tag{7-12}$$

where $(V^{-1})_{ij} V_{jk} = \delta_{ik}$. Thus a perturbation, represented by a small change in the control-parameter values, results in a small displacement of the critical point.

It is an easy matter to show that the value of the potential at the displaced critical point is

$$V(x_i^0 - (V^{-1})_{ij} V_{j\alpha} \delta c_\alpha; \delta c_\beta) = V(x^0; c^0) + V_\alpha \delta c_\vartheta$$
$$+ \tfrac{1}{2} [V_{\alpha\beta} - V_{\alpha i}(V^{-1})_{ij} V_{j\beta}] \, \delta c_\alpha \, \delta c_\beta. \tag{7-13}$$

It is also relatively straightforward to show that the stability matrix at the displaced critical point is determined from the original stability matrix and is, to lowest order,

$$V'_{ij} = V_{ij}(x^0; c^0) + P_{ij\alpha}\delta c_\alpha,$$

$$P_{ij\alpha} = V_{ij\alpha} - V_{ijk}(V^{-1})_{kl}V_{l\alpha}. \qquad (7\text{-}14)$$

As a result, the Morse saddle type cannot change under a sufficiently small perturbation as long as V_{ij} remains nonsingular.

All these results disappear when the stability matrix V_{ij} becomes singular. But at such points (bifurcation set in control-parameter space), the canonical form for the potential $V(x; c)$ can be determined (cf. Table 7-1). Rather than use (7-10) through (7-14) to determine the rearrangement of the critical points of the potential, it is possible to determine the spectrum of bifurcation possibilities and the canonical properties for each possibility, following the methods described in Section 7-5.

The simple analytic methods (7-10) through (7-14) are useful for studying the qualitative behavior of gradient dynamical systems in regions of control-parameter space where no qualitative changes occur. When the stability matrix V_{ij} becomes singular, these analytic methods fail. More powerful computational techniques are then called for. We are thus naturally forced to leave the realm of the Morse lemma and enter into a dialog with the Thom classification theorem.

Catastrophe Theory: Why It Exists

7-7   TRANSFORMATION TO CANONICAL FORM: A_3

What general procedures are involved in transforming a function with a degenerate critical point into canonical form? What additional procedures are involved in determining the most general perturbation of the resulting catastrophe germ? In this section we illustrate these procedures for a simple case: a function of a single state variable and two control parameters.

Assume $f(x; c_1, c_2)$ is a family of functions depending on a single state variable, x, and two control parameters, c_1 and c_2. This function might arise through consideration of a physical system, or it might result from the application of the Thom splitting lemma (Section 7-2) to a family of potentials depending on several state variables. In either case, we wish to determine the qualitative properties of $f(x; c_1, c_2)$, and in particular we wish to determine how these qualitative properties change as the control parameters change.

To this end, we expand the function in a Taylor series about a point x_0. This expansion is shown in line 1 of Table 7-3, to which the rest of this discussion is keyed. All Taylor series coefficients $f_i(x_0; c_1, c_2)$ depend on the expansion point, x_0, and the control parameters c_1, c_2.

At points where $f_1 \neq 0$, the implicit function theorem is applicable, and nothing interesting is happening. We therefore search through values of the expansion point, x_0, to locate a critical point: $f_1(x_0; c_1, c_2) = 0$. Since local qualitative properties are independent of the value of the function, we can choose our coordinate system so that $f_0(x_0; c_1, c_2) = 0$.

We have used up one degree of freedom searching along the x-axis, and another degree of freedom displacing the y-axis, to put the Taylor series expansion into the form shown in line 2 of Table 7-3.

There still remain two degrees of freedom to play with: c_1 and c_2. As we change these control-parameter values, we can sit on the critical point $x_0 = x_0(c_1, c_2)$, whose location depends on the control-parameter values as described in Section 7-6. All nonzero Taylor series coefficients $f_j(x_0(c_1, c_2); c_1, c_2)$, $j = 2, 3, \ldots$, depend on the two control parameters. Now it is possible that special choices of these control-parameter values annihilate one, or possibly even two, of the Taylor series coefficients f_j, $j \geq 2$. However, it is generally not possible ("nongeneric") that more than two of these coefficients are annihilated by the two control parameters.[24-26]

Table 7-3
Taylor Series Coefficients

Object	Line/Procedure	Coefficients of x^n						
		x^0	x^1	x^2	x^3	x^4	x^5	x^6
Find canonical germ	1. Taylor expansion	f_0	f_1	f_2	f_3	f_4	f_5	f_6
	2. Adjust ordinate; locate critical point	0	0	f_2	f_3	f_4	f_5	f_6
	3. Exploit control parameter degrees of freedom			0	0	f_4	f_5	f_6
	4. Smooth change of coordinates					± 1	0	0
Find canonical perturbation	5. Add arbitrary perturbation	0	ϵ_1	ϵ_2	ϵ_3	$\pm 1 + \epsilon_4$	ϵ_5	ϵ_6
	6. Smooth change of coordinates		ϵ_1	ϵ_2	ϵ_3	± 1	0	0
	7. Shift origin		ϵ_1	ϵ_2	0	± 1		

$$f(x: c_1, c_2) \doteq \underbrace{\pm x^4}_{} + \underbrace{\epsilon_1 x^1 + \epsilon_2 x^2}_{}$$

$$\text{CGl} + \text{Pert(1,2)}$$

If, say, f_6 and f_{10} are annihilated, then $f_2 \neq 0$ and the Morse lemma is applicable, and no qualitatively interesting things happen. The biggest qualitative changes occur when the earliest terms in the Taylor series expansion are annihilated. There are then two possibilities: $f_2 = 0, f_3 \neq 0$ and $f_2 = 0$, $f_3 = 0, f_4 \neq 0$. We will treat the latter case here; the former is less interesting, and is treated in exactly the same way.

Line 3 of Table 7-3 shows the Taylor series expansion after the two remaining control-parameter degrees of freedom have been used to annihilate the Taylor series coefficients f_2 and f_3.

We have not yet exploited another set of degrees of freedom. This possibility is that of finding a smooth transformation from one coordinate system to another. We look for a transformation, $x' = x'(x)$, that will transform the Taylor series into a simple canonical form. Since the leading term is x^4, we may suspect that in a suitable coordinate system, all terms higher than the quartic may be transformed away. To explore this possibility, we set

$$x' = A_1 x + A_2 x^2 + \cdots \qquad (7\text{-}15)$$

$$\pm (x')^4 = f_4 x^4 + f_5 x^5 + \cdots \qquad (7\text{-}16)$$

By plugging (7-15) into (7-16) and expanding, it is possible to formulate a simple recursive linear algorithm for computing each A_j in terms of the Taylor series coefficients $f_4, f_5, \ldots$. For example, the first two values of A_j are $A_1 = |f_4|^{1/4}$ and $A_2 = \pm f_5/4|f_4|^{3/4}$.

The canonical form for the function $f(x; c_1, c_2)$ at its most degenerate critical point is shown in line 4 of Table 7-3. To get this form, we have exploited the following degrees of freedom:

1. choice of ordinate ($f_0 = 0$);
2. choice of critical point ($f_1 = 0$);
3. choice of control parameters ($f_2 = 0, f_3 = 0$);
4. choice of nonlinear coordinate transformation ($f_4 = \pm 1, f_5 = f_6 = \cdots = 0$).

This procedure illustrates the solution to the first question posed at the beginning of this section. To illustrate the solution to the second question (most general perturbation) we proceed as follows.

An arbitrary perturbation

$$\epsilon(x) = \epsilon_1 x + \epsilon_2 x^2 + \cdots$$

is added to the canonical form. Since we are primarily concerned with the qualitative effects of the perturbation, we have dropped the constant term ϵ_0.

We have also dropped the prime on the coordinate x for clarity of presentation. Line 5 of Table 7-3 shows the Taylor series expansion of the canonical germ, x^4, after an arbitrary perturbation.

We now exploit again the degree of freedom represented by a smooth change of coordinate system. It is a rule of thumb (i.e., a theorem) that the Taylor series coefficients that could be transformed on canonical values before (in passing from line 3 to line 4 of Table 7-3) can be given the same canonical values once again. Attempts to eliminate other Taylor series coefficients (coefficients of x, x^2, x^3) will result in transformations not smoothly depending on some of these coefficients (ϵ_1, ϵ_2, ϵ_3). The form of the perturbed function, after a smooth change of coordinates, is shown in line 6 of Table 7-3.

There is one additional degree of freedom that can be exploited. Since we are interested in a family of functions rather than a single function, it is not really necessary that the critical point remain at the origin. We can therefore shift the origin of coordinates (one degree of freedom) to eliminate one of the three remaining perturbation coefficients: ϵ_1, ϵ_2, or ϵ_3. It is always possible to eliminate the coefficient ϵ_3 in this way, but not always possible to eliminate ϵ_1 or ϵ_2. For this reason, a shift along the (transformed) x-axis is chosen to eliminate the term ϵ_3. The resulting canonical form is shown in line 7 of Table 7-3.

Our results, in responding to the two questions posed in the opening paragraph of this section, are as follows.

1. We have used the several degrees of freedom at our disposal to transform the function $f(x; c_1, c_2)$, at its most degenerate critical point, to the canonical form

$$f(x; c_1, c_2) \doteq \pm \tfrac{1}{4} x^4,$$

   where we have rescaled the length for convenience.
2. We have used these degrees of freedom, in roughly reverse order, to determine that the most general perturbation of the non-Morse function x^4 is the two-parameter perturbation $\epsilon_1 x + \epsilon_2 x^2$.

Therefore, in the neighborhood of its most degenerate critical point, we have shown that $f(x; c_1, c_2)$ is represented by the canonical cusp catastrophe:

$$f(x; c_1, c_2) \doteq \pm \tfrac{1}{4} x^4 + \tfrac{1}{2} a x^2 + b x.$$

The canonical mathematical coefficients a, b are, of course, functions of the two physical control parameters c_1, c_2:

$$a = a(c_1, c_2),$$

$$b = b(c_1, c_2). \tag{7-17}$$

As is to be expected, often the most interesting and difficult part of the application of catastrophe theory to physical problems is the determination of the transformation (7-17).

The example treated in this section should not be dismissed because of its relative simplicity. The cusp catastrophe is the most widely used of all the catastrophes in physical applications.

7-8 THE GENERAL PROCEDURE

In the previous section we have illustrated, by particular example, the general mathematical procedures used to transform a non-Morse function into specific canonical form at a degenerate critical point, and the methods used to determine the most general perturbation of that canonical form of lowest dimension. In this section we shall summarize these general procedures. In the following section we will apply these procedures to a more complicated, more interesting, and less physically useful example than treated in the previous section.

The summary is presented, in schematic flow diagram form, in Figure 7-6. Catastrophe theory is in some sense like classical field theory. The near-field approximation is often tractable. So also is the far-field approximation. It is in the intermediate field regime that complexities arise and interesting things happen. So also in the case of catastrophe theory. The leading terms in the Taylor series expansion can be eliminated by choice of ordinate, critical-point, and control-parameter values. The higher-degree terms can often be eliminated by an appropriate smooth change of variables. What is left over, in the middle between the leading terms and the higher degree terms which have all vanished, are the intermediate terms that constitute the catastrophe germ. The catastrophe germ may be vividly described as that mathematical creature caught between the Rock and the Hard Place, or between Scylla and Charybdis.

The most general perturbation, of lowest dimension, of a catastrophe germ is determined by the reverse process. An arbitrary perturbation is laid on the catastrophe germ, the higher-degree terms are removed by a smooth change of coordinates, and the germ is returned to its canonical form by suitable renormalization. Then the origin of coordinates is chosen to remove

The Catastrophe Germ is that Mathematical Creature Between—

The Rock	and	The Hard Place
	or	
Scylla	and	Charybdis

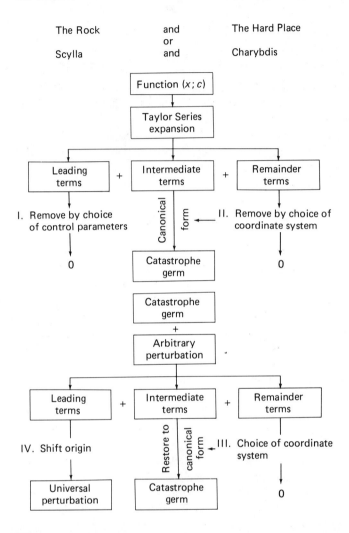

Figure 7-6
This schematic flow diagram summarizes the mathematical procedures involved in determining the canonical form for a non-Morse function in the neighborhood of a degenerate critical point, and for determining the most general perturbation, of lowest distribution, of the resulting canonical catastrophe germ.

as many as possible (exactly l) of the Taylor series coefficients characterizing the perturbation. What is left over, in addition to the canonical catastrophe germ, is its most general perturbation of lowest dimension.

7-9 TRANSFORMATION TO CANONICAL FORM $T_{3,3,3}$

As a second concrete example of the methodology described in Section 7-8, and summarized in Figure 7-6 we will consider a function of three state variables and six control parameters:

$$f(\mathbf{x}, \mathbf{c}) = f(x, y, z; c_1, c_2, c_3, c_4, c_5, c_6). \qquad \textbf{(7-18)}$$

The objectives are to determine the canonical form for this function at its degenerate critical points, and to determine the canonical form for an arbitrary perturbation of each canonical germ. This example is more difficult than the one treated in Section 7-7. It has been chosen both to illustrate the general methods and to make a point. The point will be made at the end of this section.

Construction of the Canonical Germ[27]

We begin by constructing a Taylor series expansion of the function (7-18):

$$f(\mathbf{x}, \mathbf{c}) = \sum_{\substack{i \geq 0 \\ j \geq 0 \\ k \geq 0}} \frac{1}{i!j!k!} f_{ijk}(x - x_0)^i (y - y_0)^j (z - z_0)^k.$$

This Taylor series expansion has been shown schematically in Figure 7-7. In this figure, the Taylor series coefficients are grouped according to the degree of the monomials to which they couple. We now invoke the procedures described in Section 7-7 and Section 7-8.

I. We adjust the ordinate so that the value of the function at the point studied is zero: $f_{000} = 0$.

II. We use up the three degrees of freedom associated with the three-dimensional state variable space, searching for a critical point of $f(x; c)$: $f_{100} = f_{010} = f_{001} = 0$.

III. There remain six degrees of freedom associated with the six control parameters. The control-parameter values can be chosen to annihilate no more than six of the remaining Taylor series coefficients f_{ijk}, $i + j + k \geq 2$. If at least one of the six terms of degree two (i.e., x^2, xy, etc.) remain after

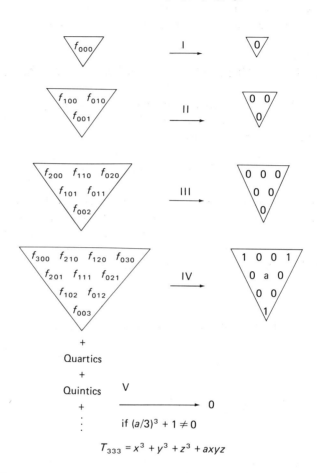

$$f(x, y, z; c_1, c_2, c_3, c_4, c_5, c_6)$$

$$T_{333} = x^3 + y^3 + z^3 + axyz$$

Figure 7-7

A simple series of steps can be used to transform a function of three state variables and six control parameters to canonical form at a degenerate critical point. These steps, which are keyed to the text, are

I. Change ordinate.

II. Locate critical point.

III. Annihilate the six quadratic coefficients using the six control parameter degrees of freedom.

IV. Linear transformation.

V. Smooth change of variables to kill off the "Taylor tail" of the function.

this step, the resulting canonical germ is among those listed in Table 7-1 or Table 7-2. Something new is obtained only if all six of the Taylor series coefficients for terms of the second degree are annihilated by appropriate choice of control-parameter values. We therefore assume that some choice of the control-parameter values will cause this mass extinction.

IV. It is possible to carry out a homogeneous linear transformation

$$\begin{pmatrix} x' \\ y' \\ z' \end{pmatrix} = \begin{pmatrix} \cdot & \cdot & \cdot \\ \cdot & \cdot & \cdot \\ \cdot & \cdot & \cdot \end{pmatrix} \begin{pmatrix} x \\ y \\ z \end{pmatrix} \tag{7-19}$$

in an attempt to put the degree-three terms into some canonical form. This linear transformation sends degree-three terms to degree-three terms, degree-four to degree-four, and so on. Since there are ten Taylor series coefficients for degree-three terms, but only nine degrees of freedom to play with in the 3×3 matrix in (7-18), nine of the ten terms of degree-three can be put into canonical form, but the tenth cannot:

$$\sum_{i+j+k=3} f_{ijk} x^i y^j z^k \doteq x^3 + y^3 + z^3 + axyz.$$

The term a, over which we have no control, is called a modulus.[27]

V. We can now attempt to find a smooth change of coordinates that will transform away all terms of degree four, five, ... This change can always be done, provided $(a/3)^3 \neq -1$.

This simple methodology shows that, under the procedures indicated in steps I, II, IV, and V and the assumption made in step III,

$$f(x; c) \doteq T_{3,3,3},$$

$$T_{3,3,3} = x^3 + y^3 + z^3 + axyz.$$

Construction of the Canonical Perturbation

Figure 7-8 summarizes the steps involved in computing the canonical perturbation of the catastrophe germ $T_{3,3,3}$. This figure represents the Taylor series expansion of the function[17]

$$F(x) = T_{3,3,3} + \text{arbitrary perturbation},$$

$$\text{arbitrary perturbation} = \sum_{i+j+k<0} \epsilon_{ijk} x^i y^j z^k.$$

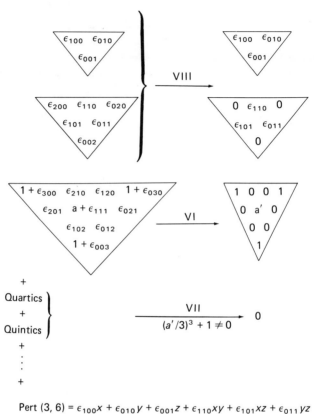

$T_{3,3,3}$ + arbitrary perturbation

$$\text{Pert (3, 6)} = \epsilon_{100}x + \epsilon_{010}y + \epsilon_{001}z + \epsilon_{110}xy + \epsilon_{101}xz + \epsilon_{011}yz$$

Figure 7-8
A simple series of steps can be used to determine the universal perturbation of the canonical catastrophe germ $T_{3,3,3}$. These steps, which are keyed to the text, are

VI. Renormalize the germ to its original canonical form.
VII. Perform a smooth change of variables to kill off the "Taylor tail" of the perturbation.
VIII. Displace the origin in state variable space.

We now carry out the steps necessary to compute the universal perturbation of $T_{3,3,3}$.

VI. A linear transformation is carried out that transforms the catastrophe germ back to its original canonical form.

VII. A nonlinear transformation is carried out that eliminates all terms of degree higher than three. This elimination can be done provided $(a'/3)^3 \neq -1$. Since a is assumed to obey this condition (otherwise the function is not determinate) a' also obeys this condition for sufficiently small perturbations.

VIII. A displacement of the origin in state variable space can be used to eliminate three of the remaining nine Taylor series coefficients of the perturbation $\epsilon(x)$. The coefficients of x^2, y^2 and z^2 can always be so eliminated.

This simple methodology shows that under the procedures indicated in steps VI, VII, and VIII, the most general perturbation of the catastrophe germ $T_{3,3,3}$ is the six-parameter perturbation

$$\text{Pert } (3, 6) = \epsilon_{100}x + \epsilon_{010}y + \epsilon_{001}z$$
$$+ \epsilon_{110}xy + \epsilon_{101}xz + \epsilon_{011}yz.$$

Taken together, these results show that in a neighborhood of a degenerate critical point of $f(x; c)$ (7-18) involving a stability matrix with three vanishing eigenvalues

$$f(x, y, z; c_1, c_2, c_3, c_4, c_5, c_6) \doteq x^3 + y^3 + z^3 + axyz$$
$$+ (\epsilon_{100}x + \epsilon_{010}y + \epsilon_{001}z$$
$$+ \epsilon_{110}xy + \epsilon_{101}z + \epsilon_{011}yz).$$

Point

When a function of three or more state variables depends on six or more control parameters, it is possible to encounter a catastrophe germ, such as $T_{3,3,3}$, depending on one or more uncontrollable parameters, called moduli.[18,28] Germs depending on no moduli are called simple, or elementary. Elementary catastrophes can be encountered for any number of dimensions (number of control parameters), as shown in Table 7-2. However, for $k > 5$, it is possible to encounter nonelementary catastrophes as well as elementary catastrophes. For $k < 6$, there are only elementary catastrophes. This is why Table 7-1 terminates at $k = 5$.

Catastrophe Theory: How It Works

7-10 CATASTROPHE CONVENTIONS

Strictly speaking, the mathematics of elementary catastrophe theory is useless for the purpose of physical applications because the mathematics has been developed for systems of mathematical equations from which all time dependence has been eliminated [cf. (7-3)]. Since the applications we have in mind involve the response of a system to time-dependent changes in the control-parameter values, the mathematics we require lies outside the mathematics called elementary catastrophe theory.

This attitude is at variance with the pragmatic needs of physicists. Therefore, two working assumptions have been widely used to make the time-independent mathematics of elementary catastrophe theory applicable to physical systems with time-dependent control parameters. These are the delay convention and the Maxwell convention, illustrated in Figure 7-9.[29-32]

The Delay Convention

If a physical system is determined by the minimum of a potential, then the system sits at the local minimum of the potential while the control parameters are changed, even if that local minimum becomes metastable with respect to other distant minima. The system jumps from one state to another only when the local minimum determining the initial state disappears in a catastrophe. This convention is illustrated in Figure 7-9(a).

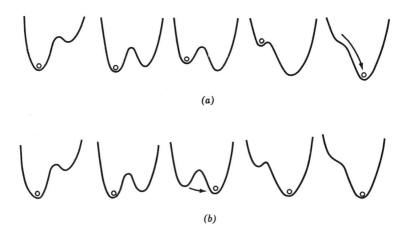

(a)

(b)

Figure 7-9

(a): Delay convention. The system state remains at a local minimum even when that minimum becomes metastable with respect to another minimum. It only jumps to a new state when the initial metastable minimum disappears. (b): Maxwell convention. The system state always occurs at the global minimum of the potential. As the initial minimum becomes metastable with respect to another minimum, the system state jumps from the former to the latter. (*From R. Gilmore, Catastrophe Theory for Scientists and Engineers [New York: Wiley & Sons, 1981], p. 143; © 1981 by John Wiley & Sons, Inc.*)

The set of control-parameter values at which Morse critical points, including minima, coalesce and disappear is called the bifurcation set. The bifurcation set, $\mathcal{S}_B$, for the cusp catastrophe is shown in Figure 7-5(*b*).

Caution

For complicated systems, the local minimum to which the system state jumps may not be well defined unless an explicit set of dynamical equations of motion are available.[33]

The Maxwell Convention

If a physical system is determined by the minimum of a potential, then the system sits at the global minimum for all values of the control parameters. If, as the control parameters are changed, one minimum becomes metastable with respect to another, the system state jumps from the initial minimum to the new global minimum. The regime where two or more minima are equally deep is called a phase transition. This convention is illustrated in Figure 7-9(*b*).

The set of control-parameter values at which two or more global minima are equally deep is called the Maxwell set. The Maxwell set, $\mathcal{S}_M$, for the cusp catastrophe is shown in Figure 7-5(*b*).

Equations that determine the Maxwell set in the space of control parameters are called Clausius-Clapeyron equations.

Neither the delay nor the Maxwell convention is a theorem—they are just useful ad hoc conventions. As seen in Figure 7-9, they are the "boundary cases" in a whole spectrum of possibilities. If the physical system described by the potential $V(x; c)$ obeys an equation of Fokker-Planck type, then it is possible to interpolate between these two conventions. The choice of convention is determined by the time scale on which the control parameters are changed and the ratio of noise level in the system (the diffusion constant in the Fokker-Planck equation) to the height of the barrier separating metastable from stable minimum. This situation is illustrated in Figure 7-10.[34]

7-11 CATASTROPHE FLAGS

How do we know when a catastrophe is present? There are two erudite ways to determine when a family of potentials hides a catastrophe. These ways follow the methods outlined in Sections 7-1 and 7-2. In the first method, one locates the critical points of the potential describing a gradient dynamical system. The critical points are located in the space of state variables.

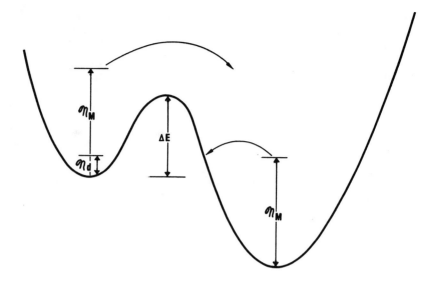

Figure 7-10
When the noise level in the system, $\mathfrak{N}_d$, is small compared to the height of the barrier, ΔE, separating metastable from stable minimum, the delay convention is applicable. When the noise level, $\mathfrak{N}_M$, is comparable with the excitation energy, ΔE, the Maxwell convention is applicable. (*From R. Gilmore, Catastrophe Theory for Scientists and Engineers* [*New York: Wiley & Sons, 1981*], *p. 144;* © *1981 by John Wiley & Sons, Inc.*)

They move about and may even coalesce as the control parameters are changed. Their degeneracy locates catastrophes for us. This method is a little like studying the locus of the roots of a linear system of equations in the complex plane as some linear system parametesr are varied.

The second method, following Section 7-2, is to carry out a Taylor series expansion of the potential about an arbitrary point for arbitrary values of the control parameters, and then to search in control-parameter space for solutions of the equation $\det[\partial^2 V/\partial x_i \partial x_j] = 0$. Sometimes this method is even useful.

But in fact we seldom use these erudite methods to learn that a catastrophe is present. More often than not, we simply blunder into it. This being the case, it is useful to know what to look for when a catastrophe is present. In this sense, catastrophes are friendly creatures. They wave flags to gain our attention. Once one of these flags has been spotted, the others may be looked for and, more often than not, found. Finding these flags confirms the presence of a catastrophe. It may then be ferreted out in detail using the erudite methods described at the beginning of this section, if desired. In many cases, it is only necessary to determine the qualitative behavior of the system (represented by the catastrophe flags), so that the hard work just described can even be avoided.

In this section we describe the five classical catastrophe flags[35, 36] and three additional catastrophe flags.[37]

Modality

In the neighborhood of a catastrophe, the potential function exhibits multiple equilibria. Each equilibrium represents a system mode, or mode of behavior (not all are stable). Multiple modality for the cusp catastrophe is illustrated in Figure 7-11.

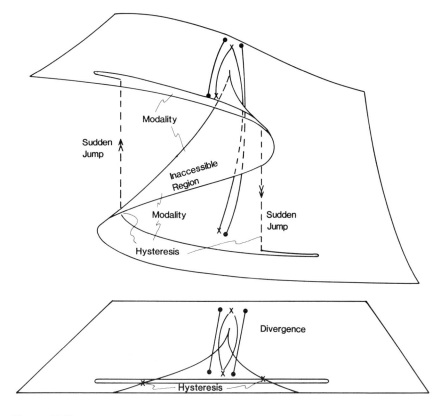

Figure 7-11
Five classic catastrophe flags are illustrated: modality, inaccessibility, sudden jumps, hysteresis, and divergence.

Inaccessibility

Two or more locally stable equilibria must be separated by at least one saddle M_i^n, $i > 0$, which is unstable. These additional critical points are therefore physically inaccessible. Therefore, multiple stable modes are always accompanied by inaccessible modes. The folded-over middle part of the cusp catastrophe manifold represents an inaccessible mode, the local intermediate maximum (Fig. 7-11).

Sudden Jumps

As the control-parameter values are changed, the system state may jump from one locally stable state to another. Such changes in physical state are represented by large changes in the values of the state variables representing the system. These large changes are called sudden jumps. They occur no matter which convention is used, and even when neither convention is applicable. Sudden jumps on the cusp catastrophe manifold are illustrated in Figure 7-11.

Hysteresis

A sudden jump from one state to another may occur as the control-parameter values are changed from one set of values to another. If the control-parameter values are changed in the reverse direction, the system may revert to its initial state. If it does, it may make the reverse jump at the same control-parameter values where the initial jump occurred (Maxwell convention) or it may not (any other convention). This lack of reversibility in the latter case is called hysteresis. Hysteresis is shown in Figure 7-11 for a situation intermediate between the Maxwell convention and the delay convention.

Divergence

The final state of the system may depend sensitively on either the initial state of the system or the process used to transform the initial state to the final state. These two ideas are illustrated in Figure 7-11.

In one case, an initial state (x) is brought to a final state with the same control-parameter values by two slightly different paths. The paths pass on different sides of the cusp point, and therefore wind up on different sheets of the cusp catastrophe manifold. In this case, one initial state gives rise to very different final states through a slight perturbation of the process.

In another case, two nearby initial states ($\bullet$) are brought, by the same process (represented by parallel lines in the control parameter plane), to final states. The identical processes pass on either side of the cusp point, and therefore the final states wind up on different sheets of the cusp catastrophe manifold.

Divergence of Linear Response

Critical points are displaced slightly when the control parameters are changed by a small amount. A linear response relation can be written [cf. (7-12)], expressing the change in the components of the jth critical point, $\delta x_i^{(j)}$, in terms of small changes in the control-parameter values

$$\delta x_i^{(j)} = \chi_{i\alpha}^{(j)} \, \delta c_\alpha.$$

The susceptibility tensor diverges when the jth critical point nears another critical point and undergoes a catastrophe. This situation is illustrated in Figure 7-12 for the cusp catastrophe.

Time Dilation

Since the gradient of the potential represents a force [cf. (7-3)], in the neighborhood of a stable Morse critical point x^0, the gradient $-\nabla V(x; c)$ describes the force that tries to return the system to x^0. In the neighborhood of a catastrophe, the force $-\nabla V$ becomes smaller, and the system takes longer and longer to return to equilibrium. If δx_i represents the coordinates of a displacement from x^0, then these coordinates obey the dynamical equations of motion

$$\frac{d}{dt} \delta x_i = -V_{ij} \, \delta x_j + \cdots.$$

The normal modes have time dependence of the form $e^{-\lambda_j t}$, where λ_j are the eigenvalues of the stability matrix V_{ij}. All eigenvalues are positive, since x^0 is a stable critical point. The time scale for relaxation to equilibrium is the reciprocal of the smallest eigenvalue of V_{ij}. In the neighborhood of a catastrophe, one or more of the eigenvalues approaches zero from above, so the system takes longer and longer to relax to equilibrium as the control parameters approach a bifurcation set.

Slowing down of relaxations is illustrated in Figure 7-13 for three cases. In case (a), a system state initially out of equilibrium rapidly approaches the equation of state manifold. This case is represented by the heavy line and

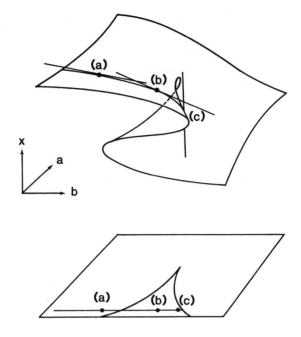

Figure 7-12
The right-hand minimum of the cusp catastrophe function depends on the control parameters a and b: $x_r = x_r(a, b)$. As the bifurcation set for the right-hand minimum is approached along any path s: $(a, b) \rightarrow (a(s), b(s))$, the derivative

$$\frac{dx}{ds} = \frac{\partial x_r}{\partial a}\frac{da}{ds} + \frac{\partial x_r}{\partial b}\frac{db}{ds}$$

becomes steeper. Shown above is the steepening of the slope along the path $(a, b) =$ (negative, increasing) far from the bifurcation set (a), closer to the bifurcation set (b), and almost at the bifurcation set (c).

double arrow. In case (b), the system approaches the cusp catastrophe manifold much more slowly, since it is moving in response to a much reduced force. This case is represented by a thin line and a single arrow. In case (c), the trajectory slows down dramatically as it passes near the folded-over part of the manifold, then speeds up as it approaches the lower sheet. The system slows down near the fold because it senses the two complex critical points that lie slightly off the real axis in the complex plane.

For initial conditions near the middle sheet representing the unstable critical point, there is also time dilation in departure from the equilibrium as well as sensitive dependence of final state on initial conditions. That is, an initial condition above (below) the middle sheet will go to the upper (lower) sheet.

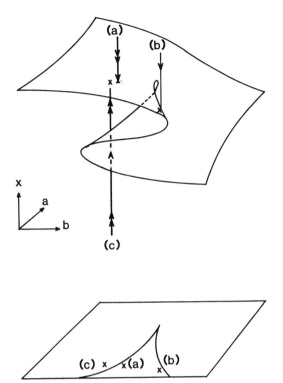

Figure 7-13
The equilibrium manifold of a gradient dynamical system is defined by $\nabla V(x; c) = 0$, here represented by the cusp catastrophe manifold. An initial state off the equilibrium manifold will return to the manifold with a speed proportional to the force acting on it. The relaxation time is rapid for a typical stable Morse critical point (*a*). The relaxation time increases dramatically as the critical point approaches degeneracy (i.e., the appropriate component of the bifurcation set is approached), whether the critical points are real (*b*) or complex conjugate pairs (*c*).

Anomalous Variance

Although purely classical systems may be described by equations of motion of gradient dynamical system type, many other kinds of systems are described by probability densities, $P(x, t)$, whose equation of motion may be of Fokker-Planck type

$$\frac{\partial}{\partial t} P(x, t) = D\nabla^2 P(x, t) + \nabla[P(x, t)\nabla V(x; c)].$$

The steady-state probabillity distribution function then has the form

$$P(x, \infty) \simeq e^{-V(x;c)/D},$$

where D, the diffusion coefficient, is a physical parameter of the appropriate dimensions that measures thermal noise, quantum fluctuations, both, or otherwise.[34,37]

In the single-mode regime, the steady-state probability distribution function is essentially a Gaussian. At the cusp point, the potential becomes a quartic, and the distribution becomes very broad and flat. In the bimodal regime, the distribution function has two equally high peaks surrounding each stable minimum when the control parameters lie on the Maxwell set $(a, b) = $ (negative, 0). Sufficiently far from the Maxwell set, as measured by the yardstick D, one or the other of the two peaks becomes dominant and the distribution is a Gaussian for all practical purposes.

The steady-state probability distribution function is illustrated for three sets of control-parameter values in Figure 7-14. As the control-parameter a decreases from positive (a) through zero (b) to negative (c), the distribution assumes bimodal character. The mean value of x for each distribution is zero, but the variance $\langle (x - \bar{x})^2 \rangle$ increases dramatically from (a) to (b) to (c).

Destructive and Nondestructive Flags

In many applications, engineering in particular, by the time a catastrophe flag has been encountered, it is too late: The bridge has fallen down or the building has collapsed. In these disciplines, it is useful to determine flags that wave in the unimodal regime.

The first five catastrophe flags, the classical diagnostic tests, are applicable in the multimodal regime.[35,36] In this sense, they are destructive flags.

The last three catastrophe flags can be applied in the unimodal regime.[37] Careful monitoring of the linear susceptibility tensor components, the normal mode frequencies, the admixture and intensities of overtones in nonharmonic oscillations, and the scatter in specific physical measurements can be used to determine, by extrapolation, the location of critical-point degeneracies and the bifurcation set in control-parameter space, while carefully avoiding dangerous multimodal regimes.

7-12   APPLICATIONS

To illustrate how catastrophe theory can be used to organize our understanding of the qualitative properties of systems in a systematic way, we consider six applications in this section. For each application, we consider the list of catastrophe flags and briefly describe the counterpart/pertinent process for the system considered. These six treatments are given briefly

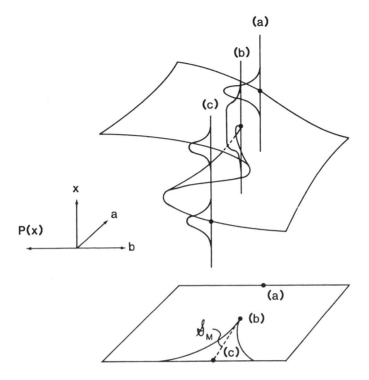

Figure 7-14
The probability distribution function $P(x) \simeq \exp\,[-V(x; a, b)/D]$ becomes increasingly broad along the line $b = 0$ as a decreases from positive to negative. For a sufficiently positive, the distribution is essentially Gaussian with the usual variance. For $a = 0$, the distribution becomes broad and the variance becomes increasingly large. For a sufficiently negative, the variance becomes macroscopic (independent of D), its root mean square approaching half the separation between the two equally deep minima.

and in parallel; the results are summarized in Figures 7-15–7-20, which all have the same structure. All applications are illustrated using the cusp catastrophe potential.

Gradient Dynamical Systems

These systems, which are dissipative, are governed by equations of the form

$$\frac{dx}{dt} = -\boldsymbol{\nabla} V\,(x; c).$$

The equations of motion require the system state to move to the nearest minimum. The eight catastrophe flags described here are summarized in Figure 7-15.

Modality.   The system can have two stable equilibria.

Inaccessibility.   The intermediate local maximum is unstable. The system will not sit long in this state if perturbations are present.

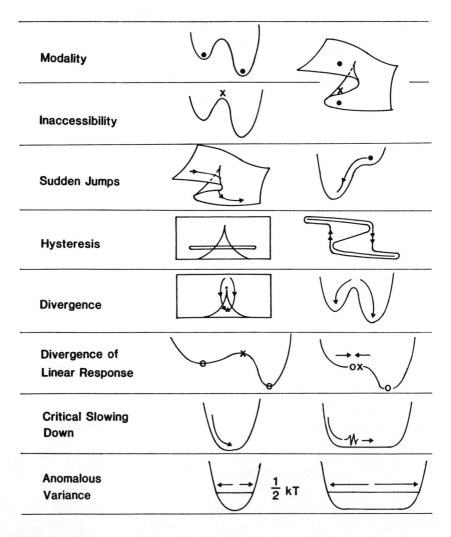

| Modality |
| Inaccessibility |
| Sudden Jumps |
| Hysteresis |
| Divergence |
| Divergence of Linear Response |
| Critical Slowing Down |
| Anomalous Variance |

Figure 7-15
Gradient dynamical systems exhibit the eight catastrophe flags as shown.

Sudden jumps. If the control parameters are slowly varied in such a way that the minimum describing the system state becomes metastable and disappears, the system will jump to the other minimum.

Hysteresis. A cyclic process cutting both bifurcation sets will induce the classic hysteresis pattern.

Divergence. If the bottom of a single well, roughly parabolic, potential is pushed upward to form a bimodal shape, the system state will wind up in the left-hand or right-hand minimum, depending sensitively on several factors.

Divergence of linear response. As two critical points approach degeneracy, they move toward each other very quickly.

Time dilation. At an isolated stable critical point, the system relaxes to equilibrium after perturbation following a Hooke's law restoring force. At the cusp point, the restoring force is the much weaker cubic, and the relaxation consequently slower. This phenomenon is called critical slowing down.

Anomalous variance. If this classical system is subjected to thermal noise, then an equipartition argument can be used to show that fluctuations become larger as the potential becomes flatter.

Newtonian Dynamical Systems

These systems, which are conservative, are governed by equations of the form

$$\frac{d^2 x}{dt^2} = -\nabla V(x; c). \tag{7-20}$$

Energy is a constant of motion for these systems. The eight catastrophe flags are summarized here and illustrated in Figure 7-16.

Modality. For energies less than the barrier height, the system may oscillate either about the left-hand minimum or the right-hand minimum.

Inaccessibility. If the energy of a system is such that the orbit contains the intermediate local maximum, the system state will take forever to reach it. More generally, trajectories approach Morse saddles only asymptotically.

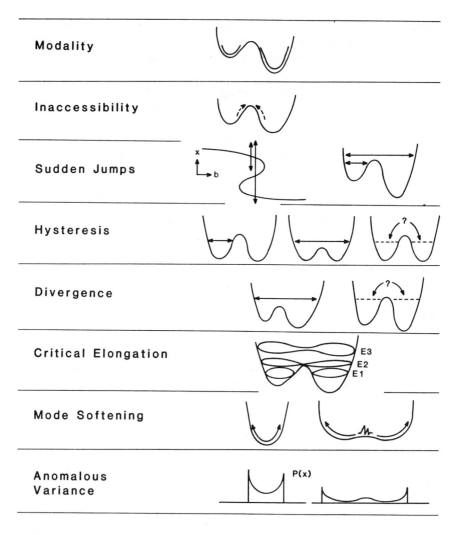

Modality	
Inaccessibility	
Sudden Jumps	
Hysteresis	
Divergence	
Critical Elongation	
Mode Softening	
Anomalous Variance	

Figure 7-16
Newtonian dynamical systems exhibit the eight catastrophe flags as shown.

Sudden jumps. If the motion is about the left-hand minimum and the barrier height is slowly decreased, the orbital range will suddenly jump so that oscillations encompass both minima.

Hysteresis. If oscillations are localized around one of the equilibria and the barrier height is first lowered, so that the oscillation encompasses both equilibria, and then raised again to its original height, the oscillation may be localized either about the original equilibrium or about the other equilibrium. Hysteresis in the dissipative and convervative cases is in some sense dual. In the dissipative case, the qualitative changes occur predictably in state variable space at different control-parameter values. In the conservative case, the qualitative changes occur unpredictably in state variable space, but at the same control-parameter values.

Divergence.   Oscillations encompassing both minima are suddenly localized to one well or the other as the barrier height is increased above the energy. The final state depends sensitively on initial conditions and process.

Divergence of linear response.   This situation is most easily illustrated for the dynamical system, depending on two state variables, x and y. It is assumed the potential is a cusp in the x-direction and quadratic in the y-direction. System orbits are intersections of the potential $V(x, y) = (\frac{1}{4} x^4 + \frac{1}{2} ax^2 + bx) + \frac{1}{2} y^2$ with the plane $V = E = $ const. These orbits become dramatically elongated as the barrier height approaches the conserved energy. This process is called critical elongation.

Time dilation.   The normal mode frequencies of (7-20) about a stable critical point are proportional to the square roots of the eigenvalues of the stability matrix evaluated at the critical point.[37] As a catastrophe is approached, one or more eigenvalues of the stability matrix goes to zero. This dramatic decrease in normal mode frequency is called mode softening.

Anomalous variance.   If the probability density, $P(x, E)$, is computed, then this density widens dramatically in the neighborhood of a catastrophe.

Statistical Mechanical Systems

These systems are governed by equations of motion for the density operator, ρ, of the form

$$ i\hbar \frac{\partial}{\partial t} \rho = [H, \rho]. $$

The system state is determined by the global minimum of the potential appearing in the Hamiltonian, H. The Maxwell convention is generally applicable to such systems. A qualitatively valid, but quantitatively incorrect description of a fluid near its critical point is provided by the cusp catastrophe.[38-40] The catastrophe flags with respect to this application are described briefly here and summarized in Figure 7-17.

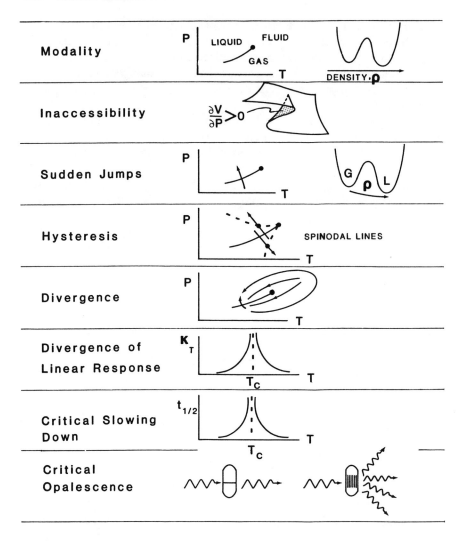

Figure 7-17
Statistical mechanical systems exhibit the eight catastrophe flags as shown.

Modality. In the neighborhood of the critical point, a fluid can exist in two states, gas and liquid. The critical point corresponds to the cusp point; the line of first-order phase transitions corresponds to its Maxwell set.

Inaccessibility. On the folded-over part of the cusp catastrophe manifold, the standard thermodynamic linear response functions assume unphysical values, signifying lack of thermodynamic stability.

Sudden jumps. As the control-parameter values (temperature, pressure) cross the gas-liquid coexistence curve, there is a sudden change in the state variable value (density).

Hysteresis. Since the Maxwell convention is generally applicable to systems in thermodynamic equilibrium, hysteresis is not generally seen. However, if an experiment is carried out very carefully, the coexistence curve can be crossed without a phase transition taking place. The resulting system state (e.g., superheated) is explosively sensitive to perturbations. The system can be driven into the metastable regime only so far. The limits of meta-stability are called spinodal lines, which are the fold lines of the cusp catastrophe.

Divergence. If the system is cooled from its fluid state, either the gas or the liquid state may result, depending sensitively on process and/or initial conditions. Conversely, a gas state can be transformed into a liquid state with nearby temperature and pressure using a path involving a first-order phase transition or one that does not involve this phase transition.

Divergence of linear response. The linear response functions for a system in thermodynamic equilibrium are the thermodynamic partial derivatives, for example

$$\kappa_T = -\frac{1}{V}\left(\frac{\partial V}{\partial P}\right)_T.$$

These functions diverge at the critical point.

Time dilation. As the critical point is approached, the system takes an increasingly long time, $t_{1/2}$, to approach equilibrium after a perturbation. This is one reason that experiments near a critical point are difficult.

Anomalous variance. In thermodynamic systems, anomalous variance is the phenomenon called critical opalescence.

Quantum Mechanical Systems

These systems are governed by equations of motion for the wavefunction, ψ, of the form

$$i\hbar\frac{\partial\psi}{\partial t} = H\psi. \tag{7-21}$$

We shall consider a laser: a system of atoms interacting with a small number of modes of the radiation field in a cavity designed to enhance the lasing modes. The systems we will consider are nonequilibrium. In fact, they are on the steady-state manifold far from the equilibrium branch.

A low-intensity signal, I_{in}, is input to the laser cavity and the output intensity, I_{out}, is monitored. The input signal sits on the shoulder of one of the cavity resonance frequencies. The output signal, I_{out}, exhibits hysteresis as a function of changing input signal.[41] This result suggested[42] that a cusp catastrophe was lurking in the mathematical description of this system.[43,44] The catastrophe flags for this bistable system are described briefly here and summarized in Figure 7-18.

Modality.   This system can exhibit two modes. In one mode the incident

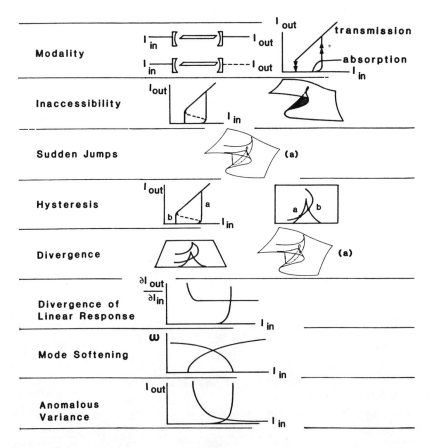

Figure 7-18
Quantum mechanical systems exhibit the eight catastrophe flags as shown.

light is absorbed by the system; in the other mode, it is transmitted by the system.

Inaccessibility. One of the steady states exhibited by the equations of motion, derived from (7-21), in unstable. The corresponding physical state is difficult to see.

Sudden jumps. The light intensity, I_{out}, exhibits sudden increases or decreases as a function of the control parameter I_{in}.

Hysteresis. The laser system, together with the experimentally observed I_{out} vs. I_{in} curves, are sketched for the optically bistable system.

Divergence. The physical parameters controlling the system can be varied experimentally. If they are chosen so that the control-parameter trajectory passes to the right of the cusp point as I_{in} is varied, then the system passes through the bistable regime. If the trajectory passes to the left of the cusp point in the control-parameter space, then the system remains in the single mode regime.

Divergence of linear response. The susceptibility, $\partial I_{out}/\partial I_{in}$, is essentially constant in both the absorption and the transmission regimes. However, near the bifurcation set, this susceptibility increases dramatically.

Time dilation. The lowest normal-mode frequency dives toward zero at both bifurcation sets.

Anomalous variance. Fluctuations $(I_{out}-\bar{I}_{out})^2$ increase noticeably as the potential flattens out when the bifurcation set is approached.

Probability Distributions

Probability densities and catastrophes are related by a logarithmic-exponential mapping:

$$P(x;\ c) \simeq e^{-V(x;\ c)/D}.$$

Since a probability density must be normalizable, only certain catastrophes ($A_{+(2k-1)}$ of Table 7-2) can be associated with probability distribution functions. However, it is possible to add an appropriate term to a catastrophe germ that "compactifies" it, that is, which allows the resulting exponential to be normalizeable.

These probability densities can be studied, either abstractly from a mathematical point of view or physically from the point of view of the systems giving rise to them. For example, we can consider a bimodal probability distribution function arising from the cusp catastrophe, or we can study the probability distribution for the projection of a magnetic moment onto the direction of an external magnetic field, or we can study the individual components of a pollster's results (i.e., people) when subjected to a divisive issue (politics, sex).[45] The catastrophe flags for probability distributions are described briefly here and summarized in Figure 7-19.

Modality. Bimodal distributions are not usually discussed in the literature of statistics. If required by data, they are usually approximated by a superposition of two Gaussians. The cusp catastrophe has finally provided a simple canonical form for a bimodal distribution.[46] The humps in the distribution represent modes of behavior (spin up or down, for or against sex).

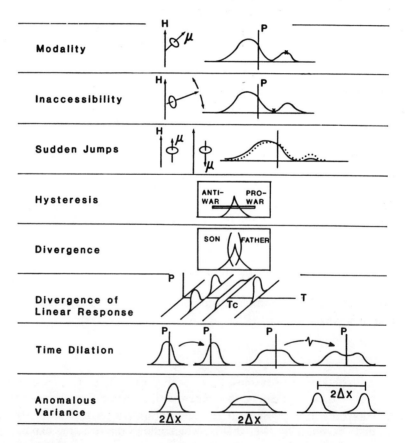

Figure 7-19
Probability densities exhibit the eight catastrophe flags as shown.

Inaccessibility. The intermediate minimum between the two peaks represents a mode of behavior that is difficult to sustain for long times (spin sideways, not knowing/don't care about sex).

Sudden jumps. In response to a changing magnetic field, each individual spin will at some point jump fairly rapidly from one spin state to another. The effect of the sudden transition of one spin on the probability density is not great. Similar sudden changes in attitudes towards politics and sex often occur under stress.

Hysteresis. Under a reversal of the control parameters the transition of a spin back to its original state will typically not occur for the same control-parameter values as the original spin-transition.

Divergence. Two persons with similar backgrounds often wind up with opposite opinions on specific issues (politics), even though they are exposed to similar experiences.

Divergence of linear response. Under stressful conditions a unimodal distribution may rapidly become bimodal. This bimodality can happen, for example, when an antiferromagnet is cooled below the Neél temperature, as well as when taxes are increased to pay for a war. This induces a polarization in the underlying sample from which the probability density is constructed.

Time dilation. Under perturbation of the control-parameter values, a bimodal distribution will adjust fairly rapidly to its new steady-state value. However, a distribution with a broad peak characteristic of the function $\exp(-x^4)$ will take a great deal longer to adjust to a perturbation of the control-parameter values.

Anomalous variance. As the stress on a system is increased either by decreasing the temperature of a ferromagnetic system or increasing the taxation level in a political system, the greater the range of responses from the individual components (projection of spins, opinions of taxpayers) when polled.

Diffraction Patterns

The equations that describe diffraction patterns have the form

$$A(c) \simeq \int e^{iV(x;\ c)/\lambda} dx, \qquad (7\text{-}22)$$

where $V(x; c)$ is a catastrophe function, λ is the wavelength of the scattered signal, and $A(c)$ is the amplitude of the scattered signal at an observation point.[47-49] The first diffraction pattern, corresponding to the fold catastrophe, was studied systematically by Airy in 1838.[50] The second diffraction pattern, corresponding to the cusp catastrophe, was discovered during World War II and described in 1946 by Pearcey.[51] Catastrophe theory provides a systematic method for treating diffraction patterns.[52-55]

To obtain a diffraction pattern, the state variables are systematically integrated out of the rapidly oscillating integral in (7-22). The result is a rapidly oscillating amplitude, dependent on the control parameters. Since the catastrophe flags depend on state variables, which here are dummy variables, we will use a poet's license to identify catastrophe flags with diffraction pattern properties. The catastrophe flags associated with diffraction patterns are described briefly here and summarized in Figure 7-20.

Modality.   The amplitude at a point depends on the number of scattering centers that reflect the signal from source to image. This, in turn, is the number of critical points of the catastrophe $V(x; c)$. For the cusp, this number is three inside the cusp-shaped region, one outside. Degenerate critical points contribute anomalously to the intensity, so the bifurcation set clearly delimits more from less brightly illuminated regions.

Inaccessibility.   All critical points contribute to the integral (7-22), independently (up to phase factor) of stability properties. Therefore, inaccessibility is not a flag for diffraction catastrophes.

Sudden jumps.   On passage through the bifurcation set, different numbers of scattering centers contribute to the intensity. There is, therefore, a relatively fast change in the average ambient illumination on crossing the bifurcation set. This change is accentuated in the short-wavelength approximation ($\lambda \to 0$) called geometrical optics, where the bifurcation set becomes a caustic.

Hysteresis.   Since the amplitude is a state function, independent of process or history, hysteresis is not a flag for diffraction catastrophes.

Divergence.   Again, the amplitude is a state function independent of starting point or process. Therefore divergence also fails to be a flag for diffraction catastrophes.

Divergence of linear response.   The amplitude changes rapidly as the bifurcation set is approached. Therefore, the complex quantities ∂A

Modality	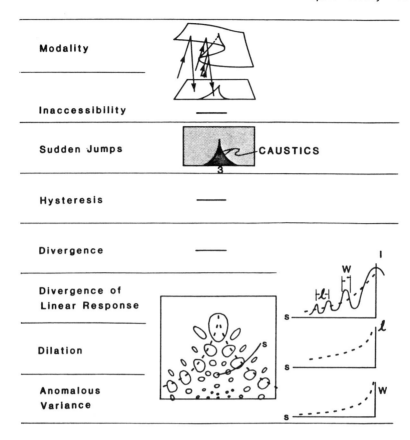
Inaccessibility	
Sudden Jumps	CAUSTICS
Hysteresis	
Divergence	
Divergence of Linear Response	
Dilation	
Anomalous Variance	

Figure 7-20
Diffraction patterns exhibit the eight catastrophe flags as shown.

$(a, b)/\partial a$ and $\partial A(a, b)/\partial b$ diverge as the cusp catastrophe bifurcation set is approached, in the $\lambda \to 0$ limit.

Time dilation. Time, of course, has no place in the discussion of diffraction patterns. What diverges as the bifurcation set is approached is the distance between relative maxima of intensity along any reasonable path cutting the bifurcation set. The intensity distribution consists of a number of peaks arranged in a symmetric way. These peaks increase in diameter and height as the bifurcation set is approached from within the cusp-shaped region. The increase in the distance between intensity peaks corresponds to time dilation.

Anomalous variance. The increase in diameter of peaks in the intensity pattern corresponds to anomalous variance.

7-13 CONCLUSIONS

Elementary catastrophe theory lies at the intersection of two branches of mathematical endeavor. It is the starting point of the program called catastrophe theory, whose objective is to study the qualitative properties of solutions of equations, and in particular to determine how these qualitative properties change as the parameters that occur in the equations change. It is also the final (at present) development in a series of theorems dealing with the qualitative shapes of functions: these theorems are the implicit function theorem, the Morse lemma, and the Thom classification theorem. The first two results typically are treated in courses on elementary or intermediate calculus. There is no good reason to exclude this third result from courses that include the first two.

The mathematical machinery required to provide a rigorous underpinning for this subject has stimulated a great deal of development of a purely mathematical nature. The qualitative features that occur when a catastrophe is present (catastrophe flags) provide a unifying framework for the observation of many phenomena in which qualitative changes occur. These catastrophe flags have been discussed in detail, and the application of catastrophe theory to a selected set of physical problems has been indicated (Figures 7-15–7-20). Although the applications have all been to physical systems, we have intimated how catastrophe theory can be applied to the softer sciences (polling, sociology, politics).

The immediate beneficiaries of catastrophe theory are the harder disciplines: mathematics, physics, chemistry, and the engineering sciences. Mathematics has benefited from the opening of new research areas. Physics and chemistry have benefited from the unification of phenomena observed when a catastrophe is present. The engineering disciplines have benefited from a systematic method for carrying out imperfection analyses.

In the longer run, the softer disciplines may be yet larger beneficiaries. The harder disciplines have equations to work with: If a catastrophe is present, it will eventually come out of these equations. The softer disciplines often have no equations to work with. However, it is not strictly quantitative results that are sought, either. Often it is the ability to predict qualitative changes, their types, and the location of their occurrences that are the goals of such disciplines. Catastrophe theory is the ideal tool for these goals.

ACKNOWLEDGMENTS

I would like to thank Professors T. Poston and I. N. Stewart for enlightening discussions. This work is supported in part by NSF grants PHY 810-2977 and 830-4846.

REFERENCES

1. R. Thom, *Structural Stability and Morphogenesis,* trans. D. H. Fowler (New York: Benjamin-Addison-Wesley, 1975).
2. E. C. Zeeman, *Catastrophe Theory: Selected Papers (1972-1977)* (Reading, Mass: Addison-Wesley, 1977).
3. T. Poston and I. N. Stewart, *Catastrophe Theory and Its Applications* (London: Pitman, 1978).
4. R. Gilmore, *Catastrophe Theory for Scientists and Engineers* (New York: John Wiley & Sons, 1981).
5. V. I. Arnol'd, *Singularity Theory* (Cambridge: University Press, 1981).
6. J. M. T. Thompson, *Instabilities and Catastrophes in Science and Engineering* (London: John Wiley & Sons, 1982).
7. Poston and Stewart, *Catastrophe Theory,* pp. 34–74.
8. Gilmore, *Catastrophe Theory,* pp.6–14.
9. M. Morse, "The Critical Points of a Function of n Variables," *Transactions of the American Mathematical Society* 33 (1931): 72–91.
10. Poston and Stewart, *Catastrophe Theory,* pp. 90–171.
11. Gilmore, *Catastrophe Theory,* pp. 15–50.
12. Zeeman, *Catastrophe Theory,* pp. 1–64.
13. Poston and Stewart, *Catastrophe Theory,* pp. 99–112.
14. Gilmore, *Catastrophe Theory,* pp. 6–14.
15. Thompson, *Instabilities and Catastrophes,* pp. 1–34.
16. Poston and Stewart, *Catastrophe Theory,* pp. 99–112.
17. Gilmore, *Catastrophe Theory,* pp. 33–50.
18. Arnol'd, *Singularity Theory,* pp. 61–90.
19. Gilmore, *Catastrophe Theory,* pp. 15–50.
20. Ibid., pp. 6–14.
21. Zeeman, *Catastrophe Theory,* pp. 1–64.
22. Poston and Stewart, *Catastrophe Theory,* pp. 172–193.
23. Gilmore, *Catastrophe Theory,* pp. 94–106.
24. Zeeman, *Catastrophe Theory,* pp. 497–561.
25. Poston and Stewart, *Catastrophe Theory,* pp. 90–98 and 123–171.
26. Gilmore, *Catastrophe Theory,* pp. 15–32 and 597–614.
27. V. I. Arnol'd, "Classification of Unimodal Critical Points of Functions," *Functional Analysis and Applications* 7 (1973):230–231.
28. Gilmore, *Catastrophe Theory,* pp. 451–467.
29. Thom, *Structural Stability,* pp. 55–100.
30. Zeeman, *Catastrophe Theory,* pp. 1–64.
31. Poston and Stewart, *Catastrophe Theory,* pp. 75–89.
32. Gilmore, *Catastrophe Theory,* pp. 141–156.
33. Ibid., p. 155.
34. R. Gilmore, "Catastrophe Time Scales and Conventions," *Physical Review* A20 (1979): 2510–2515.
35. Zeeman, *Catastrophe Theory,* pp. 1–64.
36. Poston and Stewart, *Catastrophe Theory,* pp. 75–89.
37. Gilmore, *Catastrophe Theory,* pp. 157–183.

38. Zeeman, *Catastrophe Theory,* pp. 1–64.
39. Poston and Stewart, *Catastrophe Theory,* pp. 327–359.
40. Gilmore, *Catastrophe Theory,* pp. 187–253.
41. H. M. Gibbs, S. L. McCall, and T. N. C. Venkatesan, "Differential Gain and Bistability Using a Sodium Filled Interferometer," *Physical Review Letters* 36 (1976): 1135–1138.
42. R. Gilmore and L. M. Narducci, "Relation between the Equilibrium and Non-equilibrium Critical Properties of the Dicke Model," *Physical Review* A17 (1978): 1747–1760.
43. Poston and Stewart, *Catastrophe Theory,* pp. 360–383.
44. Gilmore, *Catastrophe Theory,* pp. 367–427.
45. Zeeman, *Catastrophe Theory,* pp. 303–406.
46. L. Cobb, "The Multimodal Exponential Families of Statistical Catastrophe Theory," in *Statistical Distributions in Scientific Work,* Vol. 4., ed. C. Taille *et al.* (Boston: Reidl, 1981), pp. 67–90.
47. M. V. Berry, "Waves and Thom's Theorem," *Advances in Physics* 25 (1976): 1–26.
48. Poston and Stewart, *Catastrophe Theory,* pp. 246–283.
49. Gilmore, *Catastrophe Theory,* pp. 319–344.
50. G. B. Airy, "On the Intensity of Light in the Neighborhood of a Caustic," *Transactions of the Cambridge Philosophical Society* 6 (1838): 379–403.
51. T. Pearcey, "The Structure of an Electromagnetic Field in the Neighborhood of a Cusp of a Caustic," *Philosophical Magazine* 37 (1946): 311–317.
52. J. J. Duistermaat, "Oscillatory Integrals, Lagrange Immersions and Unfolding of Singularities," *Communications in Pure and Applied Mathematics* 27 (1974): 207–281.
53. K. Jänich, "Caustics and Catastrophes," *Mathematische Annalen* 209 (1974): 161–180.
54. V. I. Arnol'd, "Wavefront Evolution and Equivariant Morse Lemma," *Communications in Pure and Applied Mathematics* 29 (1976): 557–582.
55. H. Trinkhaus and F. Drepper, "On the Analysis of Diffraction Catastrophes," *Journal of Physics* A10 (1977): L11–L16.

About the Editor

RONALD E. MICKENS was born in Petersburg, Virginia in 1943. He received the B.A. degree in physics from Fisk University in 1964 and the Ph.D. in theoretical physics from Vanderbilt University in 1968. Since 1968 he has held a number of postdoctoral fellowships at the Massachusetts Institute of Technology, Vanderbilt University, and the Joint Institute for Laboratory Astrophysics. He has also been a visiting professor at Massachusetts Institute of Technology and Morehouse College. Currently Dr. Mickens is chairperson and professor of physics at Atlanta University. His research has covered the areas of elementary particle physics and nonlinear differential and difference equations. In addition to over seventy research papers, he is the author of the book *Nonlinear Oscillations* (Cambridge University Press, 1981).